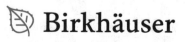

Foundations for Undergraduate Research in Mathematics

Series Editor

Aaron Wootton

Department of Mathematics, University of Portland, Portland, USA

More information about this series at http://www.springer.com/series/15561

Pamela E. Harris • Erik Insko • Aaron Wootton
Editors

A Project-Based Guide to Undergraduate Research in Mathematics

Starting and Sustaining Accessible Undergraduate Research

Editors
Pamela E. Harris
Mathematics
Williams College
Williamstown, MA, USA

Erik Insko
Mathematics
Florida Gulf Coast University
Fort Myers, FL, USA

Aaron Wootton
Mathematics
University of Portland
Portland, OR, USA

ISSN 2520-1212   ISSN 2520-1220   (electronic)
Foundations for Undergraduate Research in Mathematics
ISBN 978-3-030-37855-4   ISBN 978-3-030-37853-0   (eBook)
https://doi.org/10.1007/978-3-030-37853-0

Mathematics Subject Classification: 00A05, 00A07, 00A08, 00B10

This book is published under the imprint Birkhäuser, www.birkhauser-science.com by the registered company Springer Nature Switzerland AG.
The registered company address is: Gewerbestrasse 11, 6330 Cham, Switzerland

Series Preface

Research experience has become an increasingly important aspect of undergraduate programs in mathematics. Students fortunate enough to take part in such research, either through their home institution or via an external program, are exposed to the heart of the discipline. These students learn valuable skills and habits of mind that reach beyond what is typically addressed by the undergraduate curriculum, and are often more attractive to graduate programs and future employers than peers without research experience.

Despite their growing value in the community, research experiences for undergraduate students in mathematics are still the exception rather than the rule. The time commitment required to run or partake in a successful program can be prohibitive, and support for students and mentors is limited. For the faculty member, establishing such a program often requires taking a sophisticated topic outside the scope of the typical undergraduate curriculum and translating it in order to make it accessible to undergraduates with limited backgrounds. It also requires identifying problems or projects that are amenable to undergraduate exploration yet still relevant and interesting to the wider mathematical community. For the undergraduate, pursuing research can often mean reviewing extensive articles and technical texts while meeting regularly with a faculty member. This is no easy feat for the modern undergraduate with a heavy class load, or one who counts on a summer job to help offset academic costs.

The primary goal of the *Foundations for Undergraduate Research in Mathematics* series is to provide faculty and undergraduates with the tools they need to pursue collaborative or independent undergraduate research without the burden it often requires. In order to attain this goal, each volume in the series is a collection of chapters by researchers who have worked extensively with undergraduates to a great degree of success. Each chapter will typically include the following:

- A list of classes from the standard undergraduate curriculum that serve as prerequisites for a full understanding of the chapter
- An expository treatment of a topic in mathematics, written so a student with the stated prerequisites can understand it with limited guidance
- Exercises and Challenge Problems to help students grasp the technical concepts introduced

- A list of specific open problems that could serve as projects for interested undergraduates
- An extensive bibliography and carefully chosen citations intended to provide undergraduates and faculty mentors with a keen interest in the topic with appropriate further reading

On reading a chapter and doing the recommended exercises, the intention is that a student is now ready to pursue research, either collaboratively or independently, in that area. Moreover, that student has a number of open research problems at hand on which they can immediately get to work.

Undergraduate research programs are prevalent in the sciences, technology and engineering and their tremendous benefits are well-documented. Though far less common, the benefits of undergraduate research in mathematics are equally valuable as their scientific counterparts, and increased participation is strongly supported by all the major professional societies in mathematics. As the pioneering series of its type, *Foundations for Undergraduate Research in Mathematics* will take the lead in making undergraduate research in mathematics significantly more accessible to students from all types of backgrounds and with a wide range of interests and abilities.

Preface

The *Foundations for Undergraduate Research in Mathematics* series was created to support and promote the pursuit of undergraduate research in the mathematical sciences. To achieve this goal, the typical chapter in each volume is written with two primary objectives. First, to introduce an area of mathematics, often outside of the standard undergraduate mathematics curriculum, written at a level understandable to an undergraduate. Second, to provide a number of specific research problems or projects that are accessible to undergraduates. The goal of each chapter being that on reading that chapter a student is now ready and able to pursue research.

The current volume of the *Foundations for Undergraduate Research in Mathematics* series extends the focus of many of its chapters to a third objective: to promote development of sustainable research programs. While the chapters in the *Foundations for Undergraduate Research in Mathematics* series can initiate individual research projects, it is the development of sustainable, self-perpetuating programs that will allow undergraduate research in mathematics to thrive. Accordingly, authors in the current volume have been invited to contribute chapters given their significant achievement in developing successful research programs, which heavily include undergraduate students. While a majority of the chapters follow the traditional *Foundations for Undergraduate Research in Mathematics* series structure of introducing a topic in mathematics and providing explicit research projects, in many of the chapters, authors provide tips and tricks for how to turn initial projects into a long-term research program. For example, in the chapter "Lateral Movement in Undergraduate Research," S. Garcia provides over 20 guiding principles on sustaining undergraduate research. Likewise, in the chapter "Researching in Undergraduate Mathematics Education: Possible Directions for Both Undergraduate Students and Faculty," M. Savic provides a detailed account of how to direct undergraduates to do research in Undergraduate Mathematics Education, the first such account appearing in a Foundations for Undergraduate Research in Mathematics series volume.

Though the theme of this volume is building a sustainable program, the specific chapters span a broad spectrum of disciplines in mathematics. To provide readers with a cohesive volume, we have grouped chapters together based on the subdisciplines of the mathematics presented. The first nine chapters lie in the general area of pure mathematics with Chaps. 1–5 falling into the area of Combinatorics, Chaps. 6 and 7 Number Theory, Chap. 8 Graph Theory, and Chap. 9 Analysis. Chapter 10 falls into the area of Mathematics Education, and Chap. 11 Applied Mathematics, or more specifically, Mathematical Biology.

Chapter 1, "Folding Words Around Trees: Models Inspired by RNA," lies at the intersection of mathematics and biology, an area where mathematical models are built to address biological questions and new mathematical theories are inspired by biological structures. In this chapter, Drellich and Smith explore a combinatorial model for folding words around plane trees, which is inspired by the bonds that form between nucleotides in a single-stranded RNA molecule.

In Chap. 2, "Phylogenetic Networks," Gross, Long, and Rusinko introduce the mathematics of phylogenetic networks, which are a class of graphs sufficient for modeling a large range of evolutionary events. Phylogenetic trees have traditionally been used to describe and model the evolution of species, but trees are not sufficient to model more complex evolutionary events such as hybridization. The authors explore both the combinatorics of these graphs and an algebraic statistical model of evolution whose structure depends on the networks in this chapter.

Chapter 3, "Tropical Geometry," introduces tropical mathematics and more specifically tropical curves and surfaces. In tropical mathematics, the rules of arithmetic are redesigned so that "plus" means "take a maximum" and "times" means "plus." Tropical geometry asks the following question: what geometric shapes are defined by equations that use these new rules of arithmetic? These shapes, called tropical curves and tropical surfaces, are piecewise-linear objects, which are related to triangulations of polygons and polyhedra through a beautiful duality. Not only do these combinatorial objects mirror the behavior of the traditional curves and surfaces we are used to: they can be used to study solutions to good old-fashioned polynomial equations, through a process called tropicalization. In this chapter, Morrison investigates the structure and properties of tropical curves and tropical surfaces in their own right, as well as their connections to objects from algebraic geometry.

In Chap. 4, "Chip Firing Games and Critical Groups," Glass and Kaplan describe some of the main properties of critical groups and outline open problems that can be approached by students with a variety of backgrounds. For any finite connected graph, one can associate a finite abelian group known as the critical group. This group can be defined either in terms of a matrix associated with the graph known as the Laplacian or in terms of an elementary combinatorial operation known as chip-firing. Critical groups have been studied from a number of different perspectives, using techniques from linear algebra, combinatorics, number theory, group theory, and other areas of mathematics.

In Chap. 5, "Counting Tilings by Taking Walks in a Graph," Butler, Ekstrand, and Osborne consider the problem of tiling regions. Given a collection of shapes (tiles) and a given region, a classical question to ask is whether it is possible to cover that region with those tiles. More generally, we can ask in how many different ways a region can be covered with these tiles. As the size of the region starts to grow and the tiles become more complex, this task soon becomes daunting as the number of configurations starts to grow dramatically (usually exponentially). However, in this chapter the authors show that if we do not focus on the tiles but focus on the way that the tiles cross between consecutive layers (the zen approach to tiling), then the problem reduces to one which can be counted with basic tools of linear algebra (mixed with a bit of patience and a dash of computer programming). This opens up large sets of problems as the tiles and regions can be chosen to be much more interesting, the only limitation being imagination.

In Chap. 6, "Beyond Coins, Stamps, and Chicken McNuggets: An Invitation to Numerical Semigroups," Chapman, Garcia, and O'Neill provide several interesting projects for undergraduates whose roots lie in the intersection of number theory and factorization, linear algebra, and discrete mathematics. They start in the introductory sections by working through various concepts and examples, beginning with the historically intriguing Frobenius Coin Exchange Problem and advancing through related concepts used to study numerical semigroups. With additional insight on computational tools used in this field, the interested reader will be able to hit the ground computing examples, creating conjectures, and contributing to this very active field of research.

In Chap. 7, "Lateral Movement in Undergraduate Research: Case Studies in Number Theory," Garcia explores the thought processes, strategies, and pitfalls involved in entering new territory, developing novel projects, and seeing them through to publication. Twenty-one guiding principles for developing a sustainable undergraduate research pipeline are proposed. These ideas are then illustrated in three detailed case studies that show these guidelines in action.

Next, in Chap. 8, "Projects in (t, r) Broadcast Domination," Harris, Insko, and Johnson introduce the reader to a game played on graphs that was first explored in 2015. The goal is to place broadcasting towers in an efficient manner so that every location has good reception. The authors summarize their past work with students, describe how this game generalizes other graph domination problems, and present many remaining open problems and variations to consider. They end the chapter by providing some advice on how to continue to develop new research projects with and for students; although the mathematical content of the chapter is in domination theory, the ending suggestions can be implemented in any area.

In Chap. 9, "Squigonometry: Trigonometry in the p-Norm," Wood and Poodiack consider the effects on the classical trigonometric functions when the definition of distance is changed. The classical trigonometric functions sine and cosine parameterize the unit circle $x^2 + y^2 = 1$ in a natural way. Information like the area of a sector, arc length, and angle measure subtended follows organically; the circle is a perfect shape. A good way to see this perfection is to introduce some imperfection. What happens if the definition of distance is changed, the exponent of 2 being changed to a generic $p \geq 1$? It results in less perfect square-like circles, and with them new generalized trigonometric functions and new values of π. This chapter investigates these functions and how they capture the newly broken symmetries, exploring new connections between geometry and special functions along with a fresh appreciation of the familiar trigonometric functions.

In Chap. 10, "Researching in Undergraduate Mathematics Education: Possible Directions for Both Undergraduate Students and Faculty," Savic describes the many methods, questions, and examples of undergraduates researching in undergraduate mathematics education (RUME). For faculty, it is a chapter that can assist with the advising process, giving out examples, references, and open research questions in RUME. For undergraduate students, it has a template of writing in RUME and undergraduate stories of success. Overall, the chapter could be a resource for anyone looking to dive into the new field of RUME.

Finally, in Chap. 11, "Undergraduate Research in Mathematical Epidemiology," Bañuelos, Bush, Martinez, and Prieto-Langarica provide an introduction to using mathematics to track the spread of diseases, or anything else that travels in space and time. It is written as an inquiry-based study which the authors hope will help faculty mentors allow students to take lead on a given project. Different modeling techniques are used to answer biological/conservation questions, such as how a disease can be controlled. The authors have written the chapter with the view that any calculus student can attempt the research questions posed.

Williamstown, MA, USA   Pamela E. Harris
Fort Myers, FL, USA   Erik Insko
Portland, OR, USA   Aaron Wootton

Contents

Folding Words Around Trees: Models Inspired by RNA 1
Elizabeth Drellich and Heather C. Smith

Phylogenetic Networks 29
Elizabeth Gross, Colby Long, and Joseph Rusinko

Tropical Geometry 63
Ralph Morrison

Chip-Firing Games and Critical Groups 107
Darren Glass and Nathan Kaplan

Counting Tilings by Taking Walks in a Graph 153
Steve Butler, Jason Ekstrand, and Steven Osborne

**Beyond Coins, Stamps, and Chicken McNuggets: An Invitation
to Numerical Semigroups** 177
Scott Chapman, Rebecca Garcia, and Christopher O'Neill

**Lateral Movement in Undergraduate Research: Case Studies
in Number Theory** 203
Stephan Ramon Garcia

Projects in (t, r) Broadcast Domination 235
Pamela E. Harris, Erik Insko, and Katie Johnson

Squigonometry: Trigonometry in the p-Norm 263
William E. Wood and Robert D. Poodiack

**Researching in Undergraduate Mathematics Education: Possible
Directions for Both Undergraduate Students and Faculty** 287
Milos Savic

Undergraduate Research in Mathematical Epidemiology 303
Selenne Bañuelos, Mathew Bush, Marco V. Martinez,
and Alicia Prieto-Langarica

Folding Words Around Trees: Models Inspired by RNA

Elizabeth Drellich and Heather C. Smith

Abstract

At the intersection of mathematics and biology, we find mathematical models built to address biological questions as well as new mathematical theories inspired by biological structures. In this chapter, we explore a combinatorial model for folding words around plane trees which is inspired by the bonds that form between nucleotides in a single-stranded RNA molecule.

This chapter walks the reader through the construction of valid plane trees, structures formed by folding a word in a complementary alphabet around a plane tree, and enumerates the class of words with exactly one such folding. Valid plane trees are relatively unexplored combinatorial objects, and while we present several potential research projects, a careful reader can come up with many additional directions for further study.

Suggested Prerequisites Familiarity with proofs is sufficient for a student to get started. No additional prerequisites are required, though prior exposure to graph theory, combinatorics, and algorithms is helpful.

This work was completed while ED was affiliated with Swarthmore College.

E. Drellich
Haverford College, Haverford, PA, USA
e-mail: edrellich@haverford.edu

H. C. Smith (✉)
Davidson College, Davidson, NC, USA
e-mail: hcsmith@davidson.edu

© Springer Nature Switzerland AG 2020
P. E. Harris et al. (eds.), *A Project-Based Guide to Undergraduate Research in Mathematics*, Foundations for Undergraduate Research in Mathematics,
https://doi.org/10.1007/978-3-030-37853-0_1

1

1   Introduction

Though they may sometimes seem quite different, pure and applied fields of mathematics are in constant communication with each other. Mathematicians model real-world phenomena, which in turn can lead to new theorems and new mathematical objects. This interplay is a two way street: science inspiring mathematics and mathematics furthering science. This chapter will present a family of combinatorial questions that arise from the simplest models of RNA folding.

Nucleotides in a strand of RNA attach to others in the sequence like a length of sticky tape. Unlike DNA which forms the familiar double helix and stores biological information, single-stranded RNA molecules fold themselves into all kinds of shapes which are essential to their function.

Consider the word $w = ACGCAUGCGUUA$ with twelve nucleotides in which bonds can form between G and C and also between A and U, the Watson–Crick pairs. Figure 1 shows one way this word could be wrapped around a rooted tree with letters on opposite sides of an edge forming a bond. This combinatorial object, consisting of a word folded around a plane tree, is a simplified model of an RNA molecule.

For the first seven sections of this chapter we concern ourselves only with the combinatorial objects. We begin in Sect. 2 with the necessary background in the combinatorics of Catalan objects, particularly non-crossing matchings and plane trees, which we will use heavily throughout. Sections 3 through 6 are based on work originally published in [1] and [2]. Biologically inspired variations are discussed in Sect. 7. The RNA folding and inverse folding problems are described in Sect. 8 where we also give a number of resources for readers interested in this applied problem.

Fig. 1 A folding of
$w = ACGCAUGCGUUA$

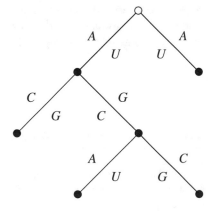

2 Catalan Numbers

Consider the following expression:

$$((9 - 7) - (4 - 6)) - ((5 - 2) - 1).$$

The parentheses tell us in which order the operations should be performed. To do this, the parentheses come in pairs, where each open parenthesis has a particular corresponding closed parenthesis, isolating a part of the expression. In fact, when we drop the numbers and the minus signs, we are left with the following arrangement:

$$(()())().$$

Even here, you can still identify the matched pairs of parentheses, and so we call such arrangements *correctly matched*. In contrast, the arrangement $())((())()$ is not correctly matched. While we would suspect that the first two parentheses are paired, this leave the third parenthesis, which is closed, without a corresponding open parenthesis to its left.

Now take 2 open parentheses and 2 closed parentheses. How many ways can you order these parentheses so that they are correctly matched? There are two ways which are correctly matched: (()) and ()(). Now try this for 3 open parentheses and 3 closed parentheses. Here are the 5 arrangements:

$$((())) \quad (()()) \quad (())() \quad ()(()) \quad ()()().$$

Exercise 1 List all 14 correctly matched arrangements of 4 pairs of parentheses.

We will use the notation C_k, where k is a natural number, to denote the number of ways to arrange k open parentheses and k closed parentheses so that they are correctly matched. Here is what we know so far:

$$C_1 = 1, \quad C_2 = 2, \quad C_3 = 5, \quad C_4 = 14.$$

These counts create an infinite sequence $(C_1, C_2, C_3, C_4, \ldots)$, but currently we only know the first 4 values. You may be hesitant to start arranging 5 sets of parentheses and just looking at the numbers, it is not entirely clear what C_5 should be. This is where the Online Encyclopedia of Integer Sequences (OEIS) [8] comes to the rescue! When we enter our sequence $(1, 2, 5, 14)$, we are quickly linked to the *Catalan numbers* which is sequence A000108. You will even notice in the comments section that there is an entry about arrangements of parentheses.

The Catalan numbers are an infinite sequence of numbers that enumerate many different types of objects including the number of ways one can arrange n pairs of parentheses which are correctly matched. The sequence begins like this:

$$C_1 = 1, \quad C_2 = 2, \quad C_3 = 5, \quad C_4 = 14, \quad C_5 = 42, \quad C_6 = 132, \quad C_7 = 429.$$

Stopping the list here, there still does not seem to be an obvious pattern to indicate what comes next, and you certainly do not want to start listing all arrangements of 8 pairs of parentheses. However, there is a closed form for computing these numbers:

$$C_n = \frac{1}{n+1}\binom{2n}{n} = \frac{1}{n+1}\frac{(2n)!}{n!\,n!}. \tag{1}$$

The Catalan numbers are so ubiquitous that many textbooks contain a whole chapter on them. A good introduction can be found in the chapter titled *Catalan Numbers* in [7]. For a much more in depth treatment of them, see [17].

The Catalan numbers can also be computed using this *recursive definition* for computing the Catalan numbers.

$$C_0 = 1, \quad C_1 = 1, \quad \text{and} \quad C_{n+1} = \sum_{i=0}^{n} C_i C_{n-i} \text{ for } n \geq 1. \tag{2}$$

The initial conditions C_0 and C_1 are given. We have already seen that $C_1 = 1$. For C_0, there is only one empty word and it vacuously meets the condition that pairs of parentheses must be correctly matching, so $C_0 = 1$. The recursive formula then tells us how to calculate C_n for any $n \geq 2$. First, $C_2 = C_0 C_1 + C_1 C_0 = 2$. Then we can use C_0, C_1, and C_2 and the recursive formula to compute C_3. Continuing in this way, we can find C_n for any natural number n.

To convince ourselves of the recursive formula, let us see how it connects with enumerating arrangements of matched parenthesis. Consider the recursive formula for C_5:

$$C_5 = C_0 C_4 + C_1 C_3 + C_2 C_2 + C_3 C_1 + C_4 C_0. \tag{3}$$

We know that C_5 counts the number of ways to arrange 5 sets of parentheses which are correctly matched. Now we need to see that the right side of the equation counts the same thing. Since there are 5 terms being added, we need to sort our arrangements into 5 different categories where the arrangements in each category share some common characteristic. We know that every arrangement starts with an open parenthesis. Now how far away is its matching closed parenthesis? We could sort by this. For example, in (()())()(), there are two pairs of parentheses between the first open parenthesis and its mate, so let us put this arrangement into Category 2. On the other hand, ((()))() goes in Category 3 while ()()()(()) goes in Category 0. There are now five categories, and every arrangement is in exactly one of them, forming a *partition*.

Now our task is to determine how many arrangements are in each category. Let us start with Category 0. In this category, each arrangement begins with () and is followed by some valid arrangement of 4 sets of parentheses. But we know this quantity to be C_4 which is equal to $C_0 C_4$ in Eq. (3). The arrangements in Category 1 have the form $(--)----$ where the first and fourth parentheses are matched. In between, we find a single pair of parentheses that must be matched and the number of possible arrangements is C_1. After the closed parenthesis in the fourth position, we must find three more pairs of parentheses that are correctly matched and there are C_3 arrangements possible. Thus the total number of arrangements in Category 1 is $C_1 C_3$. Following this logic Category 2 has $C_2 C_2$ arrangements, Category 3 has $C_3 C_1$, and Category 4 has $C_4 C_0$.

Exercise 2 Generalize this argument using strong induction to prove that the recursive formula (2) enumerates the matched arrangements of n sets of parentheses for each natural number n.

The following inequality will be needed for a proof in Sect. 6. One possible proof can be obtained by counting arrangements of parentheses.

Exercise 3 For any $n \in \mathbb{N}$ and any integer $0 \leq i \leq n$,

$$C_n > C_i C_{n-i}.$$

The Catalan numbers enumerate many other sets of objects. We focus on two here: non-crossing perfect matchings and plane trees.

Definition 1 A *perfect matching* of $\{1, \ldots, 2n\}$ for some natural number n, is a set of n ordered pairs $M = \{(i, j) : i, j \in \{1, \ldots, 2n\}, i < j\}$ such that for each $k \in \{1, \ldots, 2n\}$, there is exactly one $(i, j) \in M$ with $k \in \{i, j\}$.

Definition 2 A *non-crossing perfect matching* of size n is a perfect matching M of $\{1, \ldots, 2n\}$ which can be represented as a series of n non-intersecting arcs drawn in the plane above the line $y = 0$ where each $(i, j) \in M$ corresponds with the arc having endpoints on the x-axis with x-coordinates i and j. The set of non-crossing perfect matchings of size n is denoted by M_n.

Figure 2 shows all non-crossing perfect matchings in M_3.

Definition 3 A *plane tree* is a straight line drawing of a rooted tree embedded in the plane with the root higher than all other vertices. The set of plane trees with n edges is denoted by T_n.

Fig. 2 Non-crossing perfect matchings of size 3

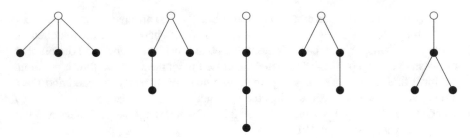

Fig. 3 All 5 distinct plane trees with 3 edges

All plane trees in T_3 are found in Fig. 3, where the root is identified by an un-filled vertex.

Before we move on, we would like a more compact way to describe a plane tree. For a drawing of a plane tree, start at the root and follow your way around the tree, counterclockwise, until you return to the root. As you go, you will walk down the left side of an edge and later you will walk up the right side of the same edge. The full path creates an ordering of the half edges of the plane tree, so we can label the half edges with the integers $1, 2, \ldots, 2n$ in the order they are visited. As a result, each edge has a pair of labels (i, j) with $i < j$ and we name the edge $e(i, j)$. For example, the edges in the second plane tree in Fig. 3 are $e(1, 4), e(2, 3), e(5, 6)$, while the edges in fourth plane tree in Fig. 3 are $e(1, 2), e(3, 6), e(4, 5)$, making these two distinct plane trees.

Not every set of pairs corresponds to a plane tree, for example, there is no plane tree with edges $e(1, 3), e(2, 4), e(5, 6)$. However, each plane tree has a unique set of edge labels that can be used to reconstruct the plane tree.

Exercise 4 Given a set of edge labels, give an algorithm to find the corresponding plane tree if one exists.

Now back to the Catalan numbers. In Exercise 2, you proved that C_n enumerates the matched arrangements of n pairs of parentheses, using strong induction and the recursive definition (2) for the Catalan numbers.

Exercise 5 Use strong induction and the recursive definition (2) to prove that, for each natural number n, the Catalan number C_n enumerates the non-crossing perfect matchings of size n and also the plane trees with n edges. (As a hint, think about what remains when the arc $(1, j)$ is removed.)

Let P_n be the set of matched arrangements of n pairs of parentheses. In Exercise 2, you proved that $|P_n| = C_n$. In Exercise 5, you used a similar method to prove $|M_n| = C_n$ and $|T_n| = C_n$, however, there is an alternative method using bijections since we already know $|P_n| = C_n$.

Definition 4 Let A and B be non-empty sets. A mapping $\tau : A \to B$ is ...

- a *function* provided, for each $a \in A$, there is exactly one $b \in B$ such that $\tau(a) = b$.
- an *injection* provided τ is a function and, for any $a, a' \in A$, if $\tau(a) = \tau(a')$, then $a = a'$.
- a *surjection* provided τ is a function and, for any $b \in B$, there exists $a \in A$ with the property $\tau(a) = b$.
- a *bijection* provided τ is both an injection and a surjection.

For two sets, A and B, if there exists a bijection $\phi : A \to B$, then $|A| = |B|$. So if we can define a bijection $\tau : M_n \to P_n$, this will imply $|M_n| = |P_n|$ and, since $|P_n| = C_n$, we will have $|M_n| = C_n$.

Define a mapping $\tau : M_n \to P_n$ as follows: For an arbitrary $M \in M_n$, we will define $\tau(M)$ to be the sequence $w = w_1 w_2 \ldots w_{2n}$ established by the following rules: For each $i \in \{1, 2, \ldots, 2n\}$

- If M contains an arc with its left endpoint at x-coordinate i, set w_i equal to (.
- Otherwise, set w_i equal to).

For an example, let M_1 be the leftmost non-crossing perfect matching in Fig. 2 and let M_5 be the rightmost one. Then $\tau(M_1)$ is ()()() and $\tau(M_5)$ is (()()). Exercise 6 asks you to prove that our mapping $\tau : M_n \to P_n$ is a bijection.

Exercise 6 To see that $\tau : M_n \to P_n$ is a bijection, verify the following three properties:

1. Function: For each M in M_n, verify that $\tau(M)$ is in P_n.
2. Injection: If $\tau(M) = \tau(N)$ for some M, N in M_n, show that $M = N$.
3. Surjection: For each P in P_n, prove that there exists M in M_n such that $\tau(M) = P$.

Now we leave it to you to give a similar proof that plane trees are counted by the Catalan numbers in the first part of Exercise 7. In upcoming sections, you will find that the connection between plane trees and non-crossing perfect matchings is essential, so the second part of Exercise 7 asks you to define a bijection directly between M_n and T_n.

Exercise 7 Give two alternate proofs for the fact $|T_n| = C_n$ as follows:

1. Find a bijection between T_n and P_n.
2. Find a bijection between T_n and M_n. Hint: Consider mapping the edge $e(i, j)$ to the pair (i, j).

3   Valid Plane Trees

We are not, in general, going to want all C_n plane trees with n edges, only those that admit a given word in a complementary alphabet.

Definition 5 A set $\mathscr{A}$ is a *complementary alphabet* if for every letter $B \in \mathscr{A}$ there is a unique complement $\overline{B} \in \mathscr{A}$ that is distinct from B and taking the complement is an involution on $\mathscr{A}$, in other words $\overline{\overline{B}} = B$ for each $B \in \mathscr{A}$. We use the notation $\mathscr{A}_m$ to denote an alphabet with $2m$ letters and we typically use $\mathscr{A}_m = \{A_1, \ldots, A_m, \overline{A}_1, \ldots, \overline{A}_m\}$ when $m \geq 3$.

Definition 6 For $n \in \mathbb{N}$, given a word $w = w_1 w_2 \ldots w_{2n}$ in a complementary alphabet $\mathscr{A}$, a plane tree T is called w-*valid* if whenever $e(i, j)$ is an edge in T, $w_i = \overline{w}_j$. A word w is called *foldable* if at least one w-valid plane tree exists. A word w is called k-*foldable* when there are exactly k plane trees which are w-valid.

Definition 6 is equivalent to the requirement that if the word w is wrapped around the plane tree T, starting at the root and traversing counterclockwise, one letter per half edge, then every edge of T has a pair of complementary letters labeling its two sides. Alternatively we could say a plane tree T is w-valid if the non-crossing matching of $\{1, \ldots, 2n\}$ corresponding to T has the property that i is matched with j only if $w_i = \overline{w}_j$.

Example 1 Consider the word $w = A\,\overline{A}\,\overline{A}\,A\overline{A}A$ with alphabet $\mathscr{A}_1$. Figure 4 shows one plane tree which is w-valid, showing that w is foldable. Figure 4 also gives a plane tree which is not w-valid, so w is k-foldable for some k with $1 \leq k < C_3$ since the total number of plane trees with 3 edges is C_3.

Exercise 8 Not all words in a complementary alphabet have a valid plane tree. Find necessary conditions for the word w to have a valid plane tree. Are your conditions sufficient?

Fig. 4 Only the plane tree on the right is valid for $w = A\,\overline{A}\,\overline{A}\,A\overline{A}A$

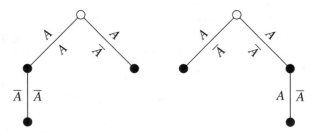

Though most words in complementary alphabets do not admit even one valid plane tree, if a sequence has a valid plane tree, one valid plane tree can be created via a simple greedy algorithm.

Definition 7 Given a word $w = w_1 w_2 \ldots w_{2n}$ in a complementary alphabet, the pseudocode for a *greedy algorithm* is given next which returns a list of edges $e(i, j)$ and a (possibly empty) stack of unused letters.

Pseudocode for the Greedy Algorithm
1. Consider the word $w_1 w_2 \ldots w_{2n}$ as stack IN with w_1 on the top and w_{2n} on the bottom. Let the stack OUT be empty.

2. Move the letter on top of stack IN to the top of stack OUT.

3. If the letters w_i and w_j, respectively, on the top of stacks IN and OUT are a complementary pair in the alphabet, pop both of them and record edge $e(i, j)$.

4. Repeat step 3 until the letters on top of the stacks IN and OUT are not a complementary pair (or one stack is empty).

5. If stack IN is non-empty, return to step 2. Else, output the recorded edges and the stack OUT.

An example of the greedy algorithm for the word $AB\overline{B}A\overline{A}B\overline{B}A$ is given in Fig. 5.

Theorem 1 ([2]) *A w-valid plane tree exists for word w whenever the stack OUT is empty at the end of the greedy algorithm.*

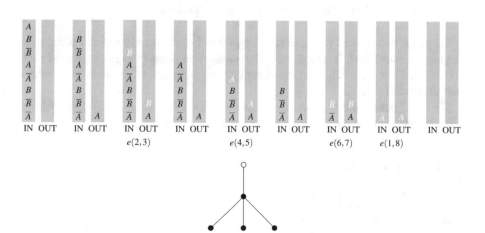

Fig. 5 The greedy algorithm is run on the word $w = AB\overline{B}A\overline{A}B\overline{B}A$ and the output edges define the given w-valid plane tree

In particular, the collection of edges returned by the greedy algorithm form a w-valid plane tree.

Proof If stack OUT is empty at the end of the greedy algorithm, consider the collection of edges $e(i, j)$ recorded in step 3 of the algorithm. Using your bijection from Exercise 7, those edges form a plane tree if and only if they determine a non-crossing perfect matching of $\{1, 2, \ldots, 2n\}$.

Given that the greedy algorithm terminated leaving both the IN and OUT stacks empty, every letter w_i must be paired with a w_j and thus

$$\{(i, j) : e(i, j) \text{ is output by the greedy algorithm}\}$$

is a perfect matching on $\{1, 2, \ldots, 2n\}$.

Furthermore an edge $e(i, j)$ could only have been produced in step 3 if all of the letters $w_{i+1}, w_{i+2}, \ldots, w_{j-1}$ had already be paired by the algorithm. None of those letters could have been paired with w_k for any $k > j$ since the algorithm has yet to encounter w_k. Nor can any of those letters been paired with w_ℓ for $\ell < i$ since w_ℓ, if it is still in the OUT stack, appears below w_i in the OUT stack from the time w_i was moved from the IN to the OUT stack. Therefore the matching created by the greedy algorithm is non-crossing and the edges produced by the greedy algorithm form a plane tree. That plane tree is w-valid by construction so a w-valid plane tree exists.   □

The converse, that a w-valid plane tree exists *only* if the OUT stack is empty at the end of the greedy algorithm, will require some tools from the next section.

4   Local Moves and the State Space Graph

If the plane tree formed by the greedy algorithm is not the only w-valid plane tree, the next consideration is how those different w-valid plane trees are related to each other. Similarity between two plane trees can be quantified by the number of edges that they share (Fig. 6).

To move between two plane trees, Condon, Heitsch, and Hoos [5] defined *local moves* to be the following operation on pairs of edges in a plane tree.

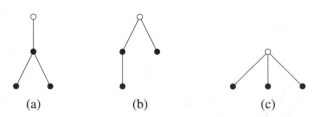

Fig. 6 Plane tree (**b**) shares edge $e(2, 3)$ with (**a**), whereas plane tree (**c**) has no edges in common with (**a**) but shares edge $e(5, 6)$ with (**b**)

(a)   (b)   (c)

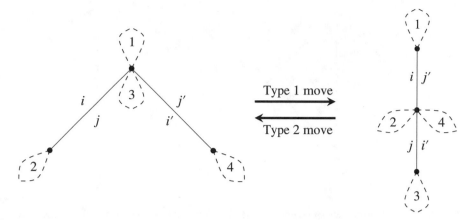

Fig. 7 Local moves between valid trees

Definition 8 Let $i < j < i' < j'$ and fix a plane tree T. The two local moves are

Type 1: if $e(i, j)$ and $e(i', j')$ are incident edges (share a common endpoint) in T, then $e(i, j)$, $e(i', j')$ are replaced by $e(i, j')$, $e(j, i')$.
Type 2: if $e(i, j')$ and $e(j, i')$ are incident edges in T, then $e(i, j')$, $e(j, i')$ are replaced by $e(i, j)$, $e(i', j')$.

A local move results in a new plane tree T'.

A local move is modeled after an unfolding-and-refolding operation on nearby base pairs in a strand of RNA. It can be thought of as "unzipping" two adjacent bonds, and "rezipping" them in the opposite orientation.

Note that local moves can be performed on edges without any successively labeled half edges, namely with $i < j - 1$ and $j < i' - 1$ and $i' < j' - 1$. In other words there can be many other edges incident to the vertices in Fig. 7, including edges that come between those sketched in the schematic.

Exercise 9 Find all 7 local moves on the plane tree with edges $e(1, 10)$, $e(2, 3)$, $e(4, 5)$, $e(6, 9)$, $e(7, 8)$ and verify that each results in a new plane tree.

The existence of local moves allows us to now define the *state space graph*.

Definition 9 Fix $n \in \mathbb{N}$. Let G_n be the graph with a vertex for each plane tree with n edges and an edge between two vertices when the corresponding plane trees have the property that one can be transformed into the other by a single local move. For simplicity, vertices are named by their corresponding plane trees.

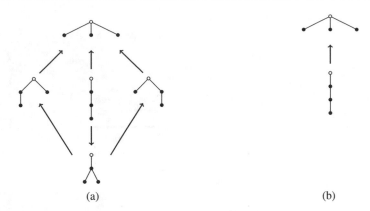

(a)   (b)

Fig. 8 (**a**) The directed state space graph $\overrightarrow{G}_3$ (**b**) The directed state space graph $\overrightarrow{G}_w$ where $w = A\overline{A}\,B\,\overline{B}\,A\,\overline{A}$

Figure 8a shows G_3, but with direction on the edges, which will be introduced later.

You may have observed that performing a local move on a w-valid tree might result in a tree that is no longer w-valid. This leads us to our next collection of definitions.

Definition 10 Fix a word w and let T be a w-valid plane tree. A local move on adjacent edges in T is a *valid local move* if the tree T' obtained by performing the local move is also w-valid.

Exercise 10 Given a word w with at least one valid plane tree, prove the following statements that will be used in the proofs of Theorems 2 and 3:

1. A local move on half edges $i < j < i' < j'$ is valid if and only if $w_i = w_{i'} = \overline{w}_j = \overline{w}_{j'}$.
2. If $w_i = \overline{w_{i+1}}$ and $e(i, i + 1)$ not an edge of w-valid plane tree T, then a single local move can be performed to create edge $e(i, i + 1)$.

We now have the machinery to prove the converse of Theorem 1.

Theorem 2 ([2]) *If a w-valid plane tree exists, the greedy algorithm will result in an empty OUT stack.*

Proof Suppose it is possible for a w-valid plane tree to exist but for the greedy algorithm to return a non-empty stack OUT. Without loss of generality assume that w is a word of shortest length for which this occurs. Since the greedy algorithm returns a valid plane tree whenever one exists for words of length 2, our word w must have length at least 4.

Even though the greedy algorithm fails to produce a w-valid plane tree, we claim that it must produce at least one edge of the form $e(j, j + 1)$. First note that if the greedy algorithm produces any edges at all, the first one produced will have the desired form. Further, every plane tree has an edge of the form $e(j, j + 1)$, so w must have two letters w_i, w_{i+1} that are complementary because it is foldable. If the greedy algorithm has not yet created an edge of the form $e(j, j + 1)$ when it gets to letter w_{i+1}, it will form the edge $e(i, i + 1)$.

Let $e(j, j+1)$ be the first edge formed by the greedy algorithm on w and let T be a w-valid plane tree. Then either $e(j, j + 1)$ is an edge in T, or a single valid local move can transform T into a w-valid tree T' containing $e(j, j + 1)$. If $e(j, j + 1)$ is not in T, then T contains edges corresponding to $\{w_j, w_k\}$ and $\{w_{j+1}, w_\ell\}$ for some w_k, w_ℓ in w. These edges are necessarily adjacent in T and, by Exercise 10, there is a valid local move transforming T into a new tree T' containing $e(j, j + 1)$ and the edge corresponding to pair $\{w_k, w_\ell\}$. So let T_j be a w-valid plane tree which contains the edge $e(j, j + 1)$.

Now let $\hat{w}$ be the word $w_1 w_2 \ldots w_{j-1} w_{j+2} \ldots w_{2n}$ and consider the plane tree $\hat{T}_j$ formed by removing the edge $e(j, j+1)$ from T_j. Then $\hat{T}_j$ is a $\hat{w}$-valid plane tree, but the greedy algorithm will produce the same non-empty OUT stack for $\hat{w}$ that it did for w making $\hat{w}$ a shorter word than w that has a valid plane tree but for which the greedy algorithm fails to produce a plane tree, contradicting the minimality of w.

$\square$

Definition 11 For a word w of length $2n$, the *restricted state space graph* G_w is the subgraph of G_n consisting of only those vertices which correspond to w-valid plane trees and edges that represent w-valid local moves.

To test your understanding of the definitions we have discussed so far, try the following problem for which no answer has been published:

Challenge Problem 1 Determine whether G_w, the subgraph of G_n, is necessarily an induced subgraph for any word w of length $2n$.

Exercise 11 Let $w = A\overline{A}B\overline{B}B\overline{B}A\overline{A}B\overline{B}A\overline{A}$, so a w-valid plane tree has 6 edges. Note that $C_6 = 132$ and thus there are 132 plane trees with 6 edges. The number of w-valid plane trees, however, is significantly smaller. How many vertices does G_w have? Draw (or program a computer to draw) G_w.

Theorem 3 ([2]) *The restricted state space graph G_w is connected for any foldable word w.*

Proof We prove this by induction on the length of the word w. If w has length 2, then the state space graph is trivially connected since it contains only the single w-valid plane tree given by the greedy algorithm. Suppose that for any word of length at most $2n - 2$, the restricted state space graph of valid plane trees is connected.

Let w be a word of length $2n$ with w-valid plane trees S and T. If S and T both have the edge $e(i, j)$, delete that edge to create new plane trees: S' and T' both valid plane trees for $w' = w_1 \ldots w_{i-1} w_{j+1} \ldots w_{2n}$ and S'' and T'' valid plane trees for $w'' = w_{i+1} \ldots w_{j-1}$. By the inductive hypothesis there is a sequence of valid local moves from S' to T' and a sequence of valid local moves from S'' to T''. Concatenating these sequences gives a sequence of valid local moves from S to T.

If S and T do not agree on any edges, then with a single local move on T, we will create a w-valid plane tree $\tilde{T}$ that shares an edge with S. Observe that S must have an edge $e(i, i + 1)$ for some $i \in \{1, \ldots, 2n - 1\}$. By Exercise 10 there must be a valid local move on T creating w-valid plane tree $\tilde{T}$ with edge $e(i, i + 1)$. Thus by the earlier argument, there is a sequence of valid local moves from S to $\tilde{T}$ which can then be extended to a sequence of valid local moves from S to T.   □

Type 1 and Type 2 local moves are inverses of each other, so we can create a directed version of a restricted state space graph G_w by replacing each edge between vertices S and T by a directed edge from S to T indicating that there is a Type 2 move which transforms S into T. Two examples are shown in Fig. 8.

Theorem 4 ([2]) *The directed state space graph $\overrightarrow{G_w}$ is acyclic and has a unique sink (a vertex with no edges pointing away from it), namely the w-valid plane tree produced by the greedy algorithm.*

Exercise 12 Prove Theorem 4. You can do this in three steps:

1. Let T_0 be the w-valid plane tree produced by the greedy algorithm. Prove that no valid local moves of Type 2 can be performed on T_0.
2. Prove that if T is not T_0, a valid local move of Type 2 can be performed on T.
3. Prove that the directed state space graph G_w has no cycles. Think about what a Type 2 move does to the sum

$$\sum_{v \in V(T)} d(v, r),$$

where $d(v, r)$ is the distance from vertex v to the root vertex r in plane tree T.

5   Enumerating Words with Only One Valid Plane Tree

Some words have no valid plane trees, such as $AAA\overline{B}\overline{B}\overline{B}$. On the other hand, there are C_n plane trees with n edges, so any word of length $2n$ can have at most C_n valid plane trees. Try the next exercise to discover words which are C_n-foldable.

Exercise 13 For any $n \in \mathbb{N}$, let $w = (A\overline{A})^n = A\overline{A}\ldots A\overline{A}$ be a word of length $2n$. Prove that every plane tree with n edges is w-valid. Are there other words of length $2n$ which have C_n valid plane trees?

There are also words which have exactly one valid plane tree, such as $AAA\overline{AAA}$, which are called one-foldable. These one-foldable words are the words that we will focus on in this section. Our goal is to characterize and enumerate the words of length $2n$ from alphabet $\mathscr{A}_m$ which have exactly one valid plane tree.

Given a word w, we saw previously that the greedy algorithm will produce a w-valid plane tree if one exists. Based on our knowledge of local moves and connectivity of the state space graph, there is only one w-valid plane tree if and only if the tree produced by the greedy algorithm does not have any valid local moves which can be made.

For a foldable word w, let $T_0(w)$ be the w-valid plane tree produced by the greedy algorithm. In Exercise 12 (proving Theorem 4), we saw that no Type 2 moves are possible in the tree produced by the greedy algorithm. So a local move is available only if we see two edges $e(i, j)$ and $e(i', j')$ of $T_0(w)$, with $i < j < i' < j'$ and $w_i = w_{i'}$ and $w_j = w_{j'}$, which share a common vertex in the plane tree.

Let us look closer at the conditions required for a local move to exist. Notice the distinction between the greedy trees for the words $A\overline{A}AA$ and $A\overline{A}A\overline{A}$. While $\{w_1, w_2\} = \{w_3, w_4\}$ in both cases, a local move is only possible for the second word. The parity of the location of the letters comes into play. Further, for any edge $e(i, j)$ in a plane tree, there is a perfect non-crossing matching of the indices $i + 1, \ldots, j - 1$ which implies there are an even number of values between i and j. As a result, one value in $\{i, j\}$ is even while the other is odd. Stated another way, if i and j have the same parity, even if w_i and w_j are complementary letters, we immediately know that $e(i, j)$ can never be an edge in a w-valid plane tree.

To take advantage of the constraints that result from parity, we introduce the notion of a doubled alphabet:

Definition 12 Let w be a word of length $2n$ from alphabet $\mathscr{A}_m$. Transform w into a new word $\hat{w}$ of length $2n$ and form the *doubled alphabet* $\mathscr{A}_{2m}$ by updating each letter of w independently using the following rules:

- If j is even and $w_j = \overline{A}_i$, then $\hat{w}_j = \overline{A}_i$.
- If j is even and $w_j = A_i$, then $\hat{w}_j = \overline{A}_{m+i}$.
- If j is odd and $w_j = A_i$, then $\hat{w}_j = A_i$.
- If j is odd and $w_j = \overline{A}_i$, then $\hat{w}_j = A_{m+i}$.

The resulting word $\hat{w}$ has the property that $w_i \in \{A_1, \ldots, A_{2m}\}$ if and only if i is odd.

Here is an example of a word w from alphabet $\mathscr{A}_2$ and the resulting word $\hat{w}$ after the transformation using the doubled alphabet:

$$w = A_1 A_1 \overline{A}_1 A_1 A_1 \overline{A}_1 \overline{A}_1 A_1 \overline{A}_1 \overline{A}_1 A_2 \overline{A}_2,$$
$$\hat{w} = A_1 \overline{A}_3 A_3 \overline{A}_3 A_1 \overline{A}_1 A_3 \overline{A}_3 A_3 \overline{A}_1 A_2 \overline{A}_2.$$

Following the transformation, we can clearly identify that the first and third letters cannot form a bond despite being complementary letters in w. Observe that w can be recovered from $\hat{w}$ given the value of m, so the transformation is injective.

Exercise 14 Prove that the number of w-valid plane trees is precisely the number of $\hat{w}$-valid plane trees.

Now consider the edges in the greedy tree $T_0(\hat{w})$. If two edges $e(i, j)$ and $e(i', j')$ share a common vertex, then j and i' have opposite parities because there must be an even number of letters in between corresponding to any edges of the tree between these edges. So if $\{\hat{w}_i, \hat{w}_j\} = \{\hat{w}_{i'}, \hat{w}_{j'}\}$, then the structure of $\hat{w}$ introduced by the doubled alphabet implies that $w_i = w_{i'}$ and $w_j = w_{j'}$. Consequently, we must have an available local move, a statement which was not true before the transformation using the doubled alphabet.

Now let us take it one step further. For T, a valid plane tree for w and for $\hat{w}$, the doubled alphabet induces a $2m$-coloring of the edges of T (which is not necessarily proper). In particular, if edge $e(i, j)$ corresponds to $\{w_i, w_j\} = \{A_k, \overline{A}_k\}$, then use color k for the edge $e(i, j)$. Notice that the tree T together with its edge coloring uniquely determines the word $\hat{w}$ (and w). We give an example in Fig. 9.

Exercise 15 Let T be a plane tree with n edges. Trace back through the transformations to prove that the number of colorings $c : E(T) \to [2m]$ of the edges of T with $2m$ colors is precisely the number of words of length $2n$ from alphabet $\mathscr{A}_m$ for which T is a valid plane tree.

We proved in Sect. 2 that the number of plane trees with n edges is C_n. With $2m$ possible colors for each edge, we see that there are $(2m)^n C_n$ pairs (T, c) where c is an edge coloring of T. Of these, we would like to enumerate the pairs (T, c) which correspond to words with exactly one valid plane tree. We start with an exercise.

Exercise 16 For a foldable word w of length $2n$ from the alphabet $\mathscr{A}_m$, let T be a w-valid plane tree. Consider the corresponding edge coloring of T induced by $\hat{w}$.

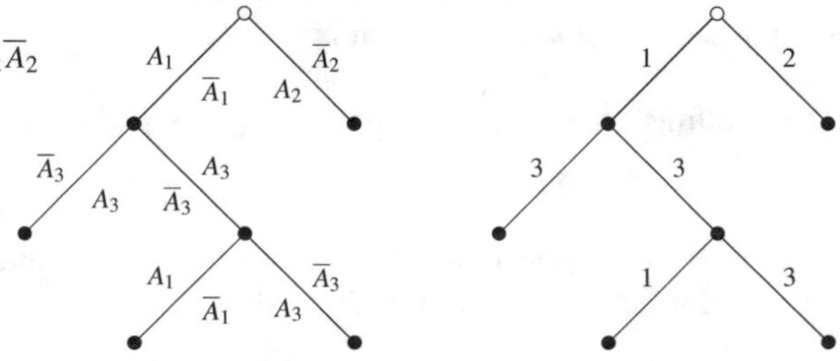

Fig. 9 A folding of $\hat{w} = A_1 \overline{A}_3 A_3 \overline{A}_3 A_1 \overline{A}_1 A_3 \overline{A}_3 A_3 \overline{A}_1 A_2 \overline{A}_2$ and the corresponding edge coloring of the tree

How do valid local moves correspond to the edge coloring? Investigate for several short words.

Setting aside the words w and $\hat{w}$, now we can identify local moves just by looking at the tree and its edge coloring. If two edges share a common vertex and have the same color, then there is a local move on these edges. These lead to our next lemma.

Lemma 1 ([1]) *The words of length $2n$ from alphabet $\mathscr{A}_m$ which are one-foldable are in bijection with the proper $2m$-edge colorings of plane trees with n edges.*

Exact enumerations and asymptotics for the number of one-foldable words for any m and n can be found in [1]. Here, let us pause to consider the case when $m = 1$. Thus we are looking for plane trees with proper 2-edge colorings. This is only possible if the degree of each vertex is at most 2. Since trees are connected, the plane tree must be a path with one vertex identified as the root. There are n such plane trees with n edges and each has exactly 2 proper edge colorings with 2 colors. So there are $2n$ words of length $2n$ from alphabet $\{A, \overline{A}\}$ which have only one valid plane tree.

Exercise 17 For each $n \in \mathbb{N}$, we claim that there are $2n$ words of length $2n$ from alphabet $\{A, \overline{A}\}$ which have only one valid plane tree. See if you can write down these words, starting with small values for n and then try generalizing your findings to larger values of n.

The connection between words and edge-colored plane trees can be useful in other contexts. Consider the following research question:

Research Project 1 Characterize the pairs (T, c) which correspond with words of length $2n$ from $\mathscr{A}_m$ which have exactly two valid plane trees. Use your characterization to enumerate these two-foldable words when $m = 1$ and when $m = 2$. Enumerate two-foldable words of length $2n$ for any m.

6 Enumeration of Valid Plane Trees

In this section, we consider the integers k for which there is a k-foldable word, a word which has exactly k valid plane trees.

Definition 13 For any natural numbers n and m, let $\mathscr{R}(n, m)$ be the set of integers k for which there is a word w of length $2n$ from alphabet $\mathscr{A}_m$ which is k-foldable.

The word $A^{2n} = AA\ldots A$ has no valid plane trees, so $0 \in \mathscr{R}(n, m)$. In Exercise 13, you proved that every plane tree with n edges is valid for the word

$(A\overline{A})^n = A\overline{A}A\overline{A}\ldots A\overline{A}$. Therefore $C_n \in \mathscr{R}(n,m)$ for any $m \in \mathbb{N}$, the maximum possible value in $\mathscr{R}(n,m)$. In the previous section, we also discovered many words which are one-foldable, such as $A^n\overline{A}^n$, so $1 \in \mathscr{R}(n,m)$.

How are the sets $\mathscr{R}(n,m)$ related for various choices of m and n? Since C_n is the maximum value in $\mathscr{R}(n,m)$ for any m, we have $\mathscr{R}(n,m) \neq \mathscr{R}(n+1,m)$. Wagner [18] further proved that $\mathscr{R}(n,m) \subsetneq \mathscr{R}(n+1,m)$. On the other hand, since the alphabet $\mathscr{A}_m$ is a subset of the alphabet $\mathscr{A}_{m+1}$, every word from alphabet $\mathscr{A}_m$ is also a word from $\mathscr{A}_{m+1}$ and we conclude $\mathscr{R}(n,m) \subseteq \mathscr{R}(n,m+1)$. However, this subset relation may be strict as demonstrated by an example when $n = 7$ given by Wagner [18].

We now focus our attention on $\mathscr{R}(n,1)$, the case when our alphabet contains just two letters $\mathscr{A}_1 = \{A, \overline{A}\}$. There are 2^{2n} words of length $2n$ and many of these words have no valid plane trees. Desiring a more thorough understanding of the set $\mathscr{R}(n,1)$, here is our primary question for this section:

Let $n \in \mathbb{N}$. Determine which integers are in the set $\mathscr{R}(n,1)$. In other words, for which $k \in \mathbb{N}$ does there exists a word w from the alphabet $\mathscr{A}_1 = \{A, \overline{A}\}$ with exactly k plane trees which are w-valid?

We approach this question from two different perspectives:

1. Which values are in $\mathscr{R}(n,1)$? In particular, what is the smallest integer not in $\mathscr{R}(n,1)$?
2. Which values are not in $\mathscr{R}(n,1)$?

Some computational work in [1] reveals the following for small values of n. Note that the ellipses indicate that all integers in the range are present in the set.

$$\mathscr{R}(1,1) = \{0,1\};$$
$$\mathscr{R}(2,1) = \{0,1,2\};$$
$$\mathscr{R}(3,1) = \{0,1,2,5\};$$
$$\mathscr{R}(4,1) = \{0,1,2,3,4,5,14\};$$
$$\mathscr{R}(5,1) = \{0,1,2,3,4,5,7,10,14,42\};$$
$$\mathscr{R}(6,1) = \{0,1,\ldots,8,10,12,14,16,18,19,25,28,42,132\}$$
$$\mathscr{R}(7,1) = \mathscr{R}(6,1) \cup \{9,15,20,30,40,43,52,56,70,84,429\}$$

6.1 Small Values in $\mathscr{R}(n,1)$

To show that an integer k appears in $\mathscr{R}(n,1)$, our primary approach is to propose a word w_k of length $2n$ from alphabet $\mathscr{A}_1$ and show that there are exactly k plane trees which are w_k-valid.

We have seen $0, 1, C_n \in \mathcal{R}(n, 1)$. The Catalan numbers are an integral part of $\mathcal{R}(n, 1)$, as first seen by the following exercise:

Challenge Problem 2 For $n \in \mathbb{N}$ with $n \geq 2$, find a word of length $2n$ from the alphabet $\mathscr{A}_1$ which shows that $C_{n-1} \in \mathcal{R}(n, 1)$. Extend your construction to show $C_t \in \mathcal{R}(n, 1)$ for each $1 \leq t \leq n$. (Wagner proves this is possible in Proposition 2.3 of [18].)

Based on the computational results given, you likely noticed that for each n, there exists $\ell \in \mathbb{N}$ such that $\{0, 1, 2, \ldots, \ell\} \subseteq \mathcal{R}(n, 1)$. For example, when $n = 6$, we have $\ell = 8$, and when $n = 7$, we have $\ell = 10$. Our first results toward discovering these values of ℓ can be summarized in the following theorem: Our first results toward discovering these values of ℓ can be summarized in the following theorem:

Theorem 5 ([1]) *For each integer $n \geq 4$, $\{0, 1, \ldots, n\} \subseteq \mathcal{R}(n, 1)$.*

This theorem is the result of a few constructions and lots of careful counting. However, there are many constructions possible and some may even lead to a stronger result, so we leave the following to you.

Challenge Problem 3 For each $i \in \{0, 1, \ldots, n\}$, find a word w_i of length $2n$ from the alphabet $\{A, \overline{A}\}$ such that there are exactly i plane trees which are w_i-valid (i.e., $i \in \mathcal{R}(n, 1)$).[1] (This appears in the proof of Proposition 4.10 in [1].)

Theorem 5 is the current status of the work and we state the following open problem:

Research Project 2 Fix $n \in \mathbb{N}$. What is the largest $\ell \in \mathbb{N}$ such that $\{0, 1, \ldots, \ell\} \subseteq \mathcal{R}(n, 1)$?

6.2 Large Values not in $\mathcal{R}(n, 1)$

You may also notice in the computational data that the largest two values in $\mathcal{R}(n, 1)$ seem to be C_{n-1} and C_n. If this is true, then any word w with more than C_{n-1} valid plane trees will have exactly C_n plane tree which are w-valid. The next theorem confirms our suspicions.

---

[1] For one possible construction, let $1 \leq \ell < n$ and consider the word $w_\ell = \overline{A} A^\ell \overline{A}{}^j A^j \overline{A}{}^\ell A$ where $j = n - 1 - \ell$. For determining the number of w_ℓ-valid plane trees, it may be useful to divide this into two cases: when $j < \ell$ and when $\ell \leq j$.

Theorem 6 ([2]) *Let $n \in \mathbb{N}$ with $n \geq 3$. If ℓ is an integer satisfying $C_{n-1} < \ell <$ C_n, then $\ell \notin \mathcal{R}(n, 1)$.*

Proof Fix $n \in \mathbb{N}$ and an integer ℓ with $C_{n-1} < \ell < C_n$. For contradiction, suppose w is a word of length $2n$ from alphabet $\mathcal{A}_m$ which has exactly ℓ valid plane trees. Transform w into $\hat{w}$ using the doubled alphabet transformation from Definition 12. By Exercise 14, $\hat{w}$ is also ℓ-foldable.

Let m_A be the number of occurrences of A in $\hat{w}$ and let m_B be the number of occurrences of B in $\hat{w}$. Therefore the number of valid plane trees for $\hat{w}$ is at most $C_{m_A} C_{m_B}$ where $m_A + m_B = n$.

In Exercise 3, you proved $C_t \geq C_i C_{t-i}$ for any $t \in \mathbb{N}$ with $0 \leq i \leq t$. Since $\ell \leq C_{m_A} C_{m_B}$, it follows that either $\ell = C_n$ or $\ell \leq C_{n-1} C_1 = C_{n-1}$ which contradicts our initial choice of ℓ.   □

A closer look at the computational data reveals many integers which are not in $\mathcal{R}(n, 1)$. The following theorem extends Theorem 6, exposing other integers that do not appear in $\mathcal{R}(n, 1)$:

Theorem 7 ([1]) *For $n \geq 8$, the largest six integers in $\mathcal{R}(n, 1)$ are*

$$C_{n-2} + C_2 \cdot C_{n-4}, \quad C_{n-2} + C_{n-3}, \quad C_3 \cdot C_{n-3}, \quad C_2 \cdot C_{n-2}, \quad C_{n-1}, \quad \text{and} \quad C_n.$$

Further, these are the largest six integers in $\mathcal{R}(n, m)$ for any integer $m \geq 1$.

This further demonstrates the fundamental connection between the Catalan numbers and the set $\mathcal{R}(n, 1)$. However, for large n, there are many more intervals of integers which are not in $\mathcal{R}(n, 1)$. This leads us to the following research project:

Research Project 3 For natural numbers n and t where $t > 6$, characterize the largest t integers in $\mathcal{R}(n, 1)$. Is there a generalized formula which would generate these?

We summarize Sect. 6 with the following research problem:

Research Project 4 For each $n \in \mathbb{N}$, completely characterize the integers in $\mathcal{R}(n, 1)$.

7 Wobble Pairs and Other Modifications to the Model

The model discussed thus far allows for matchings between complementary letters of the alphabet. This is a generalization of the *Watson–Crick pairs* (A with U and C with G) which bond when a single-strand of RNA folds onto itself. However, for the RNA molecule, there is one more type of bond which can form between nucleotides, a bond between G and U, often called a *wobble pair*. Within the categories "A," "U," "C," and "G," nucleotides can vary, resulting in dozens of additional base pairs [9]. There are even unnatural bases pairs (UBPs) created in labs [14].

Restricting our attention again to words from the alphabet $\{A, U, C, G\}$ and now allowing both Watson–Crick pairs and wobble pairs to form bonds, we have a new model to study and can ask a similar series of questions.

As a natural first question, given a word w of length $2n$ from the alphabet $\{A, U, C, G\}$, how do we decide whether there exists a w-valid plane tree? When we disallowed the $G - U$ bond, we were able to define a greedy algorithm (Definition 7) that would produce a w-valid plane tree if one existed. However, the natural extension of the greedy algorithm for our new setting no longer works. Consider the word $GUAC$. The greedy algorithm would first pair G with U and then it would halt with both A and C in the OUT stack as these cannot bond, declaring that there is no w-valid plane tree. However, we know that there is a valid plane tree for $GUAC$ by pairing G with C and U with A.

Exercise 18 In the alphabet of nucleotides $\{A, C, G, U\}$, does there exist a word that has at least one valid plane tree using only Watson–Crick pairs, but has more valid plane trees if wobble pairs $G - U$ are allowed?

One can always run a brute force algorithm, checking all possible plane trees to determine if one is w-valid, but the number of trees to examine grows exponentially with the length of w, making it an unfeasible option for long words. On the other hand, the greedy algorithm (Definition 7) for only the Watson–Crick pairs only considers each letter in the IN stack once and hence runs in linear time. This brings us to a project:

Research Project 5 For $n \in \mathbb{N}$, let w be an arbitrary word from $\{A, U, C, G\}^{2n}$. Give an algorithm, whose run-time is polynomial in n, to determine if w has at least one valid plane tree within the model where we allow bonds $A - U$, $C - G$, and $G - U$.

Ideally, an algorithm given in response to the previous project would output a w-valid plane tree if one exists, but this is not a requirement. To make small steps toward this bigger question, think about some properties which are quickly checked and may immediately reveal that a word has no valid plane tree. Let w be a word from $\{A, U, C, G\}^{2n}$ and, for each $k \in \{A, U, C, G\}$, let $n(k) = \{i : w_i = k\}$. If w is foldable, then $n(G) \leq n(C) + n(U)$, but this condition is not enough to guarantee w is foldable. Additionally, if w is foldable and $n(A) = 0$, what else must be true?

Exercise 19 Write a series of necessary conditions in order for word w in $\{A, U, C, G\}^{2n}$ to have at least one valid plane tree, allowing for wobble pairs. Are your conditions (individually or collectively) sufficient for characterizing when w is foldable?

As another observation, for each w-valid plane tree, there exists at least one $i \in \{1, \ldots, 2n - 1\}$ such that $w_i w_{i+1}$ form an edge in the tree. So, a word with at least one valid plane tree must have a consecutive pair of letters which can bond. For example, the word $w = GGUU$ has a valid plane tree in which $w_2 w_3$ forms a bond. However, for $w' = GUAC$, the consecutive pair $w'_1 w'_2$ can form a bond, but this bond does not appear in any w'-valid plane tree. The only consecutive pair which is bonded in a w'-valid plane tree is $w'_2 w'_3$. The next open-ended question seeks to quantify this.

Exercise 20 Look at some short words w and the set of w-valid plane trees for each. Are there consecutive letters that form a bond in all of the valid plane trees? Are there consecutive letters that never form a bond? Make a conjecture regarding the bonds that form among consecutive pairs in valid plane trees, then test your conjectures on longer words.

Here are some other questions about our new model which are analogous to results we discussed in for our original complementary alphabet model:

Research Project 6 Let w be a word in $\{A, C, G, U\}^{2n}$ and consider the model in which both Watson–Crick pairs and wobble pairs can form bonds.

1. Define a local move on valid plane trees analogous to the one introduced in Definition 10. Is the restricted state space graph G_w connected?
2. Characterize the words which have exactly one valid plane tree.
3. Prove that the number of w-valid plane trees is at most C_n.
4. For which $k \in \{0, 1, \ldots, C_n\}$ does there exist a word of length $2n$ which has exactly k valid plane trees?

Our exploration thus far has been motivated by bonds formed between nucleotides in RNA, bonds that have been observed in nature. However, one could consider other alphabets and rules for which pairs of letters can form bonds. Some may be more interesting than others, but you will never know until you explore. The possibilities are limitless. Here is one you might consider.

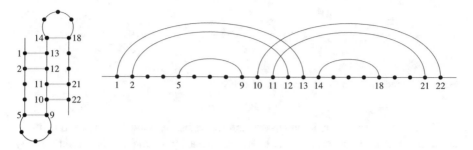

Fig. 10 A pseudoknot corresponds to a crossing matching, and the minimum distance between paired nucleotides is three base pairs

Challenge Problem 4 Consider an alphabet with 4 letters in which each letter can form a bond with each other letter (but not with itself). The greedy algorithm still does not work here. Give an example to demonstrate this. Can you write a polynomial algorithm to output a valid plane tree if one exists? The existence of such an algorithm is an open problem.

Another important physical reality of a folding RNA that we have thus far ignored is the fact that two nucleotides must be separated by at least three others to form a bond [20]. The structure formed by a bonded pair and the unbonded nucleotides between them in the sequence is called a *hairpin loop*. Because a hairpin loop is not required to form any bonds at all, it is impossible for every nucleotide to be paired off; a strand of RNA can never form a *perfect* matching. Other nucleotides might also remain unpaired as RNA folds.

The imperfect matching formed by a strand of RNA in real life might even include crossings, called *pseudoknots* like the one shown in Fig. 10. You can find more information about pseudoknots in [16].

Challenge Problem 5 Consider the word $w = A\overline{A}B\overline{B}\overline{B}A\overline{A}B\overline{B}A\overline{A}$ from Exercise 11.

1. Give all non-crossing matchings of w that have at least three unpaired letters between any two paired letters.
2. Can you arrange those matchings into a state space graph?
3. How many matchings of w have at least three unpaired letters between any two paired letters if we include those matchings with pseudoknots?

Research Project 7 Explore! Fix a set of alphabet letters and a set rules for which letters form bonds. See what you can discover about your new model.

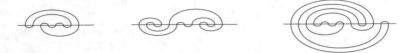

Fig. 11 Meanders of orders 4, 6, and 7 respectively

Since we are mathematicians, we cannot resist leaving you with one more combinatorial structure which has been around a long time but is surrounded by open questions which remain elusive to mathematicians. Some of the tools of this chapter, particularly local moves, may open the door for new insights.

Definition 14 For a natural number n, a *meander* of order n is a pair of size n non-crossing perfect matchings, with one drawn above the x-axis and one drawn below the x-axis, which together form a single closed curve (Fig. 11).

The study of meanders can be traced back to the work of Poincaré [15], however, the enumeration of meanders of size n remains an open problem. Recent efforts have been made via mixing time arguments for Markov chains. These stem from a study of local moves on meanders. The local move that we learned about for plane trees (Definition 8) can be restated as a local move on non-crossing perfect matchings via the bijection you found in Exercise 7. We can then define a local move on a meander by applying one local move to the non-crossing perfect matching above the x-axis and another local move to the matching below the x-axis, provided the outcome is still a meander. These are precisely the local moves studied for meanders in [6]. Could variations on these local moves, inspired by RNA, yield more insight toward a hundred-year-old question in pure math: the enumeration of meanders?

8 Numerical Models of RNA Folding

Everything we have talked about so far falls under the umbrella of mathematics *inspired by* molecular biology, but, as that phrase implies, its origins are in the realm of mathematics *to further* molecular biology. If, having read this far, you still have a nagging feeling that these mathematical abstractions are ignoring the overwhelming biological complexity of the foundational questions, this section will give you some directions to explore. If you are intrigued by the combinatorial elegance of the abstractions, this section will direct you to some additional biological properties that may inspire you to ask new questions. For a broader overview of the landscape of these numerical models, [4] is a good place to start.

$$(((((((((..((((((.........)))).(((((.......)))))......((((((.......))))))))))))))....$$

Fig. 12 The secondary structure of E. coli tRNA molecule tdbD00000067 [10]

8.1   Structures of RNA

While DNA encodes an organism's blueprint in the familiar double helix, the molecules carrying out those instructions are the single-stranded RNA. These shorter snippets fold on themselves, with base pairs bonding and shifting to make functional molecules. What any given strand of RNA will do depends not just on its sequence of nucleotides, but on the ways in which those nucleotides bond with each other and how they orient themselves in three dimensional space. The three levels of complexity are referred to as the *primary* (nucleotide sequences), *secondary*, and *ternary* structures. In this sense, a w-valid plane tree is a secondary structure for the primary structure given by word w. The secondary structure can even be represented as a sequence of matched parentheses indicating bonded pairs (similar to Sect. 2), and dots indicating unbonded nucleotides (Fig. 12).

Like a word w may have many w-valid plane trees, a particular RNA primary structure may have many secondary and ternary structures it can exhibit. These are expensive, in terms of both time and money, to observe experimentally. Moreover the secondary and ternary structures of RNA can be relatively unstable: how it is folded can change. The primary structure, on the other hand, is relatively cheap and easy to observe and therefore modeling the way that primary structures can take on secondary and ternary structures is an important and active area of combinatorial molecular biology. The text by Waterman [19] and the book edited by Patcher and Sturmfels [12] are good places for curious readers to continue exploring the mathematical techniques researchers use to study computational biology, beyond those mentioned in [4].

8.2   The RNA Folding Problem

The RNA folding problem seeks to accurately model the secondary, and eventually ternary structures of RNA strands based only on the sequence of base pairs. Many of the numerical models for predicting the secondary structure of RNA rely on the thermodynamic theory that molecules tend toward states with the least free energy. These *minimal free energy models* assign an energy value to every substructure of a folded strand of RNA, and then use numerical methods to identify structures that minimize free energy and thus are deemed most likely to occur. Even though these models use thousands of parameters, most of them experimentally determined, they can still struggle to correctly predict how even the shortest, simplest strands of RNA, transfer RNA, will fold [13, 20].

Nevertheless, transfer RNA (tRNA) is a good place to start if you want to look at some real RNA sequences. A tRNA sequence is relatively short, about 75

Fig. 13 The secondary
structure of a tRNA molecule

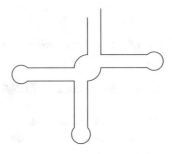

nucleotides long, and both the secondary and ternary structures are known, as is their biological function. Transfer RNA molecules carry the amino acids that are specified by the codons in DNA and other forms of RNA. To read and execute these instructions, tRNA molecules all have closely related secondary structures, illustrated in Fig. 13. Check out the over 10,000 tRNA sequences in the tRNA Database, hosted jointly by the Universities of Leipzig, Marburg, and Strasbourg [9, 10].

8.3   The Inverse RNA Folding Problem

The inverse problem to modeling how a particular strand of RNA will fold is to start with a particular secondary structure and ask what sequences of RNA can fold into that shape. Some of the work of finding these sequences has been crowdsourced. The game Eterna, which you can play at eternagame.org, lets players build their own RNA sequences and uses a folding model to determine what secondary structure the artificial sequence will form. The goal of this project is to find RNA sequences that, because of their secondary structure, can help detect and cure illness [11].

Several other programs are available to computationally find sequence that fold into a particular shape. These tend to be somewhat less user-friendly, but given a secondary structure, they can return possible sequences that could form it. A recent survey details the strengths and weaknesses of five other programs: RNAinverse, RNAiFold, AntaRNA, NUPACK, and IncaRNAfbinv [3].

The first step to investigating the numerical models for the folding and inverse folding problems is to figure out how to work with the programs that already exist.

Challenge Problem 6 Explore the capabilities of one of these folding or inverse folding programs. Necessary steps will include:

1. Download the appropriate files from github or another repository. A link to each of the five inverse folding programs is in [3]. Two folding programs are cited directly in the references of this paper [13, 20].
2. Follow the steps in the README file to install the program on your school or personal computer.
3. Determine what inputs the program takes and what outputs it can give you.

If you have successfully gotten one or more of these programs running and are interested in the intricacies of the model, you can check out the associated research papers. From there you can formulate your own questions or contact the authors of the program you are interested in to see if there are specific problems you can study.

9 Conclusion

We leave you with the broad strokes of this story: how mathematical molecular biology led to new combinatorial problems. An RNA molecule binds to itself forming a structure which determines its functionality. In modeling this bonded structure, biologists discovered that the molecule forms a tree-like structure. Mathematicians model these structures as plane trees and these simplified models led to the new combinatorial objects, valid plane trees, that we discussed in this chapter.

With the exception of the numerical models in Sect. 8, everything in this chapter could be studied purely as an extension of pre-existing combinatorial objects. Planes trees and non-crossing perfect matchings have been around a long time. But our new spin with wrapping words around plane trees has led us to focus on just those trees which are valid for a particular word. This opens up a door to a wide range of combinatorial questions, including:

1. Which words have at least one valid plane tree? (Theorem 2)
2. What does the space of all w-valid plane trees look like? (Sect. 4)
3. Which words have only one valid plane tree? (Sect. 5)
4. Given a $k \in \mathbb{N}$, is there a word which has exactly k valid plane trees? (Sect. 6)

As we saw in Sect. 7, new pure math research questions can be found by adding more of the biological constraints, but there is no reason to stop there! Follow your curiosity as a mathematician and consider other alphabets and other rules for how bonds can form. Or perhaps this chapter has served as a springboard for you to learn more about the biological study of RNA folding as you explore the numerical models in Sect. 8.

We have posed a number of research directions, but do not limit yourself to these. Explore examples, look for patterns, and ask questions. The possibilities are endless! We would love to hear about your findings as you delve deeper into the topics discussed in this chapter.

References

1. Beth Bjorkman, Garner Cochran, Wei Gao, Lauren Keough, Rachel Kirsch, Mitch Phillipson, Danny Rorabaugh, Heather Smith, and Jennifer Wise. k-foldability of words. *Discrete Applied Math.*, 259:19–30, 2019.
2. Francis Black, Elizabeth Drellich, and Julianna Tymoczko. Valid plane trees: combinatorial models for RNA secondary structures with Watson-Crick base pairs. *SIAM J. Discrete Math.*, 31(4):2586–2602, 2017.

3. Alexander Churkin, Matan D. Retwitzer, Vladimir Reinharz, Yann Ponty, Jérôme Waldispühl, and Danny Barash. Design of RNAs: comparing programs for inverse RNA folding. *Briefings in Bioinformatics*, 19(2):350–358, 2018.

4. Qijun He, Matthew Macauley, and Robin Davies. RNA Secondary Structures: Combinatorial Models and Folding Algorithms. In Raina Robeva, editor, *Algebraic and Discrete Mathematical Methods for Modern Biology*, chapter 13. Academic Press, 2015.

5. Christine E. Heitsch. Combinatorics on plane trees, motivated by RNA secondary structure configurations. *preprint*, 2006.

6. Christine E. Heitsch and Prasad Tetali. Meander Graphs. *Discrete Mathematics & Theoretical Computer Science*, DMTCS Proceedings vol. AO, 23rd International Conference on Formal Power Series and Algebraic Combinatorics (FPSAC 2011), January 2011.

7. P. Hilton, D. Holton, and J. Pedersen. *Mathematical Vistas: From a Room with Many Windows*. Undergraduate Texts in Mathematics. Springer New York, 2013.

8. OEIS Foundation Inc. The on-line encyclopedia of integer sequences, http://oeis.org, 2019.

9. Frank Jühling, Mario Mörl, Mathias Sprinzl, Peter F. Stadler, Roland K. Hartmann, and Joern Pütz. tRNAdb 2009: compilation of tRNA sequences and tRNA genes. *Nucleic Acids Research*, 37(Suppl 1):D159–D162, 10 2008.

10. Frank Jühling, Mario Mörl, Mathias Sprinzl, Peter F. Stadler, Roland K. Hartmann, and Joern Pütz. Transfer RNA database, http://trna.bioinf.uni-leipzig.de/DataOutput, 2019.

11. Jeehyung Lee, Wipapat Kladwang, Minjae Lee, Daniel Cantu, Martin Azizyan, Hanjoo Kim, Alex Limpaecher, Snehal Gaikwad, Sungroh Yoon, Adrien Treuille, Rhiju Das, and EteRNA Participants. RNA design rules from a massive open laboratory. *Proceedings of the National Academy of Sciences*, 111(6):2122–2127, 2014.

12. Bernd Sturmfels Lior Pachter, editor. *Algebraic Statistics for Computational Biology*. Cambridge University Press, 2005.

13. Amrita Mathuriya, David A. Bader, Christine E. Heitsch, and Stephen C. Harvey. Gtfold: A scalable multicore code for RNA secondary structure prediction. In *Proceedings of the 2009 ACM Symposium on Applied Computing*, SAC '09, pages 981–988, New York, NY, USA, 2009. ACM.

14. Takashi Ohtsuki, Michiko Kimoto, Masahide Ishikawa, Tsuneo Mitsui, Ichiro Hirao, and Shigeyuki Yokoyama. Unnatural base pairs for specific transcription. *Proceedings of the National Academy of Sciences*, 98(9):4922–4925, 2001.

15. H. Poincaré. Sur un théorème de géométrie. *Rendiconti del Circolo Matematico di Palermo (1884–1940)*, 33(1):375–407, 1912.

16. Christian Reidys. *Combinatorial Computational Biology of RNA: Pseudoknots and Neural Networks*. Springer, New York, NY, 2011.

17. R.P. Stanley. *Catalan Numbers*. Cambridge University Press, 2015.

18. Zsolt Adam Wagner. On some conjectures about combinatorial models for RNA secondary structures. *preprint*: https://arxiv.org/abs/1901.10238, 2019.

19. Micahel S. Waterman. *Introduction to Computational Biology*. Chapman and Hall/CRC, New York, 1995.

20. Tianbing Xia, John SantaLucia, Mark E. Burkard, Ryszard Kierzek, Susan J. Schroeder, Xiaoqi Jiao, Christopher Cox, and Douglas H. Turner. Thermodynamic parameters for an expanded nearest-neighbor model for formation of RNA duplexes with Watson-Crick base pairs. *Biochemistry*, 37(42):14719–14735, 1998. PMID: 9778347.

Phylogenetic Networks

Elizabeth Gross, Colby Long, and Joseph Rusinko

Abstract

Phylogenetics is the study of the evolutionary relationships between organisms. One of the main challenges in the field is to take biological data for a group of organisms and to infer an evolutionary tree, a graph that represents these relationships. Developing practical and efficient methods for inferring phylogenetic trees has led to a number of interesting mathematical questions across a variety of fields. However, due to hybridization and gene flow, a phylogenetic network may be a better representation of the evolutionary history of some groups of organisms. In this chapter, we introduce some of the basic concepts in phylogenetics and present related research projects on phylogenetic networks that touch on areas of graph theory and abstract algebra. In the first section, we describe several open research questions related to the combinatorics of phylogenetic networks. In the second, we describe problems related to understanding phylogenetic statistical models as algebraic varieties. These problems fit broadly in the realm of algebra, but could be more accurately classified as problems in *algebraic statistics* or *applied algebraic geometry*.

E. Gross
University of Hawai'i at Mānoa, Honolulu, HI, USA
e-mail: egross@hawaii.edu

C. Long (✉)
The College of Wooster, Wooster, OH, USA
e-mail: clong@wooster.edu

J. Rusinko
Hobart and William Smith Colleges, Geneva, NY, USA
e-mail: rusinko@hws.edu

© Springer Nature Switzerland AG 2020
P. E. Harris et al. (eds.), *A Project-Based Guide to Undergraduate Research in Mathematics*, Foundations for Undergraduate Research in Mathematics,
https://doi.org/10.1007/978-3-030-37853-0_2

Suggested Prerequisites An introductory course in graph theory or discrete mathematics for the research projects in Sect. 2. For the projects in Sect. 3, an introductory course in abstract algebra would also be helpful.

1   Introduction

The field of phylogenetics is concerned with uncovering the evolutionary relationships between species. Even before Darwin proposed evolution through variation and natural selection, people used family trees to show how individuals were related to one another. Since Darwin's theory implies that all species alive today are descended from a common ancestor, the relationships among any group of individuals, even those from different species, can similarly be displayed on a phylogenetic tree. Thus, the goal of phylogenetics is to use biological data for a collection of individuals or species, and to infer a tree that describes how they are related. In modern phylogenetics, the biological data that we consider is most often the aligned DNA sequences for the species under consideration. Understanding how species have evolved has important applications in evolutionary biology, species conservation, and epidemiology [36].

Perhaps unsurprisingly, there is a rich interplay between phylogenetics and mathematics. A tree can be viewed as a certain type of graph, and graph theory is an entire field of mathematics dedicated to understanding the structure and properties of graphs. Similarly, DNA mutation is a random process, and understanding random processes falls in the domain of probability and statistics. As such, there are many mathematical tools that have been developed for doing phylogenetic inference. Often, developing a new tool or trying to answer a novel question in phylogenetics requires solving some previously unsolved mathematical problem. It is also common for a phylogenetic problem to suggest a mathematical problem that is interesting in its own right.

The outline above, where every set of species is related by a phylogenetic tree, is a simplified description of the evolutionary process. Rarely does the evolutionary history for a set of species neatly conform to this story. Instead, species hybridize and swap genes. Moreover, genes within individuals have their own unique evolutionary histories that can differ from that of the individuals in which they reside [29, 44]. The result is that in many cases, a tree is simply insufficient to represent the evolutionary process. Recognizing this, many researchers have argued that networks can be a more appropriate way to represent evolution. While using networks might be more realistic from a biological standpoint, there are many complexities and new mathematical questions that must be solved in order to infer phylogenetic networks. In particular, understanding inference for networks requires proving results for networks analogous to those known for trees. The projects that we present in this chapter are examples of some of the new lines of inquiry inspired by using networks in phylogenetics.

The first category of problems that we describe concern the combinatorics of phylogenetic networks. Inferring phylogenies for large sets of species can often be computationally intensive regardless of the method chosen. One approach for dealing with this in the tree setting is to consider small subsets of species one at a time. Once phylogenetic trees have been built for each subset, the small trees are then assembled to construct the tree for the entire set of species. The details of actually doing this can of course become quite complicated. Thus, different heuristics and algorithms have been proposed, and understanding their performance and properties leads to a number of interesting questions about the combinatorics of trees. As a first example, one might consider if it is even possible to uniquely determine the species tree for a set of species only from knowledge of how each subset of a certain size is related. Even if this is possible, one then might like to know how to resolve contradictions between subtrees if there is error in the inference process. Adopting a similar strategy for inferring phylogenetic networks from subnetworks leads to a host of similar combinatorial questions about networks. In Sect. 2, we will explore the structure of phylogenetic networks in greater depth and formulate some of these questions more precisely for potential research projects.

The second class of problems we discuss concerns the surprising connections between abstract algebra and phylogenetics. One of the ways that researchers have sought to infer phylogenies is by building models of DNA sequence evolution on phylogenetic trees. Once the tree parameter is chosen, the numerical parameters of the model control the rates and types of mutations that can occur as evolution proceeds along the tree. Once all the parameters for the model are specified, the result is a probability distribution on DNA site-patterns. That is, the model predicts the frequency with which different DNA site-patterns will appear in the aligned DNA sequences of a set of species. For example, the model might predict that at the same DNA locus for three species, there is a 5% chance that the DNA nucleotide A is at that locus in each species. Another way to write this is to write that for this choice of parameters, $p_{AAA} = .05$. Algebra enters the picture when we start to consider the algebraic relationship between the predicted sight pattern frequencies. For example, we might find that for a particular model on a tree T, no matter how we choose the numerical parameters the probability of observing ACC under the model is always the same as the probability of observing GTT. We can express this via the polynomial relationship $p_{ACC} - p_{GTT} = 0$, and polynomials that always evaluate to zero on the model we call *phylogenetic invariants* [8] for the model on T.

The set of all phylogenetic invariants for a model is an algebraic object called an *ideal*. By studying the ideals and invariants associated to phylogenetic models, researchers have been able to prove various properties of the models, such as their dimension and whether or not they are identifiable, as well as to develop new methods for phylogenetic inference (see, e.g., [2, 7, 9, 34]). As with some of the combinatorial questions above, there are a number of papers studying these questions in the case of trees, but few in the case of networks. In Sect. 3, we show how to associate invariants and ideals to phylogenetic networks and describe several related research projects. While there are fascinating connections between these algebraic objects and statistical models of DNA sequence evolution, our presentation distills some of the background material and emphasizes the algebra. There is also a computational algebra component to some of these projects and we provide example computations with Macaulay2 [17] code.

2   Combinatorics of Phylogenetic Networks

In this section, we give the background necessary to work on the research and challenge questions related to the structure of phylogenetic networks. We begin by introducing some of the concepts from graph theory necessary to formally define a phylogenetic tree and a phylogenetic network. We then discuss some ways to encode trees and networks and common operations that we can perform on them. Much of the terminology around trees and graphs is standard in graph theory, and so we have omitted some of the basic definitions that can be found in the first chapter of any text on the subject. An affordable and helpful source for more information and standard graph theoretic results would be [10]. The terms that are specific to phylogenetic trees and networks we have adapted largely from [16, 35]. The textbook [22] provides a thorough introduction to phylogenetic networks, though the specific terminology being used in the research literature is still evolving. A broader introduction to phylogenetics from a mathematical perspective can be found in [39].

2.1   Graphs and Trees

The outcome of a phylogenetic analysis is typically a phylogenetic tree, a graph that describes the ancestry for a set of taxa. As an example, an interactive phylogenetic tree relating hundreds of different species can be accessed at https://itol.embl.de/itol.cgi.

In mathematical terms, a *tree* is a connected graph with no cycles. We refer to the degree one vertices of a tree as the *leaves* of the tree. The leaves correspond to the extant species for which we have data in a phylogenetic analysis and so we label these vertices by some label set. In theoretical applications, the label set for an n-leaf tree is often just the set $[n] := \{1, \ldots, n\}$, and we call such a tree an n *-leaf phylogenetic tree*. Note that we consider two n-leaf phylogenetic trees to be distinct even if they differ only by the labeling of the leaves. In technical terms, two n-leaf trees are the same if and only if there is a graph isomorphism between them that also preserves the leaf-labeling.

We often distinguish one special vertex of an n-leaf phylogenetic tree which we call the *root*. If the root is specified, then the tree can be regarded as a directed graph, where all edges point away from the root. The root corresponds to the common ancestor of all of the species of the tree, hence, the directed edges can be thought of as indicating the direction of time. We also often restrict the set of trees we consider to those that are *binary*. A binary tree is one in which every vertex other than the root has degree one or degree three. If the root is specified for a binary tree, then it will have degree two. We use these rooted binary phylogenetic trees as a model of

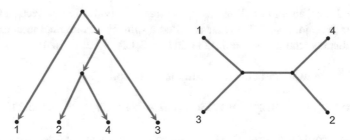

Fig. 1 A rooted 4-leaf binary phylogenetic tree and the tree obtained by unrooting this tree

evolution. The degree three internal vertices correspond to speciation events, where there is one species at the time just prior to the vertex, and two species that emerge from the vertex.

Depending on the application, it is common in phylogenetics to consider both rooted and unrooted trees. As such, we can think about *rooting* a tree, where we place a degree two vertex along an edge and direct the edges away from this vertex (so that there is a directed path from the root to every vertex in the graph). Or, we can think about *unrooting* a tree, where we suppress the degree two vertex (see the trees in Fig. 1). As an example, there is only one 3-leaf binary phylogenetic tree, however, there are three different rooted 3-leaf binary phylogenetic trees that can be obtained by rooting along the three different edges of the unrooted tree.

Example 1 Figure 1 shows a rooted 4-leaf binary phylogenetic tree and the tree obtained by unrooting this tree. Notice that the edges of the rooted tree are directed, but that this is unnecessary since the root determines the direction of each edge. Also observe that if we root the unrooted tree along the edge labeled by 1, we obtain the rooted tree at left. Finally, notice that swapping the labels 1 and 3 in the rooted tree produces a distinct rooted 4-leaf binary phylogenetic tree, whereas for the unrooted tree, swapping these labels leaves the tree unchanged.

Exercise 1 How many edges are there in an n-leaf rooted binary phylogenetic tree?

Exercise 2 Prove that there exists a unique path between any pair of vertices in a tree.

Exercise 3 Prove that the number of rooted binary phylogenetic n-trees is $(2n - 3)!!$ Here the symbol !! does not mean the factorial of the factorial, but rather multiplying by numbers decreasing by two. For example, $7!! = 7 \times 5 \times 3 \times 1 = 105$ and $10!! = 10 \times 8 \times 6 \times 4 \times 2 = 3840$.

Each edge of an unrooted phylogenetic tree subdivides the collection of leaves into a pair of disjoint sets. This pair is called a split. For example, the unrooted tree in Fig. 1 displays the splits $S = \{1|234, 2|134, 3|124, 4|123, 13|24\}$.

Exercise 4 Draw an unrooted tree with the set of splits

$S=\{1|23456, 2|13456, 3|12456, 4|12356, 5|12346, 6|12345, 13|2456, 135|246, 46|1235\}$.

Challenge Problem 1 Prove that two unrooted phylogenetic trees are isomorphic if and only if they display the same set of splits.

2.2   Phylogenetic Networks

As mentioned in the introduction, a tree might not always be sufficient to describe the history of a set of species. For example, consider the graphs depicted in Fig. 3. Notice that there are vertices in these graphs with in-degree two and out-degree one. There are a few ways that we might interpret these *reticulation events*. It could be that two distinct species entered the vertex, and only one, their hybrid, emerged. Or, it might be that one of the edges directed into the degree two vertices represents a gene flow event where species remain distinct but exchange a small amount of genetic material. If we undirect all of the edges of either of these graphs, the result is clearly not a tree since the resulting undirected graph contains a cycle. In fact, this is a phylogenetic network. A more thorough introduction to phylogenetic networks than we offer here can be found in [22, 31]. The website "Who's who in phylogenetic networks" [1] is also an excellent resource for discovering articles and authors in the field.

Definition 1 A phylogenetic network N on a set of leaves $[n]$ is a rooted acyclic directed graph with no edges in parallel (i.e., no multiple edges) and satisfying the following properties:

 (i) The root has out-degree two.
 (ii) The only vertices with out-degree zero are the leaves $[n]$ and each of these have in-degree one.
 (iii) All other vertices either have in-degree one and out-degree two, or in-degree two and out-degree one.

In the preceding definition, the term acyclic refers to the fact that the network should contain no *directed* cycles. The vertices of in-degree two are called the *reticulation vertices* of the network since they correspond to reticulation events. Likewise, the edges that are directed into reticulation vertices are called *reticulation edges*. Observe that the set of rooted binary phylogenetic trees is the subset of the set of phylogenetic networks.

Exercise 5 For what $m \in \mathbb{N}$ is it possible to draw a rooted 3-leaf phylogenetic network with exactly m edges?

Exercise 6 Show that there are an infinite number of rooted n-leaf phylogenetic networks.

The ability of phylogenetic networks to describe more complicated evolutionary histories comes at a cost in that networks can be much more difficult to analyze. Since there are infinitely many phylogenetic networks versus only finitely many phylogenetic trees, selecting the best network to describe a set of species is particularly challenging. Because there are so many networks, it is often desirable to consider only certain subclasses of phylogenetic networks depending on the particular application. One way to restrict the class of networks is by considering only networks with a certain number of reticulations or those of a certain *level*. The concept of the level of a network, introduced in [25], relies on the definition of a *biconnected component* of a graph.

Definition 2 A graph G is *biconnected* (or 2-connected) if for every vertex $v \in V(G)$, $G - \{v\}$ is a connected graph. The *biconnected components* of a graph are the maximal biconnected subgraphs.

Definition 3 The *reticulation number* of a phylogenetic network is the total number of reticulation vertices of the network. The *level* of a rooted phylogenetic network is the maximum number of reticulation vertices in a biconnected component (considered as an undirected graph) of the network.

Exercise 7 What is the reticulation number of the two phylogenetic networks pictured in Fig. 2? What are the biconnected components of each network? What are the levels of the two networks?

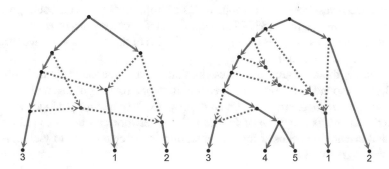

Fig. 2 Two rooted binary phylogenetic networks

Exercise 8 Suppose that you remove one reticulation edge from each pair of edges directed into a reticulation vertex in a phylogenetic network. Show that if you undirect the remaining edges, the result is a connected, acyclic graph.

Exercise 9 How many 3-leaf rooted phylogenetic networks with a single reticulation vertex are there?

2.3   Semi-Directed Networks

Whether we work with rooted or unrooted phylogenetic trees or networks depends upon the particular application. As an example, for some statistical models of DNA sequence evolution, the models for two distinct rooted trees will be the same if the trees are the same when unrooted [39, Chapter 7]. Thus, when working with such models, there is no basis for selecting one location of the root over any other, and so we only concern ourselves with unrooted trees.

Rooting a tree is one way of assigning a direction to each of its edges. When constructing evolutionary models associated to phylogenetic networks it can occur that the direction of some edges can be distinguished by the model, but that the directions associated to other edges cannot. Thus it makes sense to consider the class of unrooted networks in which some of the edges are directed which are known as *semi-directed* networks.

For certain algebraic models of evolution, the models will not necessarily be the same if the unrooted phylogenetic network parameters are the same. However, they will be if the underlying *semi-directed topology* of the networks is the same.

Definition 4 The *semi-directed topology* of a rooted phylogenetic network is the semi-directed network obtained by unrooting the network and undirecting all non-reticulation edges.

Because of the increasing importance of networks in phylogenetics, several authors have investigated the combinatorics of both rooted and unrooted phylogenetic networks (e.g., [16,23,33]). The semi-directed topology has recently appeared in some applications [18,38], but the combinatorics of these networks have received comparatively little attention.

Of course, the semi-directed networks that we are interested in are those that actually correspond to the semi-directed topology of a rooted phylogenetic network, which we call *phylogenetic semi-directed networks*. An edge in a phylogenetic semi-directed network is a *valid root location* if the network can be rooted along this edge and orientations chosen for the remaining undirected edges to yield a rooted phylogenetic network.

Example 2 Figure 3 shows three semi-directed networks. The 3-leaf semi-directed network is a phylogenetic semi-directed network, which can be seen by noting that it is the semi-directed topology of the 3-leaf network in Fig. 2. The 4-leaf semi-

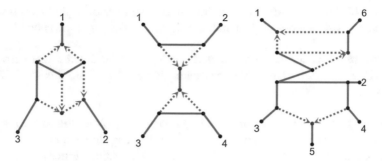

Fig. 3 A 3-leaf phylogenetic semi-directed network, a 4-leaf semi-directed network that is not phylogenetic, and a 6-leaf phylogenetic semi-directed network

Fig. 4 The semi-directed network referenced in Exercise 14

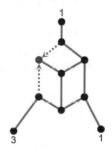

directed network is not a phylogenetic semi-directed network. Notice that there is no way to orient the edge connecting the reticulation vertices without creating vertices of in-degree 3 and out-degree 3, violating the conditions of Definition 1. The 6-leaf network is also a phylogenetic semi-directed network (Exercise 11).

Exercise 10 Find all of the valid root locations for the 3-leaf phylogenetic semi-directed network in Fig. 3.

Exercise 11 Show that the 6-leaf semi-directed network in Fig. 3 is a phylogenetic semi-directed network. Find all of the valid root locations.

Exercise 12 Draw the semi-directed topology of the 5-leaf rooted phylogenetic network in Fig. 3.

Exercise 13 How many 4-leaf semi-directed networks with a single reticulation are there?

Exercise 14 Show that there is no way to direct any of the existing undirected edges in the semi-directed network in Fig. 4 to obtain a phylogenetic semi-directed network.

Exercise 15 Find a formula for the reticulation number of a phylogenetic semi-directed network in terms of the number of leaves and edges of the network.

Exercise 16 Consider the semi-directed topology of the rooted 5-leaf network in Fig. 2. How many different rooted phylogenetic networks share this semi-directed topology?

Challenge Problem 2 Prove or provide a counterexample to the following statement. It is impossible for two distinct phylogenetic semi-directed networks to have the same unrooted topology and the same set of reticulation vertices (i.e., to differ only by which edges are the reticulation edges).

As a hint for this challenge problem, consider the 3-leaf phylogenetic semi-directed network in Fig. 3. Two of the reticulation vertices are incident to leaf edges in the network. As a first step, it may be helpful to consider whether or not there is any way to reorient the edges into one of these vertices so that it is still a reticulation vertex and so that the network remains a phylogenetic semi-directed network.

Challenge Problem 3 Find an explicit formula for the number of semi-directed networks with a single reticulation vertex and n leaves.

> **Research Project 1** Find an explicit formula for the number of level-1 semi-directed networks with n leaves and m reticulation vertices. Can you generalize this formula to level-k networks with n leaves and m reticulation vertices?

For Research Project 1, the level of a semi-directed network is defined in terms of the unrooted, undirected topology just as for phylogenetic networks. Thus, any phylogenetic network and its semi-directed topology will have the same level. A starting point would be to look for patterns in small families of trees or networks. To begin thinking about proof techniques you might examine the proofs of the number of rooted trees with n leaves, or perhaps the number of distinct unlabeled tree topologies with n leaves. Chapter three of Felsenstein's book *Inferring Phylogenies* provides some intuition about tree counting [15].

Challenge Problem 4 Find necessary and sufficient conditions for a semi-directed network to be a phylogenetic semi-directed network.

Fig. 5 The rooted phylogenetic network corresponding to the adjacency matrix in Example 3

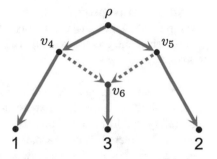

> **Research Project 2** Determine a method or algorithm for counting valid root locations in a phylogenetic semi-directed network (i.e., count the number of rooted networks corresponding to a particular semi-directed network).

By definition, a phylogenetic semi-directed network must have at least one valid root location. A simple, though extremely inefficient algorithm for finding all valid root locations would be to check all edges as root locations and then all possible orientations for the other edges. To improve on this naive algorithm, you might start by considering each reticulation vertex one at a time. Does a single pair of reticulation edges place restrictions on the possible valid root locations?

It also might be helpful to have an efficient representation of a phylogenetic network. Since a phylogenetic network is just a special type of graph, it can be represented by an *adjacency matrix*. The adjacency matrix for a directed graph with vertex set V is a square $|V| \times |V|$ matrix where the rows and columns are indexed by the elements of V. The (v_i, v_j) entry of the graph is 1 if and only if there is a directed edge from v_i to v_j and 0 otherwise. The following example shows the adjacency matrix for a rooted phylogenetic network.

Example 3 Consider the rooted 3-leaf phylogenetic network pictured in Fig. 5. The network has 8 vertices which form the row and column labels of the adjacency matrix. The 7 edges of the network correspond to the 7 non-zero entries of the adjacency matrix.

$$
\begin{array}{c@{\quad}c@{\ }c@{\ }c@{\ }c@{\ }c@{\ }c@{\ }c@{\ }c}
 & 1 & 2 & 3 & 4 & \rho & v_4 & v_5 & v_6 \\
1 & 0 & 0 & 0 & 0 & 0 & 0 & 0 & 0 \\
2 & 0 & 0 & 0 & 0 & 0 & 0 & 0 & 0 \\
3 & 0 & 0 & 0 & 0 & 0 & 0 & 0 & 0 \\
4 & 0 & 0 & 0 & 0 & 0 & 0 & 0 & 0 \\
\rho & 0 & 0 & 0 & 0 & 0 & 1 & 1 & 0 \\
v_4 & 1 & 0 & 0 & 0 & 0 & 0 & 0 & 1 \\
v_5 & 0 & 1 & 0 & 0 & 0 & 0 & 0 & 1 \\
v_6 & 0 & 0 & 1 & 0 & 0 & 0 & 0 & 0
\end{array}
$$

.

Notice that properties of the matrix correspond to properties of the graph. For example, the fact that v_6 is the only vertex with in-degree two can be seen from the matrix, since the column corresponding to v_6 is the only column with two non-zero entries.

There are some subtleties involved in constructing the adjacency matrix for a semi-directed network, as there are both directed and undirected edges. One possibility would be to encode semi-directed networks by treating the undirected edges as bidirected. In any case, it might then prove useful to construct a dictionary between properties of a network and properties of the adjacency matrix of that network.

Research Project 3 Construct a fast heuristic algorithm which will determine if a semi-directed network is a phylogenetic semi-directed network. Alternatively, determine the computational complexity of determining if a given semi-directed network with n leaves and reticulation number m is a phylogenetic semi-directed network.

These research projects may be closely related to Research Project 2 above. After all, determining if a semi-directed network is phylogenetic amounts to determining if there exist *any* valid root locations. Thus, one might consider some of the suggestions above when approaching these problems. Determining the computational complexity may prove very difficult indeed, and it may be a challenge to prove something even when $m = 1$.

One general strategy for proving computational complexity results is to find a transformation from the problem of interest into another problem with a known computational complexity. A good model for how this might work in the context of phylogenetics can be found in [6] a project which was the result of collaboration between undergraduates and faculty members.

2.4   Restrictions of Networks

In phylogenetics it is frequently necessary to pass back and forth between analyzing full datasets on a complete set of organisms $[n]$ and a more confined analysis on subset of $[n]$. For instance, you may have access to an existing data set on $[n]$ but are only interested in some subset of the organisms. Alternatively you may have information on a collection of subsets of $[n]$ and want to piece them together to determine information about the complete set of organisms.

Definition 5 Let N be an n-leaf phylogenetic network with root ρ, and let $A \subseteq [n]$. The *restriction of N to A* is the phylogenetic network $N_{|A}$ constructed by

(i) Taking the union of all directed paths from ρ to a leaf labeled by an element of A.
(ii) Deleting any vertices that lie above a vertex that is on every such path.
(iii) Suppressing all degree two vertices other than the root.
(iv) Removing all parallel edges.
(v) Applying steps (iii) and (iv) until the network is a phylogenetic network.

We say that N *displays* $N_{|A}$.

While the definition of restriction is defined in terms of a rooted phylogenetic network, we can also apply this definition to a semi-directed network. Given a phylogenetic semi-directed network, its restriction to a subset $A \subset [n]$ is found by rooting the network at a valid root location, restricting the rooted phylogenetic network to A, and then taking the semi-directed topology of the restricted phylogenetic network. The following challenge problem shows that this operation is well-defined.

Challenge Problem 5 Suppose that a valid rooting is chosen for an n-leaf semi-directed network and that the network is then restricted to a subset of the leaves of size $k \leq n$. Show that the k-leaf semi-directed network obtained by unrooting the restricted network is independent of the original rooting chosen.

In practice it can be computationally difficult to directly estimate a phylogenetic network from sequence data corresponding to the set $[n]$. One potential workaround is to infer phylogenetic networks on a collection of subsets of $[n]$, and then select a larger network N which best reflects the networks estimated on the various subsets.

Definition 6 A set of phylogenetic networks $\mathscr{A} = \{N_1, N_2, \cdots, N_k\}$ whose leaves are all contained in a set $[n]$ is called *compatible* if there exists a phylogenetic network N for which the restriction of N to the leaf set of N_i is isomorphic to N_i for all $1 \leq i \leq k$.

It is common when working with unrooted trees to restrict the trees to four element subsets of the leaves. The resulting 4-leaf trees are called *quartets*, and an n-leaf phylogenetic tree is uniquely determined by its $\binom{n}{4}$ quartets. Similarly, when working with a network, we can construct a *quarnet* by restricting the network to a four element subset of its leaves. In this paper, since we are working with semi-directed phylogenetic networks, we will use the term quarnet to mean a 4-leaf semi-directed phylogenetic network. However, note that in other sources a quarnet may refer to an unrooted 4-leaf network.

Exercise 17 Determine if the following collections of quarnets are compatible.

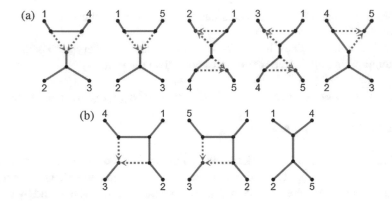

It is possible that some collections of subnetworks can be displayed by multiple phylogenetic networks. In practice we might want to know when a collection of subnetworks can be used to represent a unique network.

Definition 7 Let $\mathcal{A} = \{N_1, N_2, \cdots, N_k\}$ be a collection of phylogenetic networks for which the union of all of the corresponding leaf sets is $[n]$. The collection $\mathcal{A}$ is said to *distinguish* a phylogenetic network N, if N is the only phylogenetic network with leaf set $[n]$ such that the restriction of N to the leaf set of N_i is isomorphic to N_i for all $1 \leq i \leq k$.

Exercise 18 Find a collection of three quarnets which are displayed by the 6-leaf phylogenetic network in Fig. 6 which do not distinguish that network.

Challenge Problem 6 Show that the set of all quarnets of a level-one semi-directed network distinguishes that network.

A good strategy for proving this might be to consider two distinct level-one semi-directed networks, and then show that there must be a quarnet on which they differ.

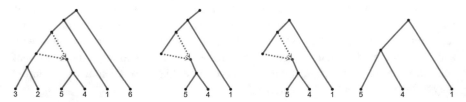

Fig. 6 The restriction of a 6-leaf phylogenetic network to the set $\{1, 4, 5\}$. The networks pictured are obtained by applying (i), (ii), and then (iii), (iv), and (iii) again to obtain the restricted phylogenetic network

Challenge Problem 7 Find all minimal sets of quarnets which distinguish the semi-directed topology of the 6-leaf phylogenetic network in Fig. 6.

Challenge Problem 8 Give criteria for determining whether or not a collection of quarnets are compatible. Hint: There are known criteria for determining if a set of quartet trees are compatible [19].

Research Project 4 Describe an algorithm that determines if a set of semi-directed networks $\mathscr{A} = \{N_1, N_2, \cdots, N_k\}$ is compatible. Bonus points if the algorithm is efficient, constructive, or determines if the collection distinguishes a unique network. This question is already interesting in the case that each of the N_i is a quarnet.

The previous research problem is based on the notion that one could computationally estimate quarnets from DNA sequence data, and then the compatible quarnets could be combined to determine a single network which describes the evolution across a broader collection of organisms. This idea has proven successful when building phylogenetic trees, thus, a number of authors have studied whether or not networks can be constructed by building up large networks from smaller structures (e.g., [21, 23, 24, 26]). Insights and techniques from these papers will likely prove valuable for attacking some of these research and challenge questions. However, it is unlikely that any results will translate directly, since each of the sources cited place different restrictions on the types of input networks and the types of networks constructed.

As a warmup to this activity one might examine similar results on trees as can be found throughout Chapters 3 and 6 in the textbook *Basic Phylogenetic Combinatorics* [12], the introductory chapters of which also provide a nice mathematical framework for working with trees and networks. However, there is a level of abstraction in this book which mandates that readers may need to keep a running list of concrete examples nearby to connect the text with their intuitive understanding of trees and networks.

Exercise 19 Construct a phylogenetic network which displays the following quarnets: Either prove this collection distinguishes the network or find the set of all networks which display this collection.

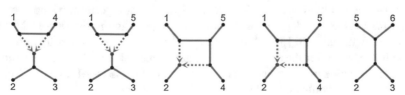

In practice, the estimation of quarnets from data is likely to be imperfect. Thus, even if we produce data from a model on an [n]-leaf semi-directed network, the collection of estimated quarnets is likely to be incompatible. The same issue applies no matter the size of the inferred subnetworks. In such cases, one would like to construct a phylogenetic network which displays the maximum number of quarnets or other semi-directed networks in a collection $\mathscr{A}$.

Exercise 20 Find a phylogenetic network which displays the maximum number of the following collection of quarnets.

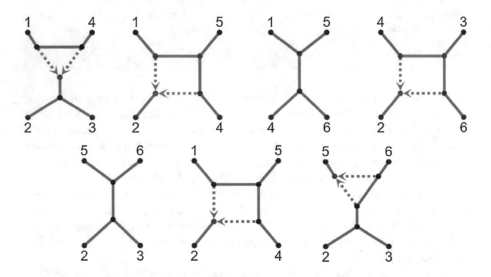

Research Project 5 Given a collection $\mathscr{A} = \{N_1, N_2, \cdots, N_k\}$ of semi-directed networks, give an efficient algorithm for computing a semi-directed network N which maximizes the number of subnetworks in $\mathscr{A}$ displayed by N.

As a warmup example, in the case of trees, there are two very popular approaches to this problem. The first is described in a series of papers describing the ASTRAL family of software [30], where the solution tree T is assumed to have certain features which appear in the trees in $\mathscr{A}$. This is a very efficient algorithm which provably solves the problem under this assumption. An alternative is the quartet MaxCut family of algorithms [37] which provide a fast heuristic for solving this problem. While it does not offer the same theoretical guarantees of the ASTRAL methods, it also removes some of the restrictive assumptions of the ASTRAL method. Both of these algorithms would be good starting points for exploration.

In moving towards networks, one might examine the recent software SNAQ [38] which builds phylogenetic networks based on input from a collection of quarnets. This research problem is very broad and allows for many types of restrictions that would still be interesting in practice. One should feel free to restrict both the types of networks in the collection $\mathscr{A}$ and the type of semi-directed network N which is allowed. Consider restrictions both on the number of leaves, level, and number of reticulation vertices.

3   Algebra of Phylogenetic Trees and Networks

In the previous section, we introduced networks as a possible explanation for the evolutionary history of a set of species and explored some combinatorial questions related to their structure. In this section we move from combinatorial to algebraic questions. In particular, we study *phylogenetic ideals*, collections of polynomials associated to models of DNA sequence evolution. Phylogenetic ideals associated to tree models have been well-studied (e.g., [4, 13, 42]) and have been used not only for model selection but also to prove theoretical results about the models. For example, they have been used to show that the tree parameters of certain models are *identifiable* (e.g., [2, 4, 9, 28]). A model parameter is identifiable if each output from the model uniquely determines the value of that parameter. This is an important consideration for using phylogenetic models for inference, since it would be undesirable to have multiple different trees explain our data equally well. In contrast to trees, the ideals associated to semi-directed networks have not been well-studied and the topic is rich enough that even the simplest networks give rise to interesting research questions. Therefore, in this section, we will work with phylogenetic semi-directed networks with only a single reticulation vertex. As an undirected graph, a phylogenetic semi-directed network with a single reticulation has a unique cycle of length k, and so we call these networks *k-cycle networks*.

Phylogenetic ideals are determined by two things: a model of DNA sequence evolution and a tree or network. In this section, we fix the model of DNA sequence evolution, and then focus on how the polynomials change based on different network attributes. The model of evolution that is quietly sitting in the background is the Cavendar–Farris–Neyman (CFN) model. While there are four DNA bases (adenine (A), guanine (G), cytosine (C), and thymine (T)), the CFN model only distinguishes between purines (A, G) and pyrimidines (C, T). Thus, it is a 2-state model of evolution where the two states are represented by 0 and 1. For the CFN model on a fixed n-leaf tree T, the mutations between purines and pyrimidines are modeled as a Markov process proceeding along the tree. The numerical parameters of the model determine the probabilities that mutations occur along each edge. Once the numerical parameters are specified, the model gives a probability distribution on the set $\{0, 1\}^n$. Put another way, the tree determines a map, or *parameterization*, that sends each choice of numerical parameters to a probability distribution. Because each coordinate of this map is a polynomial, we can consider it as a ring homomorphism. The kernel of this homomorphism is the phylogenetic ideal associated to T.

We begin in Sect. 3.1 describing in greater detail how to construct an ideal from an unrooted phylogenetic tree. In the subsequent sections, we show how a similar process can be used to associate an ideal to different classes of k-cycle networks. Phylogenetic network ideals were originally studied in [18], and it is likely they will receive increasing attention as researchers look to apply methods that have proven successful for trees to phylogenetic networks. In these sections, we also present a number of research projects related to uncovering generating sets and properties of network ideals as well as comparing the ideals for different networks.

While not essential for the projects presented below, for those interested in learning more about the CFN model and the connections to phylogenetic ideals we recommend [5]. One reason that we do not dwell on the details of the maps referenced above is that we actually work in a set of transformed coordinates called the *Fourier coordinates*, introduced in [14]. This is common when studying phylogenetic ideals, as it makes many of the computations feasible. Though the derivation and details of the transform are outside the scope of this chapter, they can be found in [13, 14, 41]. Viewing phylogenetic statistical models from an algebraic perspective fits broadly into the field of *algebraic statistics*. An overview of some of the basic concepts and significant results in this area can be found in [43, Chapter 15]. Similarly, many of the concepts below come from computational algebraic geometry, and some good first references for students are [11, 20]. If the reader has not yet had a course in abstract algebra, [3, Chapter 4] provides an excellent introduction to the algebraic viewpoint on phylogenetics which is accessible to readers who are familiar with matrices.

3.1 Ideals Associated to Trees

An ideal I of a ring R is a subset of R closed under addition and multiplication by ring elements, that is, for all $f, g \in I$, we have $f + g \in I$, and for all $r \in R$ and $f \in I$, we have $rf \in I$. In this and the proceeding sections, our rings of interest will be polynomial rings. The polynomial ring $\mathbb{Q}[x_1, \ldots, x_n]$ is the set of all polynomials in variables $x_1, \ldots, x_n$ with coefficients in $\mathbb{Q}$. The mathematical fields for studying polynomial rings and their ideals are ring theory and commutative algebra. One subfield of commutative algebra is combinatorial commutative algebra. In combinatorial commutative algebra, it is quite common to encounter ideals arising as the kernel of ring homomorphisms described by a combinatorial structure (such as a tree, graph, or simplicial complex). Phylogenetic ideals are also defined in this way. To demonstrate the process for constructing a phylogenetic ideal, we will begin by taking an in-depth look into how to construct the phylogenetic ideal associated to the 4-leaf tree T pictured in Fig. 7.

For what follows, we will use $\mathbb{Z}_2$ to denote the quotient group $\mathbb{Z}/2\mathbb{Z}$. This group has two elements, 0 and 1, with addition modulo 2. We will see that the tree T encodes a ring homomorphism between two polynomial rings R_T and S_T. Because

Fig. 7 An unrooted 4-leaf binary phylogenetic tree

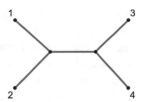

T has four leaves, the variables of R_T are indexed by all 4-tuples of the form (i_1, i_2, i_3, i_4) where $i_1, i_2, i_3, i_4 \in \mathbb{Z}_2$ and $i_1 + i_2 + i_3 + i_4 = 0$. This restriction comes from the fact that the CFN model is a *group-based model* for the group $\mathbb{Z}_2$ [41]. Using this restriction on the indices (and dropping parentheses and commas), we get

$$R_T = \mathbb{Q}[q_{0000}, q_{0011}, q_{0101}, q_{0110}, q_{1001}, q_{1010}, q_{1100}, q_{1111}].$$

The tree T has five edges, four of which are leaf edges and one of which is an internal edge. The variables of the ring S_T come in two forms corresponding to the two types of edges. Each of these variables has an upper and lower index. The variables corresponding to the leaf edges are of the form $a_0^{(j)}, a_1^{(j)}$ where j is the label of the leaf and 0 and 1 are elements of $\mathbb{Z}_2$. The variables corresponding to the one internal edge are $b_0^{(1)}$ and $b_1^{(1)}$. Thus, the ring S_T, with all its variables listed, is

$$S_T = \mathbb{Q}[a_0^{(1)}, a_1^{(1)}, a_0^{(2)}, a_1^{(2)}, a_0^{(3)}, a_1^{(3)}, a_0^{(4)}, a_1^{(4)}, b_0^{(1)}, b_1^{(1)}].$$

The structure of the tree T determines the following ring homomorphism:

$$\phi_T : R_T \to S_T$$

defined by

$$\phi_T(q_{0000}) = a_0^{(1)} a_0^{(2)} a_0^{(3)} a_0^{(4)} b_0^{(1)}$$

$$\phi_T(q_{0011}) = a_0^{(1)} a_0^{(2)} a_1^{(3)} a_1^{(4)} b_0^{(1)}$$

$$\phi_T(q_{0101}) = a_0^{(1)} a_1^{(2)} a_0^{(3)} a_1^{(4)} b_1^{(1)}$$

$$\phi_T(q_{0110}) = a_0^{(1)} a_1^{(2)} a_1^{(3)} a_0^{(4)} b_1^{(1)}$$

$$\phi_T(q_{1001}) = a_1^{(1)} a_0^{(2)} a_0^{(3)} a_1^{(4)} b_1^{(1)}$$

$$\phi_T(q_{1010}) = a_1^{(1)} a_0^{(2)} a_1^{(3)} a_0^{(4)} b_1^{(1)}$$

$$\phi_T(q_{1100}) = a_1^{(1)} a_1^{(2)} a_0^{(3)} a_0^{(4)} b_0^{(1)}$$

$$\phi_T(q_{1111}) = a_1^{(1)} a_1^{(2)} a_1^{(3)} a_1^{(4)} b_0^{(1)}.$$

Notice that the image of each variable $q_{i_1 i_2 i_3 i_4}$ under ϕ_T is a monomial that has one variable for each leaf edge, $a_0^{(j)}$ or $a_1^{(j)}$, and one variable, $b_0^{(1)}$ or $b_1^{(1)}$, for the internal edge. In order to determine the "color" (either 0 or 1) of each variable in the monomial corresponding to $q_{i_1 i_2 i_3 i_4}$, we consider the index (i_1, i_2, i_3, i_4) as a coloring of the leaves of T. The lower index of each leaf variable is just the color of that leaf. To determine the lower index of the internal edge variable, we sum up the colors (as elements of $\mathbb{Z}_2$) on either side of the split $12|34$ which is induced by the internal edge (splits are discussed in Sect. 2.1). This sum is always the same regardless of which side of the split we consider. For example, the variable q_{0101} maps to a monomial with the variable $b_1^{(1)}$. We could determine the subscript of this variable by summing up the colors of 1 and 2 on one side of the split $(0 + 1 = 1)$ or by summing up the colors of 3 and 4 on the other side of the split $(1 + 0 = 1)$.

We now can define the phylogenetic ideal of T. The phylogenetic ideal of T is the kernel of ϕ_T, that is,

$$I_T := \ker(\phi_T) = \{f \in R_T \ : \ \phi_T(f) = 0\}.$$

An example of an element in I_T is $q_{1111} q_{0000} - q_{1100} q_{0011}$. Elements of I_T are referred to as *phylogenetic invariants* as described in the introduction to this section.

The ideal $I_T \subseteq R_T$ contains an infinite number of elements, but we would prefer to have a finite way to describe I_T. This is similar to the way that we use bases in linear algebra to describe a vector space that has an infinite number of elements. Luckily, the ideal I_T is finitely generated, meaning that there exist $g_1, \ldots, g_m \in R_T$ such that for any $f \in I_T$, there exist $r_1, \ldots, r_m \in R_T$ such that $f = r_1 g_1 + r_2 g_2 + \ldots + r_m g_m$. Any set $\{g_1, \ldots, g_m\}$ that satisfies the preceding definition is called a *generating set* of I_T. The problem of finding generating sets for phylogenetic tree ideals has been solved in many cases, and for small trees, explicit lists of generators are available online at https://www.shsu.edu/~ldg005/small-trees/, the work of which is described in Chapter 15 of [32].

Exercise 21 Show that the polynomial

$$q_{1111} q_{0000} - q_{1100} q_{0011}$$

is in I_T. Find another polynomial in I_T.

Exercise 22 Show that the kernel of any ring homomorphism is an ideal.

3.2   Ideals Associated to Sunlet Networks

To begin our discussion of network ideals, we will first consider a specific type of phylogenetic semi-directed network called a *sunlet*. As an undirected graph, a phylogenetic semi-directed network with a single reticulation vertex has a unique

Fig. 8 A 4-sunlet network
and a 6-sunlet network

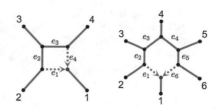

cycle of length k, and so we call such a network a *k-cycle network*. A *k-sunlet* network is a k-leaf, k-cycle network. Starting with sunlet networks will allow us to introduce network ideals in a simplified setting, before we show how to associate an ideal to a general k-cycle network in Sect. 3.3.

Exercise 23 Show that as an undirected graph, a phylogenetic semi-directed network with a single reticulation vertex has a unique cycle that contains the reticulation vertex.

To describe how to associate a polynomial ideal to a sunlet, we will begin by looking at the 4-leaf sunlet network N_4 pictured in Fig. 8. As in the previous section on trees, the sunlet network N_4 encodes a ring homomorphism between two polynomial rings. For the 4-sunlet, these two rings are called R_4 and S_4. The variables of R_4 are indexed by all 4-tuples of the form (i_1, i_2, i_3, i_4) where $i_1, i_2, i_3, i_4 \in \mathbb{Z}_2$ and $i_1 + i_2 + i_3 + i_4 = 0$. This is exactly the same as the ring R_T in the previous section because both the 4-sunlet and T have four leaves, thus,

$$R_4 = \mathbb{Q}[q_{0000}, q_{0011}, q_{0101}, q_{0110}, q_{1001}, q_{1010}, q_{1100}, q_{1111}].$$

The ring S_4 has two variables for each edge of N_4 just as the ring S_T had two variables for each edge of T. Since the sunlet network N_4 has three more internal edges than T, the polynomial ring S_4 contains six additional variables not in S_T. Thus,

$$S_4 = \mathbb{Q}[a_0^{(1)}, a_1^{(1)}, a_0^{(2)}, a_1^{(2)}, a_0^{(3)}, a_1^{(3)}, a_0^{(4)}, a_1^{(4)}, b_0^{(1)}, b_1^{(1)}, b_0^{(2)}, b_1^{(2)}, b_0^{(3)}, b_1^{(3)}, b_0^{(4)}, b_1^{(4)}].$$

The 4-sunlet N_4 defines the map

$$\phi_4 : R_4 \to S_4$$

$$q_{i_1, i_2, i_3, i_4} \mapsto a_{i_1}^{(1)} a_{i_2}^{(2)} a_{i_3}^{(3)} a_{i_4}^{(4)} \left(b_{i_1}^{(1)} b_{i_1+i_2}^{(2)} b_{i_1+i_2+i_3}^{(3)} + b_{i_2}^{(2)} b_{i_2+i_3}^{(3)} b_{i_2+i_3+i_4}^{(4)} \right).$$

The image of each variable in R_4 under the map ϕ_4 is listed in Example 4. To understand the combinatorial nature of the parameterization ϕ_4, observe that if we remove one of the reticulation edges of N_4, the result is an unrooted 4-leaf tree

Fig. 9 The 4-sunlet network N_4 and the two trees T_1 and T_2 obtained by removing each reticulation edge

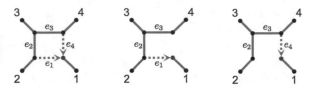

with labeled leaves (see Fig. 9). These two resulting trees, T_1 and T_2, are not binary since removing a reticulation edge in the network will leave vertices of degree two. Consider the image of $q_{i_1 i_2 i_3 i_4}$ under ϕ_4. Multiplying through by $a_{i_1}^{(1)} a_{i_2}^{(2)} a_{i_3}^{(3)} a_{i_4}^{(4)}$ gives us a binomial where the first part corresponds to the tree T_1 and the second part corresponds to the tree T_2. The indices of the a variables appearing in $\phi_4(q_{i_1, i_2, i_3, i_4})$ correspond to the coloring (i_1, i_2, i_3, i_4) of the leaves of T_1 and T_2, while the indices of the b variables are obtained by considering the splits of T_1 and T_2. In particular, the index of the $b^{(j)}$ variable in the first part of the binomial is obtained by removing the jth edge of T_1 and summing the colors (as elements of $\mathbb{Z}_2$) on either side of the induced split, and the index of the $b^{(j)}$ variable in the second part of the binomial is obtained by a similar process on T_2. Notice that T_1 and T_2 each have three internal edges, and thus, there are three b variables in each part of the binomial. As in the tree case, the phylogenetic ideal associated to N_4 is the kernel of the map ϕ_4:

$$I_4 := \ker(\phi_4).$$

As we will see in Example 4, the ideal I_4 is generated by a single polynomial, $q_{0110}q_{1001} - q_{0101}q_{1010} + q_{0011}q_{1100} - q_{0000}q_{1111}$.

Exercise 24 Verify that the polynomial $q_{0110}q_{1001} - q_{0101}q_{1010} + q_{0011}q_{1100} - q_{0000}q_{1111}$ is in I_4.

Now that we have gone through the process of constructing the phylogenetic ideal for the 4-leaf sunlet, let us generalize the procedure for arbitrary k-sunlets. Let N_k be the k-sunlet network with the leaves labeled clockwise, from 1 to k, starting from the leaf extending from the single reticulation vertex. For example, N_6 is the network pictured at right in Fig. 8. For each k, we now define two polynomials rings, R_k and S_k, a ring map ϕ_k, and the phylogenetic ideal I_k.

The first polynomial ring we consider is R_k, which is a generalization of R_4.

$$R_k := \mathbb{Q}[q_{i_1, \dots, i_k} : i_1, \dots, i_k \in \mathbb{Z}_2, \ i_1 + \dots + i_k = 0]$$

Exercise 25 Enumerate the indeterminates, i.e., variables, for R_3. In general, how many indeterminates does R_k have?

The next ring we will consider is a ring with two indeterminates associated to each edge in N_k. The k-sunlet network N_k has $2k$ edges, k of which are leaf edges and k of which are internal (non-leaf) edges. We label the leaf edges of the network from 1 to k to match the corresponding leaf labels. Similarly, we label the internal edges from 1 to k, starting with the reticulation edge clockwise from the leaf edge labeled by 1 and continuing around the sunlet (as in Fig. 8). To each edge of the sunlet, we associate two variables, one for each element of $\mathbb{Z}_2$. We denote the variables for the leaf edge labeled by j as $a_0^{(j)}$ and $a_1^{(j)}$ and the variables for the internal edge labeled by j as $b_0^{(j)}$ and $b_1^{(j)}$. The second polynomial ring of interest is

$$S_k := \mathbb{Q}[a_i^{(j)}, b_i^{(j)} : 1 \leq i \leq k, \ j \in \mathbb{Z}_2].$$

Now that we have defined the rings R_k and S_k, we can define the ring map ϕ_k and the ideal I_k. The ideal I_k associated to the phylogenetic network N_k is the kernel of the ring homomorphism:

$$\phi_k : R_k \rightarrow S_k$$

$$q_{i_1,\ldots,i_k} \mapsto a_{i_1}^{(1)} \cdots a_{i_k}^{(k)} \left(\prod_{j=1}^{k-1} b_{i_1+\ldots+i_j}^{(j)} + \prod_{j=2}^{k} b_{i_2+\ldots+i_j}^{(j)} \right).$$

In other words,

$$I_k := \ker(\phi_k) = \{f \in R_k : \phi_k(f) = 0\}.$$

Exercise 26 For $k = 3$ write down the rings R_k and S_k. Let $f = 3q_{1,1,0}q_{1,0,1}^2 + q_{0,0,0}$. Compute $\phi_3(f)$.

Exercise 27 Find a non-zero polynomial in I_3 or prove that no such polynomial exists.

As mentioned previously, when studying ideals associated to phylogenetic networks, we are interested in the polynomials in the ideal. In some cases, just knowing a few polynomials in the ideal is helpful, but we can obtain a more complete understanding of the ideal if we can determine a generating set.

There are algorithms based on the theory of Gröbner bases for determining the generating set for an ideal from its parameterization. A Gröbner basis is a special type of generating set for an ideal and we encourage curious readers to learn more about them before starting on some of the research problems in this section (see, e.g., [11, 40]). However, while these algorithms give us a means of determining a generating set for an ideal, in most cases of interest, it is infeasible to perform all the computations necessary by hand. Therefore, we will want to use a computer algebra system to do most of the tedious work for us. In this chapter, we will use the computer algebra system *Macaulay2* [17]. As a first example, we show below how to use this program to find a generating set for I_4, the ideal associated to the sunlet network N_4.

Example 4 Let us consider N_4, the 4-leaf sunlet network pictured in Fig. 8. Recall that for N_4, the two rings of interest are

$$R_4 = \mathbb{Q}[q_{0000}, q_{0011}, q_{0101}, q_{0110}, q_{1001}, q_{1010}, q_{1100}, q_{1111}], \text{ and}$$

$$S_4 = \mathbb{Q}[a_0^{(1)}, a_1^{(1)}, a_0^{(2)}, a_1^{(2)}, a_0^{(3)}, a_1^{(3)}, a_0^{(4)}, a_1^{(4)}, b_0^{(1)}, b_1^{(1)}, b_0^{(2)}, b_1^{(2)}, b_0^{(3)}, b_1^{(3)}, b_0^{(4)}, b_1^{(4)}].$$

The ring homomorphism ϕ_4 is described as follows:

$$\phi_4(q_{0000}) = a_0^{(1)} a_0^{(2)} a_0^{(3)} a_0^{(4)} \left(b_0^{(1)} b_0^{(2)} b_0^{(3)} + b_0^{(2)} b_0^{(3)} b_0^{(4)} \right),$$

$$\phi_4(q_{0011}) = a_0^{(1)} a_0^{(2)} a_1^{(3)} a_1^{(4)} \left(b_0^{(1)} b_0^{(2)} b_1^{(3)} + b_0^{(2)} b_1^{(3)} b_0^{(4)} \right),$$

$$\phi_4(q_{0101}) = a_0^{(1)} a_1^{(2)} a_0^{(3)} a_1^{(4)} \left(b_0^{(1)} b_1^{(2)} b_1^{(3)} + b_1^{(2)} b_1^{(3)} b_0^{(4)} \right),$$

$$\phi_4(q_{0110}) = a_0^{(1)} a_1^{(2)} a_1^{(3)} a_0^{(4)} \left(b_0^{(1)} b_1^{(2)} b_0^{(3)} + b_1^{(2)} b_0^{(3)} b_0^{(4)} \right),$$

$$\phi_4(q_{1001}) = a_1^{(1)} a_0^{(2)} a_0^{(3)} a_1^{(4)} \left(b_1^{(1)} b_1^{(2)} b_1^{(3)} + b_0^{(2)} b_0^{(3)} b_1^{(4)} \right),$$

$$\phi_4(q_{1010}) = a_1^{(1)} a_0^{(2)} a_1^{(3)} a_0^{(4)} \left(b_1^{(1)} b_1^{(2)} b_0^{(3)} + b_0^{(2)} b_1^{(3)} b_1^{(4)} \right),$$

$$\phi_4(q_{1100}) = a_1^{(1)} a_1^{(2)} a_0^{(3)} a_0^{(4)} \left(b_1^{(1)} b_0^{(2)} b_0^{(3)} + b_1^{(2)} b_1^{(3)} b_1^{(4)} \right),$$

$$\phi_4(q_{1111}) = a_1^{(1)} a_1^{(2)} a_1^{(3)} a_1^{(4)} \left(b_1^{(1)} b_0^{(2)} b_1^{(3)} + b_1^{(2)} b_0^{(3)} b_1^{(4)} \right).$$

Using *Macaulay2* we can compute a generating set for I_4. In the code below, we use I for this ideal, R and S for the rings R_4 and S_4, and phi for the map ϕ_4.

```
i1: R   = QQ[q_{0,0,0,0}, q_{0,0,1,1}, q_{0,1,0,1},
                                       q_{0,1,1,0},
            q_{1,0,0,1}, q_{1,0,1,0}, q_{1,1,0,0},
                                       q_{1,1,1,1}];

i2: S   = QQ[a1_0, a1_1, a2_0, a2_1, a3_0, a3_1, a4_0, a4_1,
             b1_0, b1_1, b2_0, b2_1, b3_0, b3_1, b4_0, b4_1];

i3: phi = map(S, R,
        {a1_0*a2_0*a3_0*a4_0*(b1_0*b2_0*b3_0+b2_0*b3_0*b4_0),
        a1_0*a2_0*a3_1*a4_1*(b1_0*b2_0*b3_1+b2_0*b3_1*b4_0),
        a1_0*a2_1*a3_0*a4_1*(b1_0*b2_1*b3_1+b2_1*b3_1*b4_0),
        a1_0*a2_1*a3_1*a4_0*(b1_0*b2_1*b3_0+b2_1*b3_0*b4_0),
        a1_1*a2_0*a3_0*a4_1*(b1_1*b2_1*b3_1+b2_0*b3_0*b4_1),
        a1_1*a2_0*a3_1*a4_0*(b1_1*b2_1*b3_0+b2_0*b3_1*b4_1),
        a1_1*a2_1*a3_0*a4_0*(b1_1*b2_0*b3_0+b2_1*b3_1*b4_1),
        a1_1*a2_1*a3_1*a4_1*(b1_1*b2_0*b3_1+b2_1*b3_0*b4_1)})

i4: I   = ker phi

o4: ideal(q_{0, 1, 1, 0}*q_{1, 0, 0, 1}-q_{0, 1, 0, 1}
                                      *q_{1, 0, 1,0}+
```

$$q\_\{0, \ 0, \ 1, \ 1\}*q\_\{1, \ 1, \ 0, \ 0\}-q\_\{0, \ 0, \ 0, \ 0\}$$
$$*q\_\{1, \ 1, \ 1, \ 1\}).$$

The output of the last command tells us that I_4 is generated by a single polynomial, namely

$$q_{0110}q_{1001} - q_{0101}q_{1010} + q_{0011}q_{1100} - q_{0000}q_{1111}.$$

Exercise 28 Compute the ideal I_5 for the 5-sunlet N_5 using *Macaulay2* or another computer algebra system. How many generators are returned? What are the degrees of the returned generators?

Exercise 29 Verify (computationally or by hand) that the polynomial

$$q_{01100}q_{10010} - q_{01010}q_{10100} + q_{00110}q_{11000} - q_{00000}q_{11110}$$

is in the ideal I_5.

On a standard laptop, the computation in Exercise 28 will finish, but not immediately. You may notice the difference in the time it takes to run the computation for I_4 in Example 4 and for I_5 in Exercise 28. As we increase k, computing I_k becomes even more complex, to the point that a computer may take several hours or days or may run out of memory before returning a generating set. The computer, of course, will execute an algorithm to determine a generating set for I_k. In many cases, however, executing all the steps of the algorithm is not actually necessary to obtain the information about the ideal that we are interested in. Therefore, we can use some tricks and techniques to reduce the size of the computations and extract information about the ideals.

For example, we can use some of the built-in options in *Macaulay2* such as `SubringLimit`, a command that stops the computation after a specified number of polynomials have been found. If using this strategy, we will obtain a set of polynomials in the ideal I_k, but we will not have a certification that these polynomials generate I_k. However, if we let J be the ideal, they generate then we know that $J \subseteq I_k$. We can show that $J = I_k$ if we can show that J is *prime* and that the *dimension* of J is equal to that of I_k. An ideal $I \subseteq R$ is prime if for all $f, g \in R$, if $fg \in I$, then $f \in I$ or $g \in I$. Checking whether an ideal is prime and finding its dimension can be done in *Macaulay2* using the `isPrime` and `dim` commands. Of course, we do not have a set of generators for I_k, since that is what we are trying to find, so we cannot use `dim` to find its dimension. However, we can still determine a lower bound on the dimension of I_k from the map ϕ_k using the rank of the Jacobian matrix as shown in Example 5. Since $J \subseteq I_k$, we have $\dim(I_k) \leq \dim(J)$, and so if the rank of the Jacobian is equal to $\dim(J)$, then $\dim(I_k) = \dim(J)$.

This `SubringLimit` method of determining a generating set for an ideal was used to prove Proposition 4.6 in [18]. That paper also includes supplementary *Macaulay2* code which may prove useful.

Example 5 Let $I = \langle q_{0110}q_{1001} - q_{0101}q_{1010} + q_{0011}q_{1100} - q_{0000}q_{1111}\rangle$ be the ideal returned from Example 4. The following *Macaualy2* code is used to determine whether the dimension of the ideal I is the same as the dimension of the ideal I_4 as well as whether or not I is prime. This serves as verification that I is indeed equal to I_4.

```
i5: phimatrix = matrix{{
        a1_0*a2_0*a3_0*a4_0*(b1_0*b2_0*b3_0+b2_0*b3_0*b4_0),
        a1_0*a2_0*a3_1*a4_1*(b1_0*b2_0*b3_1+b2_0*b3_1*b4_0),
        a1_0*a2_1*a3_0*a4_1*(b1_0*b2_1*b3_1+b2_1*b3_1*b4_0),
        a1_0*a2_1*a3_1*a4_0*(b1_0*b2_1*b3_0+b2_1*b3_0*b4_0),
        a1_1*a2_0*a3_0*a4_1*(b1_1*b2_1*b3_1+b2_0*b3_0*b4_1),
        a1_1*a2_0*a3_1*a4_0*(b1_1*b2_1*b3_0+b2_0*b3_1*b4_1),
        a1_1*a2_1*a3_0*a4_0*(b1_1*b2_0*b3_0+b2_1*b3_1*b4_1),
        a1_1*a2_1*a3_1*a4_1*(b1_1*b2_0*b3_1+b2_1*b3_0*b4_1)}}

i6: rank Jacobian phimatrix == dim(I)

i7: isPrime I.
```

Challenge Problem 9 Compute I_6 in *Macaulay2* by imposing a limit on the number of polynomials returned using `SubringLimit`. Verify that the ideal that is returned is indeed I_6.

One will only get so far using the strategy described above, as for larger k, there may be many polynomials required to generate I_k and they may take a very long time to find. In these cases, just being able to compute the ideal I_k becomes an interesting project on its own.

Research Project 6 Find a generating set for the ideal I_k of the k-sunlet network N_k when $k = 7, 8, 9$.

Moving from the computational to the theoretical, it is sometimes possible to give a description of a generating set for a whole class of ideals.

Research Project 7 Give a description of a set of phylogenetic invariants in the sunlet ideal I_k. Does this set of invariants generate the ideal? Does this set of invariants form a Gröbner basis for the ideal with respect to some term order?

We can envision two different approaches to Research Project 7. The first is to compute the sunlet ideals for a range of examples. As you are able to compute I_k for higher k, patterns should emerge. We see this even for $k = 4$ and $k = 5$. For example, Exercise 29 might give a hint of how we can find some invariants for larger k by doing computations for small k. Once you discover a pattern, you could then try to prove that this pattern holds in general.

The second approach would be to try to construct invariants for sunlet networks using the known invariants in the ideals of the trees that they display. As an example, consider the map ϕ_4 encoded by the sunlet N_4. Consider the following two maps. The first, $\phi_{T_1} : R_4 \to S_4$, sends $q_{i_1 i_2 i_3 i_4}$ to the term of $\phi_4(q_{i_1 i_2 i_3 i_4})$ that includes $b_{i_1}^{(1)}$ and the other, $\phi_{T_2} : R_4 \to S_4$, sends $q_{i_1 i_2 i_3 i_4}$ to the second term of $\phi_4(q_{i_1 i_2 i_3 i_4})$ that includes $b_{i_2+i_3+i_4}^{(4)}$. So, for example,

$$\phi_{T_1}(q_{0000}) = a_0^{(1)} a_0^{(2)} a_0^{(3)} a_0^{(4)} b_0^{(1)} b_0^{(2)} b_0^{(3)} \text{ and } \phi_{T_2}(q_{0000}) = a_0^{(1)} a_0^{(2)} a_0^{(3)} a_0^{(4)} b_0^{(2)} b_0^{(3)} b_0^{(4)}.$$

The ideal $I_{T_1} = \ker(\phi_{T_1})$ is the ideal of the tree created by removing the reticulation edge e_4 from the 4-sunlet in Fig. 8. The ideal $I_{T_2} = \ker(\phi_{T_2})$ is the ideal of the tree created by removing the reticulation edge e_1. The problem of finding invariants for trees is discussed briefly in Sect. 3.1.

Tree ideals are parametrized by monomials which makes it easier to find invariants. In particular, invariants for ideals parameterized by monomials can be found by examining the additive relationships between the exponents of the monomials. This means that finding invariants for these ideals can be done using only tools from linear algebra.

Example 6 Let $f : \mathbb{R}^2 \to \mathbb{R}^3$ be the map defined by $(t_1, t_2) \mapsto (t_1^2, t_1 t_2, t_2^2)$. We can represent this map by a 2×3 matrix A, where the ij-th entry is the exponent of t_i in the j-th coordinate of the image of (t_1, t_2),

$$A = \begin{pmatrix} 2 & 1 & 0 \\ 0 & 1 & 2 \end{pmatrix}.$$

Elements of the integer kernel of A encode binomial invariants in $\ker(f)$. For example, the integer vector $(1, -2, 1)^T$ is a vector of integers in $\ker(A)$. We can interpret the positive entries as the monomial $y_1 y_3$ and the negative integers as the monomial y_2^2, and conclude that $y^2 - y_1 y_3$ is in $\ker(f)$.

Notice in the preceding example that while the parameterization was in terms of monomials, the invariant we constructed is a binomial. While there are many different formal definitions, the class of ideals which are parameterized by monomials are called *toric ideals* and it is known that toric ideals are generated by binomials. This fact is proven in [40, Chapter 4], which might also serve as a good reference for learning more about the invariants of toric ideals. The following exercise shows why toric ideals may prove useful when trying to find invariants for sunlet ideals.

Exercise 30 Consider the ideals I_4, $I_{T_1} = \ker(\phi_{T_1})$, and $I_{T_2} = \ker(\phi_{T_2})$ described above.

(a) Show that if $f \in I_4$, then $f \in I_{T_1} \cap I_{T_2}$.
(b) Compute $I_{T_1} = \ker(\phi_{T_1})$ and $I_{T_2} = \ker(\phi_{T_2})$ using *Macaulay2*. You can verify that your computations are correct using the online catalog of invariants referenced in Sect. 3.1. Specifically, by looking under "Invariants in Fourier coordinates" for the "Neyman 2-state model" (another name for the CFN model).
(c) Verify that the generator for I_4 found in Example 4 is contained in I_{T_1} and in I_{T_2}. (Hint: To determine if a polynomial f is contained in an ideal I, you can verify in *Macaulay2* that f % I == 0 returns TRUE).

The previous exercise shows that $I_4 \subset I_{T_1} \cap I_{T_2}$. Put another way, invariants in I_{T_1} and I_{T_2} are candidates to be invariants in I_4. Similar statements hold for all of the ideals I_k in this section, and for the ideals I_N that we describe in the next section. Thus, exploring toric ideals might prove useful for finding network invariants.

3.3   Beyond Sunlet Networks

Sunlet networks have a very particular structure, and the ring map we described in Sect. 3.2 is specific to sunlets. In this section, we set up the ring map ϕ more generally, which will allow us to explore the algebra of general n-leaf, k-cycle networks, such as the two pictured below in Fig. 10.

Let N be an n-leaf, k-cycle network. The first ring we will consider is of the same form as that from the previous section,

$$R_n := \mathbb{Q}[q_{i_1,\ldots,i_n} : i_1, \ldots, i_n \in \mathbb{Z}_2, \; i_1 + \ldots + i_n = 0].$$

The next ring we will consider is a ring with two indeterminates associated to each edge of N. As with the sunlet, an n-leaf, k-cycle network has $2n$ edges, but unlike with sunlets, we no longer make a distinction between the leaf edges and the interior edges when labeling and so label all the edges by $\{1, \ldots, 2n\}$. As before, we associate two parameters to each edge, indexed by the edge label and the elements of $\mathbb{Z}_2$.

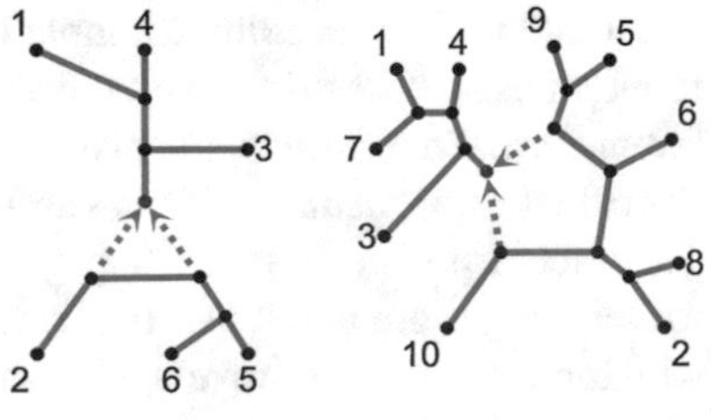

Fig. 10 A 6-leaf, 3-cycle network and a 10-leaf, 5-cycle network

$$S_n := \mathbb{Q}[a_0^{(i)}, a_1^{(i)} : 1 \le i \le 2n].$$

Our next step will be to define the map $\phi_N : R_n \to S_n$. If we remove either of the reticulation edges of N, the result is an unrooted n-leaf tree with labeled leaves. These two trees, T_1 and T_2, are not binary since removing a reticulation edge in the network will leave vertices of degree two.

The map ϕ_N sends each variable in R_n to a binomial in S_n where the two terms are determined by T_1 and T_2. For what follows, let $L_m \subset [2n]$ be the set of edge indices of T_m. For the variable $q_{i_1,\dots,i_n}$ the term that is associated to T_m will be a monomial, with one indeterminate, $a_0^{(j)}$ or $a_1^{(j)}$, for each edge of the tree. In order to determine the "color" (0 or 1) of each edge indeterminate, we consider $(i_1, \dots, i_n)$ as a labeling of the leaves of T_m by elements of $\mathbb{Z}_2$. If we remove an edge e_j of T_m, the resulting graph has two connected components which splits the leaves into two sets. Let $s_j^m(i_1, \dots, i_n)$ be the group sum of the leaf labels on either side of the split induced by removing the edge e_j from T_m. (Note the sum of leaf labels is the same on either side of the split.) The indeterminate associated to the edge e_j is then $a_{s_j^m(i_1,\dots,i_n)}^{(j)}$. Thus, we have the map

$$\phi_N : R_n \to S_n$$

$$q_{i_1,\dots,i_n} \mapsto \prod_{j \in L_1} a_{s_j^1(i_1,\dots,i_n)}^{(j)} + \prod_{j \in L_2} a_{s_j^2(i_1,\dots,i_n)}^{(j)}.$$

Now the phylogenetic ideal I_N associated to N is the kernel of ϕ_N:

$$I_N := ker(\phi_N) = \{f \in R_n : \phi_N(f) = 0\}.$$

Example 7 Let N be the 6-leaf, 3-cycle network pictured in Fig. 10. Removing the reticulation edges of N creates two trees, T_1 and T_2, with edge indices $L_1 = [12] \setminus \{9\}$ and $L_2 = [12] \setminus \{10\}$. To determine the parameterization for the coordinate q_{111100}, we color the leaves by $(1, 1, 1, 1, 0, 0)$. Here, we show vertices and edges colored by 1 as magenta.

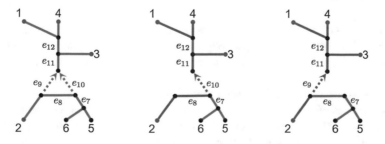

To determine the color of an edge, we sum the leaf colors on either side of the split created by removing that edge. For example, the edge e_{11} in T_1 corresponds to the split 134|256. Since

$$i_1 + i_3 + i_4 = 1 + 1 + 1 = i_2 + i_5 + i_6 = 1 + 0 + 0 = 1,$$

e_{11} is colored by 1. Thus, the indeterminate for e_{11} is $a_1^{(11)}$. Then for the map ϕ_N, we have

$$q_{111100} \mapsto a_1^{(1)} a_1^{(2)} a_1^{(3)} a_1^{(4)} a_0^{(5)} a_0^{(6)} a_0^{(7)} a_1^{(8)} a_1^{(10)} a_1^{(11)} a_0^{(12)} +$$
$$a_1^{(1)} a_1^{(2)} a_1^{(3)} a_1^{(4)} a_0^{(5)} a_0^{(6)} a_0^{(7)} a_0^{(8)} a_1^{(9)} a_1^{(11)} a_0^{(12)}.$$

Exercise 31 Let N be the quarnet with leaf label set $\{1, 2, 3, 4\}$ from Exercise 20. Write out R_n, S_n, and ϕ_N. Use Macaulay2 to compute I_N.

Exercise 32 Let N be the sunlet network N_k. Show that the ring map in this section is the same ring map as the previous section if we replace the b indeterminates with the appropriate a indeterminates.

Notice that if we swap the leaf labels 1 and 3 in the 4-sunlet N_4, we obtain a different 4-sunlet network. In the following challenge problem, we will see how changing the labeling of a network changes which polynomials are in the ideal.

Challenge Problem 10 How many labeled 4-leaf sunlets are there? Compute I_N for each of the 4-leaf sunlets. How are the generating sets of each of these ideals related to I_4?

For the sunlet graphs, we could factor all of the indeterminates corresponding to the leaf edges from the binomial but no other indeterminates. In essence, we could write the image of every variable in R_k as a monomial multiplied by a binomial. We can also do this for the k-cycle networks, and as Example 7 shows, sometimes we can factor out many more indeterminates. From that example, we could write

$$q_{111100} \mapsto a_1^{(1)} a_1^{(2)} a_1^{(3)} a_1^{(4)} a_0^{(5)} a_0^{(6)} a_0^{(7)} a_1^{(11)} a_0^{(12)} (a_1^{(8)} a_1^{(10)} + a_1^{(9)} a^{(11)}).$$

The following challenge problem is aimed to get at this general phenomenon for k-cycle networks.

Challenge Problem 11 Write out the map ϕ_N for several 4-leaf and 5-leaf k-cycle networks. For each graph, which edge indeterminates can you factor for every binomial in the map? Can you describe the general pattern for k-cycle networks?

In our explorations we have seen that different networks may induce different phylogenetic ideals. The ideals of certain networks may contain the ideals of other networks with the same leaf set. This suggests we might try to understand the relationship between ideal containment and the corresponding network structures.

Research Project 8 Draw all of the 5-leaf level-one networks. Which networks have the same ideal under the CFN model? Which networks have ideals that are contained in one another?

To explore the structure of the ideals you might use the Macaulay 2 command `isSubset(J,I)` to determine if the ideal J is contained in the ideal I. Similarly, `I == J` will tell you if two ideals are equal. In order to formalize the ideal containment structures you identify, it might be helpful to use a mathematical object called a partially ordered set (poset). The definition of a poset as well as examples can be found in Chapter 6 of [27].

Acknowledgements This material is based upon work supported by the National Science Foundation under Grant No. DMS-1616186 and Grant No. DMS-1620109.

References

1. Tushar Agarwal, Philippe Gambette, and David Morrison. Who is who in phylogenetic networks: Articles, authors, and programs, 2016.
2. Elizabeth S. Allman, Sonja Petrović, John A. Rhodes, and Seth Sullivant. Identifiability of 2-tree mixtures for group-based models. *IEEE/ACM Trans. Comp. Biol. Bioinformatics*, 8(3):710–722, 2011.
3. Elizabeth S. Allman and John A. Rhodes. *Mathematical Models in Biology, an Introduction.* Cambridge University Press, Cambridge, United Kingdom, 2004.
4. Elizabeth S. Allman and John A. Rhodes. The identifiability of tree topology for phylogenetic models, including covarion and mixture models. *J. Comp. Biol.*, 13(5):1101–1113, 2006.
5. Elizabeth S. Allman and John A. Rhodes. *Reconstructing Evolution: New Mathematical and Computational Advances*, chapter 4. Oxford University Press, UK, June 2007.
6. Maria Anaya, Olga Anipchenko-Ulaj, Aisha Ashfaq, Joyce Chiu, Mahedi Kaiser, Max Shoji Ohsawa, Megan Owen, Ella Pavlechko, Katherine St. John, Shivam Suleria, Keith Thompson, and Corinne Yap. On determining if tree-based networks contain fixed trees. *Bulletin of Mathematical Biology*, 78(5):961–969, 2016.
7. Marta Casanellas and Jesús Fernández-Sánchez. Performance of a new invariants method on homogeneous and nonhomogeneous quartet trees. *Molecular biology and evolution*, 24(1):288–293, 2006.
8. J.A. Cavender and Joseph Felsenstein. Invariants of phylogenies in a simple case with discrete states. *J. of Class.*, 4:57–71, 1987.
9. J.T. Chang. Full reconstruction of Markov models on evolutionary trees: identifiability and consistency. *Math. Biosci.*, 137(1):51–73, 1996.
10. Gary Chartrand and Ping Zhang. *A First Course in Graph Theory.* Courier Corporation, 2013.
11. David Cox, John Little, and Donal O'shea. *Ideals, Varieties, and Algorithms.* Undergraduate Texts in Mathematics. Springer Science+Business Media, third edition, 2007.
12. Andreas Dress, Katharina T. Huber, Jacobus Koolen, Vincent Moulton, and Andreas Spillner. *Basic Phylogenetic Combinatorics.* Cambridge University Press, Cambridge, United Kingdom, 2012.
13. Nicholas Eriksson. Using invariants for phylogenetic tree construction. In *Emerging applications of algebraic geometry*, pages 89–108. Springer, 2009.

14. S.N. Evans and T.P. Speed. Invariants of some probability models used in phylogenetic inference. *Ann. Statist.*, 21(1):355–377, 1993.
15. Joseph Felsenstein. *Inferring Phylogenies*. Sinauer Associates, Inc., Sunderland, UK, 2004.
16. Andrew Francis, Charles Semple, and Mike Steel. New characterisations of tree-based networks and proximity measures. *Advances in Applied Mathematics*, 93, February 2018.
17. D.R. Grayson and M.E. Stillman. Macaulay2, a software system for research in algebraic geometry. Available at http://www.math.uiuc.edu/Macaulay2/, 2002.
18. Elizabeth Gross and Colby Long. Distinguishing phylogenetic networks. *SIAM J. Appl. Algebra Geometry*, 2(1):72–93, 2018.
19. Stefan Grünewald, Peter J Humphries, and Charles Semple. Quartet compatibility and the quartet graph. *the electronic journal of combinatorics*, 15(1):103, 2008.
20. Brendan Hassett. *Introduction to Algebraic Geometry*. Cambridge University Press, New York, 2007.
21. Katharina T. Huber, Vincent Moulton, Charles Semple, and Taoyang Wu. Quarnet inference rules for level-1 networks. *Bulletin of Mathematical Biology*, 80(8):2137–2153, August 2018.
22. Daniel H. Huson, Regula Rupp, and Celine Scornavacca. *Phylogenetic Networks: Concepts, Algorithms and Applications*. Cambridge University Press, Cambridge, United Kingdom, 2010.
23. Leo Van Iersel and Vincent Moulton. Trinets encode tree-child and level-2 phylogenetic networks. *J. Math Biol.*, 68(7):1707–1729, June 2014.
24. J. Jansson, N.B. Nguyen, and W.K. Sung. Algorithms for combining rooted triplets into a galled phylogenetic network. *SIAM J. Comput.*, 35(5):1098–1121, 2006.
25. Jesper Jansson and Wing-Kin Sung. Inferring a level-1 phylogenetic network from a dense set of rooted triples. *Theoretical Computer Science*, 363(1):60–68, October 2006.
26. Judith Keijsper and R.A. Pendavingh. Reconstructing a phylogenetic level-1 network from quartets. *Bulletin of Mathematical Biology*, 76(10):2517–2541, October 2014.
27. Mitchel T Keller and William T Trotter. *Applied Combinatorics*. Mitchel T. Keller, William T. Trotter, 2016.
28. Colby Long and Seth Sullivant. Identifiability of 3-class Jukes–Cantor mixtures. *Advances in Applied Mathematics*, 64:89–110, 3 2015.
29. W.P. Maddison. Gene trees in species trees. *Syst. Biol.*, 46(523–536), 1997.
30. S. Mirarab, R. Reaz, M.S. Bayzid, T. Zimmermann, M.S. Swenson, and T. Warnow. ASTRAL: genome-scale coalescent-based species tree estimation. *Bioinformatics*, 30:i541–i548, 2014.
31. Luay Nakhleh. *Problem Solving Handbook in Computational Biology and Bioinformatics*, chapter Evolutionary Phylogenetic Networks: Models and Issues, pages 125–158. Springer Science+Business Media, LLC, 2011.
32. Lior Pachter and Bernd Sturmfels, editors. *Algebraic Statistics for Computational Biology*, page 101. Cambridge University Press, Cambridge, United Kingdom, 2005.
33. Fabio Pardi and Celine Scornavacca. Reconstructible phylogenetic networks: Do not distinguish the indistinguishable. *PLoS Comput Biol.*, 11(4), April 2015.
34. Joseph P. Ruskino and Brian Hipp. Invariant based quartet puzzling. *Algorithms Mol Biol.*, 7(35), 2012.
35. Charles Semple. Phylogenetic networks with every embedded phylogenetic tree a base tree. *Bulletin of Mathematical Biology*, 78(1):132–137, 2016.
36. Charles Semple and Mike Steel. *Phylogenetics*. Oxford University Press, Oxford, 2003.
37. S. Snir and S. Rao. Quartet MaxCut: a fast algorithm for amalgamating quartet trees. *Molecular Phylogenetics and Evolution*, 62(1):1–8, January 2012.
38. Claudia Solís-Lemus and Cécile Ané. Inferring phylogenetic networks with maximum pseudolikelihood under incomplete lineage sorting. *PLOS Genetics*, 2016.
39. Mike Steel. *Phylogeny: Discrete and Random Processes in Evolution*. CBMS-NSF Regional Conference Series in Applied Mathematics. SIAM, 2016.
40. Bernd Sturmfels. *Gröbner bases and Convex Polytopes*, volume 8. American Mathematical Soc., 1996.

41. Bernd Sturmfels and Seth Sullivant. Toric ideals of phylogenetic invariants. *J. Comp. Biol.*, 12(2):204–228, 2005.

42. Bernd Sturmfels and Seth Sullivant. Combinatorial secant varieties. *Quarterly Journal of Pure and Applied Mathematics*, 2:285–309, 2006.

43. Seth Sullivant. *Algebraic Statistics*. Graduate Studies in Mathematics. American Mathematical Society, 2018.

44. M. Syvanen. Horizontal gene transfer: evidence and possible consequences. *Annu. Rev. Genet.*, 28:237–261, 1994.

Tropical Geometry

Ralph Morrison

Abstract

Tropical mathematics redefines the rules of arithmetic by replacing addition with taking a maximum, and by replacing multiplication with addition. After briefly discussing a tropical version of linear algebra, we study polynomials built with these new operations. These equations define piecewise-linear geometric objects called tropical varieties. We explore these tropical varieties in two and three dimensions, building up discrete tools for studying them and determining their geometric properties. We then discuss the relationship between tropical geometry and algebraic geometry, which considers shapes defined by usual polynomial equations.

Suggested Prerequisites We use standard set theory notation (unions, functions, etc.) throughout this chapter. Section 1 draws on terminology and motivation from abstract algebra and linear algebra, but can be understood without them. Section 2 draws on topics from discrete geometry, although it is mostly self-contained. Section 3 includes geometry in three dimensions, which uses some notation from a standard course in multivariable calculus. Section 4 uses ring theory terminology from an abstract algebra course.

R. Morrison (✉)
Williams College, Williamstown, MA, USA
e-mail: 10rem@williams.edu

© Springer Nature Switzerland AG 2020
P. E. Harris et al. (eds.), *A Project-Based Guide to Undergraduate Research in Mathematics*, Foundations for Undergraduate Research in Mathematics,
https://doi.org/10.1007/978-3-030-37853-0_3

1   Tropical Mathematics

Take a piece of graph paper, or draw your own rectangular grid. Pick some of
the grid points, and join them up to form a polygon. Be sure it is convex, so that
all the angles are less than 180°. Now, start connecting grid points to each other
with line segments, never letting any two line segments cross. Keep going until you
cannot split things up anymore. You should end up with lots of triangles, like the
first picture in Fig. 1.

Using a different color, say purple, put a dot in every triangle. Connect two dots
with a line segment if their triangles share a side. If a triangle has a boundary edge,
just draw a little edge coming out of the dot. Your picture will now look like the
middle of Fig. 1. Now, try to draw your purple shape again, but with the following
rule: each line segment you draw should be perpendicular to the shared side of the
triangle.[1] Now you might have a picture like on the right in Fig. 1. Congratulations!
You have drawn your first *tropical curve*.[2]

Tropical curves, and more generally *tropical varieties*, are geometric shapes
that can be defined by familiar equations called *polynomials*. However, these
polynomials are interpreted using different rules of arithmetic than usual addition
and multiplication, replacing addition with taking a maximum and multiplication
with addition. The study of these shapes is called *tropical geometry*, although we
can also study other areas of mathematics with these new rules of arithmetic. In
general, we call these subjects *tropical mathematics*.

The first question most people have about tropical mathematics is why it is
called "tropical." One of the pioneers of tropical mathematics was Imre Simon,
a mathematician and computer scientist who was a professor at the University of
São Paulo Brazil. The adjective *tropical* to describe the field was coined by French
mathematicians (Dominique Perrin or Christian Choffrut, depending on who you

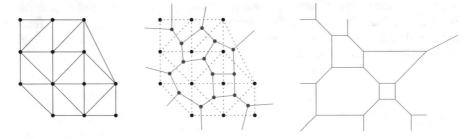

Fig. 1 Drawing a tropical curve

---

[1] You can draw this shape away from the polygon, so it is ok if your line segment between the two
dots doesn't cross the side of the triangle anymore! If it does not seem possible: go back to Step 1,
draw your triangles differently, and try again.

[2] Unless you have drawn one before. But hopefully it was fun anyway!

ask [47,52]) in Professor Simon's honor, based on the proximity of his university to the Tropic of Capricorn.

The second question most people have is why on Earth we would ever redefine our rules of arithmetic in this way. It turns out that it leads to some incredibly useful and beautiful mathematics. The first applications of this max-plus arithmetic were in the world of tropical linear algebra, where studying matrix multiplication and related problems in this setting helped solve automation and scheduling problems. More recently, tropical geometry arose as a skeletonized version of algebraic geometry, a major area of mathematics that studies solutions to polynomial equations. By "tropicalizing" solution sets to polynomial equations, we can turn algebro-geometric problems into combinatorial ones, studying more hands-on objects and then lifting that information back to the classical world. Beyond having applications to computational algebraic geometry, this has allowed for theorems, some new and some old, to be proven in a purely tropical way.

1.1   Tropical Arithmetic and Tropical Linear Algebra

The set of real numbers $\mathbb{R}$, equipped with addition $+$ and multiplication $\times$, has the algebraic structure of a *field*. This means we can add, subtract, multiply, and divide (except by zero), and that arithmetic works essentially how we expect it to. For instance, there is an additive identity 0, which does not change anything when added to it; and there is a multiplicative identity 1, which does not change anything when multiplied by it. The operations also play well together: for any $a, b, c \subset \mathbb{R}$, we have $a \times (b + c) = a \times b + a \times c$. If we forget about the fact that we can divide for a minute, all these properties (together with commutativity and associativity of our operations) mean that $(\mathbb{R}, +, \times)$ is a *commutative ring with unity*.

Let us now redefine arithmetic on the real numbers with *tropical addition* $\oplus$ and *tropical multiplication* $\odot$, where $a \oplus b = \max\{a, b\}$ and $a \odot b = a + b$. So, $2 \oplus 3 = 3$ and $2 \odot 3 = 5$. Instead of only allowing real numbers, we use the slightly larger set $\overline{\mathbb{R}} = \mathbb{R} \cup \{-\infty\}$, where $-\infty$ has the property that it is smaller than any element of $\mathbb{R}$. This means, for instance, that $-\infty \oplus 2 = 2$ and $-\infty \odot 2 = -\infty$.

The triple $(\overline{\mathbb{R}}, \oplus, \odot)$ *almost* has the structure of a commutative ring with unity, with $-\infty$ as the additive identity and 0 as the multiplicative identity. However, elements do not have additive inverses. The equation $1 \oplus x = 0$ has no solution, since we cannot "subtract" 1 from both sides. Thus, the triple $(\overline{\mathbb{R}}, \oplus, \odot)$ is a *semiring*, and in particular we call it the *tropical semiring*.[3]

---

[3] We could have just as easily defined tropical addition as taking the minimum of two numbers. (Instead of $-\infty$, we would have used ∞ as our additive identity.) Some researchers use the min convention, which is especially useful when studying connections to algebraic geometry; others use the max convention, which is more useful for highlighting certain dualities. Pay attention to the introductions of books and papers to determine which convention they are using!

Exercise 1 Verify that tropical addition and tropical multiplication satisfy the law of distributivity. That is, show that for any $a, b, c \in \overline{\mathbb{R}}$, we have $a \odot (b \oplus c) = (a \odot b) \oplus (a \odot c)$. Then explain why every element of $\overline{\mathbb{R}}$, besides the additive identity, has a multiplicative inverse. Because of this it would also be reasonable to refer to $(\overline{\mathbb{R}}, \oplus, \odot)$ as the *tropical semifield*.

Historically, the first use of these max-plus operations as an alternative to plus-times came in the world of *max-linear algebra*, which is similar to linear algebra over the real numbers except that all instances of $+$ and $\times$ are replaced with $\oplus$ and $\odot$. An example of matrix multiplication with these operations would be

$$\begin{pmatrix} 5 & 2 \\ -1 & 8 \end{pmatrix} \odot \begin{pmatrix} 1 & 0 \\ 2 & -\infty \end{pmatrix} = \begin{pmatrix} (5\odot1)\oplus(2\odot2) & (5\odot0)\oplus(2\odot-\infty) \\ (-1\odot1)\oplus(8\odot2) & (-1\odot0)\oplus(8\odot-\infty) \end{pmatrix} = \begin{pmatrix} 6 & 5 \\ 10 & -1 \end{pmatrix}. \tag{1}$$

There are many natural questions, equations, or definitions coming from usual linear algebra that, when studied tropically, boil down to a scheduling, optimization, or feasibility problem. We list a few here, and refer the reader to [8] for more details:

- Solving equations of the form $A \odot \mathbf{x} \leq \mathbf{b}$, where A and $\mathbf{b}$ are given, solves a scheduling problem.
- Finding the determinant of a matrix solves a job assignment problem. (We have to be careful what we mean by "determinant," since there are no negatives tropically!)
- Finding an eigenvalue of a matrix finds the shortest weighted cycle on the weighted graph given by the matrix. (And strangely, this matrix only has that one eigenvalue.)

Challenge Problem 1 Explain why each of the above linear algebra topics has the given interpretation when working tropically.

Research Project 1 Study the complexity of tropical matrix multiplication. For both tropical and classical matrix multiplication, the usual algorithm for multiplying two $n \times n$ matrices (namely taking the dot product of rows and columns) uses n^3 multiplications. However, an algorithm for classical matrix multiplication due to Strassen [53] has a runtime of $O(n^{2.807})$, with more recent algorithms pushing the runtime down to $O(n^{2.3728639})$ [33]. Can such improvements be made for tropical matrix multiplication?

More generally, study the computational complexity of problems in max-linear algebra.

1.2   Tropical Polynomials and Tropical Varieties

A traditional *polynomial* in n variables over $\mathbb{R}$ is a sum of terms, each of which consists of a coefficient from $\mathbb{R}$ multiplied by some product of those n variables (possibly an empty product; possibly with repeats). We study the set of points where these polynomials *vanish*; in other words, we set these polynomials equal to 0, and study the solution sets in $\mathbb{R}^n$.

Example 1 The polynomial $x^2 - 5x + 6$, the polynomial $x^2 + y^2 - 1$, and the polynomial $x^2 + y^2 + z^2 - 1$ are polynomials in one, two, and three variables, respectively. The solution sets obtained by setting these polynomials equal to 0 are the finite set $\{2, 3\}$ in $\mathbb{R}$; the unit circle in $\mathbb{R}^2$; and the unit sphere in $\mathbb{R}^3$, respectively.

Note that the solution set of $\{2, 3\}$ to $x^2 - 5x + 6 = 0$ (usually referred to as the *roots* of the polynomial) gives a factorization, namely $x^2 - 5x + 6 = (x - 2)(x - 3)$. This illustrates the fundamental theorem of algebra: that any non-constant polynomial in one variable can be factored into linear terms, each of the form $x - \alpha$ with α a root.[4]

Algebraic geometry is the field of mathematics that studies shapes defined by the vanishing of polynomials. *Tropical geometry*, in parallel, studies shapes defined by *tropical polynomials*. Tropical polynomials are the same as usual polynomials, except with all addition and multiplication replaced with tropical addition and tropical multiplication. This includes multiplication of variables, so that $x^2 y$ is interpreted as $x \odot x \odot y - x + x + y - 2x + y$.

Example 2 The tropical polynomial in one variable $x^2 \oplus (2 \odot x) \oplus (-1)$ can be written in classical notation as $\max\{2x, x + 2, -1\}$. The graph of this polynomial, interpreted as a function from $\mathbb{R}$ to $\mathbb{R}$, is illustrated in Fig. 2.

Although we could set a tropical polynomial equal to 0, the resulting solution set would not be especially meaningful: most tropical polynomials in one variable are equal to 0 at at most one point, which does not give much information about the polynomial. Instead, we study the points where the **maximum is achieved (at least) twice**. In the polynomial from Example 2, the maximum is achieved twice at two points: when $x = -3$ (where the $2 \odot x$ and -1 terms tie for the maximum), and when $x = 2$ (where the x^2 and $2 \odot x$ terms tie for the maximum).

---

[4]There is a bit more fine print: we must work over $\mathbb{C}$, the field of complex numbers, which is algebraically closed; and we may have to include multiple copies of the same term, based on the *multiplicity* of the root.

Fig. 2 The graph of the
tropical polynomial
$x^2 \oplus (2 \odot x) \oplus (-1)$

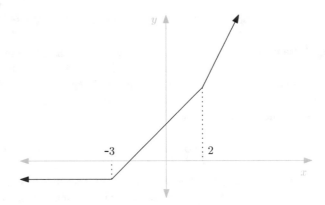

Definition 1 We say that a tropical polynomial $p(x_1, \ldots, x_n)$ *vanishes* at a point $(a_1, \ldots, a_n)$ if the maximum in $p(a_1, \ldots, a_n)$ is achieved at least twice. If $p(x)$ is a tropical polynomial in one variable that vanishes at a, we say that a is a *root* of $p(x)$.

As with classical roots, we can give tropical roots a notion of multiplicity: it is the change in slope going from one linear portion of the graph to the next at that root. So in Example 2, both roots have multiplicity 1, since the slope changes from 0 to 1, and then from 1 to 2.

Exercise 2 We say that a tropical polynomial in one variable has a *root at* $-\infty$ if the leftmost linear part of its graph does not have slope 0; the *multiplicity* of that root is defined to be the slope of that ray. With this definition, prove that any tropical polynomial in one variable of degree n has exactly n roots in $\overline{\mathbb{R}}$, counted with multiplicity. (In this sense, $\overline{\mathbb{R}}$ is "tropically algebraically closed.")

A natural question to ask is whether the tropical roots of a tropical polynomial in one variable have any real meaning. At least in our example, they give information about how to factor the polynomial: the reader can verify that $x^2 \oplus (2 \odot x) \oplus (-4) = (x \oplus -3) \odot (x \oplus 2)$. This property holds in general, if we are willing to consider factorizations that give the correct function, even if not the correct polynomial. (Check and see why $x^2 \oplus 0$ and $x^2 \oplus (-100 \odot x) \oplus 0$ define the same function, even though they are different polynomials!)

Challenge Problem 2 Prove the tropical fundamental theorem of algebra: that any tropical polynomial $p(x)$ in one variable is equal, as a function, to

$$c \odot (x \oplus \alpha_1)^{\mu_1} \odot (x \oplus \alpha_2)^{\mu_1} \odot \cdots \odot (x \oplus \alpha_k)^{\mu_k}, \tag{2}$$

where $\alpha_1, \ldots, \alpha_k$ are the tropical roots of p, with multiplicities $\mu_1, \ldots, \mu_k$, respectively, and where c is a constant.

Research Project 2 Study the factorization of tropical polynomials in more than one variable. Work in this direction has been done in [37], who provide efficient algorithms for certain classes of polynomials, even though in general this is an NP-complete problem.

Moving beyond polynomials in just one variable, we obtain tropical vanishing sets more complex than finite collections of points. In Sect. 2 we study tropical polynomials in two variables in depth, as well as the *tropical curves* they define in $\mathbb{R}^2$. In Sect. 3 we consider tropical polynomials in three variables, which define tropical surfaces. We also describe how intersecting such surfaces can give rise to tropical curves in three dimensions. In Sect. 4 we discuss the connection between algebraic geometry and tropical geometry through the tool of tropicalization.

1.3   Some Tropical Resources

Throughout this chapter we provide many references to books and articles on tropical geometry, both as sources for results and as great places to find ideas for research projects. We will frequently reference *An Introduction to Tropical Geometry* by Maclagan and Sturmfels [39], a graduate text that thoroughly develops the structure of tropical varieties and their connection to algebraic geometry. That book uses the min convention, while we use the max convention, so we adapt their results as necessary.

The material presented in this chapter, as well as in [39], looks at tropical geometry from an *embedded* perspective, where tropical varieties are subsets of Euclidean space. Another fruitful avenue is to look at tropical varieties, especially tropical curves, from an *abstract* perspective, under which tropical curves are thought of as graphs, possibly with lengths assigned to the edges. In the case of graphs without edge lengths, this theory is thoroughly explored in [16]. We also refer the reader to [3, 13, 15, 22, 41] for research articles incorporating this perspective.

Finally, there are many fantastic computational tools that help in exploring tropical geometry, both for computing examples and for implementing algorithms. Here are a few that we will reference in this chapter, all free to download:

- Gfan [29], a software package for computing Gröbner fans and tropical varieties.
- Macaulay2 [25], a computer algebra system. Especially useful for us are the Polyhedra and Tropical packages.
- polymake [23], which is open source software for research in polyhedral geometry. Among many other things, it can deal with polytopes and tropical hypersurfaces.

- TOPCOM [50], a package for computing Triangulations Of Point Configurations and Oriented Matroids. As we will see in Sects. 2 and 3, being able to find triangulations of polygons and polytopes goes hand in hand with researching tropical varieties.

2 Tropical Curves in the Plane

Let $p(x, y)$ be a tropical polynomial in two variables with at least two terms. Let S be the set of all pairs $(i, j) \in \mathbb{Z}^2$ such that a term of the form $c_{ij} \odot x^i \odot y^j$ appears in $p(x, y)$ with $c_{ij} \neq -\infty$; in other words, S is the set of all exponent pairs that actually show up in $p(x, y)$. We can then write our polynomial as

$$p(x, y) = \bigoplus_{(i,j) \in S} c_{ij} \odot x^i \odot y^j, \tag{3}$$

or in classical notation as

$$p(x, y) = \max_{(i,j) \in S} \{c_{ij} + ix + jy\}. \tag{4}$$

As established in Definition 1, we say $p(x, y)$ vanishes at a point if this maximum is achieved at least twice at that point. We call the set of points in $\mathbb{R}^2$ where p vanishes the *tropical curve* defined by p. Let $\mathcal{T}(p)$ denote this tropical curve.

Example 3 Let $p(x, y) = x \oplus y \oplus 0$. Written in classical notation, $p(x, y) = \max\{x, y, 0\}$. The maximum in this expression is achieved at least twice if two of the terms are equal, and greater than or equal to the third. This occurs at the point $(0, 0)$,[5] and along three rays emanating from this point: when $x = y \geq 0$, when $x = 0 \geq y$, and when $y = 0 \geq x$. The tropical curve $\mathcal{T}(p)$ is illustrated in Fig. 3. As mentioned in Exercise 3, we call this tropical curve a tropical line.

Fig. 3 The tropical line defined by $x \oplus y \oplus 0$

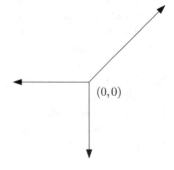

(0,0)

---

[5]In fact, the maximum occurs three times at this point.

Exercise 3 Any tropical curve defined by a tropical polynomial of the form $a \odot x \oplus b \odot y \oplus c$, with $a, b, c \in \mathbb{R}$, is called a *tropical line*. Determine all the possibilities for what a tropical line can look like. What if we allow one of a, b, or c to be $-\infty$?

2.1   Convex Hulls and Newton Polygons

A set in $\mathbb{R}^2$ (or more generally in $\mathbb{R}^n$) is called *convex* if any line segment connecting two points in the set is also contained in the set. The *convex hull* of a collection of points is the "smallest" convex set containing all the points.[6] The *Newton polygon* of $p(x, y)$, written Newt(p), is the convex hull of all the points in S. That is,

$$\text{Newt}(p) = \text{conv}\left(\{(i, j) \in \mathbb{Z}^2 \mid x^i \odot y^j \text{ appears in } p(x, y) \text{ with } c_{ij} \neq -\infty\}\right).$$
(5)

As the convex hull of finitely many points in $\mathbb{R}^2$, Newt(p) is either empty, a point, a line segment, or a two-dimensional polygon. To avoid certain trivial cases, we will assume that we have chosen p such that Newt(p) is a two-dimensional polygon. It is a *lattice polygon*, meaning that all vertices are *lattice points*, which are points with integer coordinates. In the special case that Newt$(p) = \text{conv}\{(0, 0), (d, 0), (0, d)\}$ for some positive integer d, we say that the polynomial has degree d, and we call the Newton polygon the *triangle of degree d*, denoted T_d.

Example 4 Let $p(x, y) = (1 \odot x^2) \oplus (1 \odot y^2) \oplus (2 \odot xy) \oplus (2 \odot x) \oplus (2 \odot y) \oplus 1$. Then we have that $S = \{(2, 0), (0, 2), (1, 1), (1, 0), (0, 1), (0, 0)\}$, so Newt$(p)$ is the triangle of degree 2, and $p(x, y)$ is a polynomial of degree 2. The Newton polygon, along with the tropical curve $\mathcal{T}(p)$, is illustrated in Fig. 4. Some preliminary connections between Newt(p) and $\mathcal{T}(p)$ can already be observed: the rays in $\mathcal{T}(p)$ point in directions that are perpendicular and outward relative to the edges of Newt(p). However, there are other features of the tropical curve not visible from

Fig. 4 The Newton polygon of $(1 \odot x^2) \oplus (1 \odot y^2) \oplus (2 \odot xy) \oplus (2 \odot x) \oplus (2 \odot y) \oplus 1$, along with the tropical curve the polynomial defines

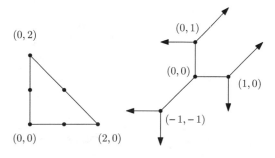

---

[6]More formally, it is the intersection of all convex sets containing the points. See if you can prove that such an intersection is still convex!

the Newton polygon; for instance, there are three bounded edges, and there are four *vertices*, where multiple edges or rays come together.

2.2   Subdivisions and the Duality Theorem

Since it was presented without justification, the reader might wonder: how did we determine $\mathscr{T}(p)$ in Example 4? One brute force way could be to take every possible pair among the 6 terms in $p(x, y)$ (there are 15 ways to do this), set them equal to each other, and try to determine whether those two terms ever tie for the maximum, and if so, where. It turns out that studying the Newton polygon of p leads to a much more elegant approach.

Let P be a lattice polygon, and $S = P \cap \mathbb{Z}^2$ be the set of integer coordinate points in P. Let $h : S \to \mathbb{R}$ be any function assigning real number values[7] to each element of S; we refer to h as a *height function*. We then define a set A of points in $\mathbb{R}^3$ by "lifting" the points of S to the heights prescribed by h:

$$A = \{(i, j, h(i, j)) \mid (i, j) \in S\}. \tag{6}$$

Take the convex hull of A in $\mathbb{R}^3$. Unless all the points of A lie on a plane, this convex hull is a three-dimensional *polytope*, the three-dimensional analog of a polygon, whose boundary consists of two-dimensional polygonal faces meeting along edges. Viewed from above, conv(A) looks like P, except subdivided by these upper polygonal faces. We call this subdivision of P the *subdivision induced by h*. The faces of conv(A) that are visible from above form the *upper convex hull* of A.

Example 5 Let $p(x, y)$ be as in Example 4. Let $P = \text{Newt}(p)$, and $S = P \cap \mathbb{Z}^2$. Define $h : S \to \mathbb{R}$ using the coefficients of $p(x, y)$, so that $h(i, j) = c_{i,j}$. Then the set A consists of the six points $\{(0, 0, 1), (1, 0, 2), (2, 0, 1), (0, 1, 2), (1, 1, 2), (0, 2, 1)\}$, illustrated on the left in Fig. 5. Their convex hull is then a polytope with 8

Fig. 5 The points of A labelled as $\times$'s, their convex hull, the induced subdivision of the triangle, and the dual tropical curve

---

[7]This definition will still work even if we define $h : S \to \mathbb{R} \cup \{-\infty\}$, as long as h does not map any vertices of P to $-\infty$.

triangular faces, illustrated in the middle of the figure. Of these faces, the 4 that are colored are visible from above, giving the induced subdivision of P shown towards the right. The tropical curve $\mathcal{T}(p)$ is reproduced, with vertices colored the same as their corresponding triangles, as described in Theorem 1 below.

The subdivision of the Newton polygon induced by the coefficients of the tropical polynomial gives us almost all the information regarding how to draw the tropical curve in the plane. Although this result holds in much more generality, we spell it out explicitly in the case of two variables.

Theorem 1 (The Duality Theorem, [39, Proposition 3.1.6]) *Let $p(x, y)$ be a tropical polynomial with $P = Newt(p)$ two-dimensional. Then the tropical curve $\mathcal{T}(p)$ is* dual *to the subdivision of P induced by the coefficients of $p(x, y)$ in the following sense:*

- *Vertices of $\mathcal{T}(p)$ correspond to polygons in the subdivision of P.*
- *Edges of $\mathcal{T}(p)$ correspond to interior edges in the subdivision of P.*
- *Rays of $\mathcal{T}(p)$ correspond to boundary edges in the subdivision of P.*
- *Regions of $\mathbb{R}^2$ separated by $\mathcal{T}(p)$ correspond to lattice points of P used in the subdivision.*

Moreover, two vertices of $\mathcal{T}(p)$ are connected by an edge if and only if their corresponding polygons in the subdivision share an edge, and the edge in the Newton polygon is perpendicular to the edge in the subdivision; and the rays emanating from a vertex in $\mathcal{T}(p)$ correspond to boundary edges of the corresponding polygon in the subdivision, with the rays in the outward perpendicular directions to the boundary edges of P.

So once we have found the subdivision of our Newton polygon, we know exactly what the tropical curve will look like, up to scaling edge lengths and up to translation. If we find the subdivision from Example 5, then our tropical curve could be either of the ones illustrated in Fig. 6 (or infinitely many others!). However, we can nail down the exact coordinates of the vertices by solving for the relevant three-way-ties. For instance, the top-most vertex of the tropical curve corresponds to the triangle with vertices at $(0, 2)$, $(0, 1)$, and $(1, 1)$ in the subdivision, so the coordinates of the vertex are located at the (unique) three-way tie between the y^2, the y, and the xy terms.

Sometimes there is information present in the polynomial or in the subdivision of the Newton polygon that is lost in the tropical curve. For instance, if $p(x, y) = x^2 \oplus y^2 \oplus 0$, then $\mathcal{T}(p)$ is, as a set, the tropical line from Fig. 3. By only considering this tropical curve as a set, we thus lose information about the starting polynomial. This leads us to decorate the edges and rays of our tropical curves with *weights*. In particular, each edge or ray is given a positive integer weight m, where m is equal to one less than the number of lattice points on the dual edge of the subdivision. Several tropical curves with the same Newton polygon are illustrated in Fig. 7, with

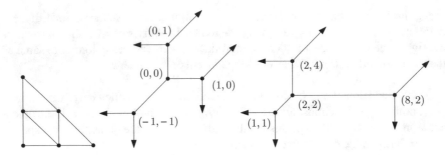

Fig. 6 A subdivision of a Newton polygon, and two possible tropical curves dual to it

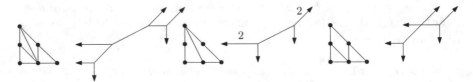

Fig. 7 Three tropical curves with the same Newton polygon, dual to different subdivisions. The first tropical curve is smooth, and the other two are not

all weights above 1 labelled. If a tropical curve has all weights equal to 1, and each vertex has a total of three edges and rays emanating from it, then we call the tropical curve *smooth*. Equivalently, a tropical curve is smooth if its dual subdivision is a *unimodular triangulation*, meaning that every polygon in the subdivision is a triangle with no lattice points besides its vertices.[8]

Exercise 4 Let $p(x, y)$ be a tropical polynomial of degree d such that $\mathscr{T}(p)$ is smooth. Determine the number of edges, rays, and vertices of $\mathscr{T}(p)$. (**Hint**: count up the corresponding objects in a unimodular triangulation of the triangle T_d. You can use the fact that any triangle in such a triangulation has area $1/2$.)

Challenge Problem 3 Show that any tropical curve satisfies the following *balancing condition*[9]: choose a vertex, and let $\langle a_1, b_1 \rangle, \langle a_2, b_2 \rangle, \ldots, \langle a_\ell, b_\ell \rangle$ be the outgoing directions of the rays and edges emanating from the vertex, where $a_i, b_i \in \mathbb{Z}$ and $\gcd(a_i, b_i) = 1$ for all i. Let m_i denote the weight of the ith edge/ray. Show that $m_1 \times \langle a_1, b_1 \rangle + m_2 \times \langle a_2, b_2 \rangle + \cdots + m_\ell \times \langle a_\ell, b_\ell \rangle = \langle 0, 0 \rangle$.

---

[8]By Pick's Theorem [46], this in turn is equivalent to every polygon in the subdivision being a triangle with area $1/2$.

[9]This is a special case of a much more general result called the *Structure Theorem*, which says that any tropical variety has the structure of a weighted, balanced polyhedral fan of pure dimension. See [39, Theorem 3.3.5].

Fig. 8 A set that turns out to
be a tropical curve

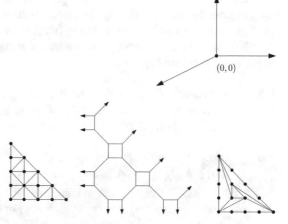

Fig. 9 A regular subdivision
with a dual tropical curve, and
a non-regular subdivision

Exercise 5 Consider the subset C of $\mathbb{R}^2$ illustrated in Fig. 8. It consists of three rays, all emanating from the origin, in the directions $\langle 1, 0 \rangle$, $\langle 0, 1 \rangle$, and $\langle -2, -1 \rangle$. Show that C is a tropical curve by finding a tropical polynomial $p(x, y)$ such that $C = \mathscr{T}(p)$. (**Hint**: the previous Challenge Problem might be useful!)

Armed with our Duality Theorem, one way to study tropical curves is the following: choose a tropical polynomial, find the induced subdivision of its Newton polygon, and draw it, solving for the exact coordinates of the vertices. Perhaps the most challenging step is finding the induced subdivision; this can be accomplished with such computational tools as `polymake`, `TOPCOM`, and `Macaulay2`.

Here we take another approach, similar to the very start of this chapter. Rather than starting with a tropical polynomial, choose the Newton polygon, and simply draw a subdivision, perhaps a unimodular triangulation. Then try to draw a tropical curve dual to it. (This is exactly the method from the start of Sect. 1.) An example of a triangulation of the triangle of degree 4 is illustrated in Fig. 9, along with a tropical curve that is dual to it. Note that to draw this tropical curve, we never needed to find a tropical polynomial defining it!

Sadly, this approach does not always work. A tropical curve can be drawn dual to a subdivision if and only if the subdivision is *regular*, meaning that it is induced by some height function.

Exercise 6 Consider the subdivision on the right in Fig. 9. Show that it is not a regular triangulation. (You might argue that no height function could have induced that triangulation; or you could argue that it is impossible to draw a tropical curve dual to it.)

It turns out that there are 1279 unimodular triangulations of the triangle of degree 4 up to symmetry [2, 7], and only one of them is non-regular: it is the unique unimodular triangulation that completes the non-regular subdivision from Fig. 9.

Similar phenomena occur for "small" polygons, whereby most triangulations end up being regular, so that drawing dual tropical curves is usually possible. For larger polygons, regular subdivisions seem to become rarer and rarer. See [30] for many results in the case that the polygon is a lattice rectangle, as well as [19] for results in a more general setting.

Challenge Problem 4 Let n be a positive integer, and let P be a $1 \times n$ lattice rectangle. Prove that any subdivision of P is regular. How many unimodular triangulations are there of P?

Research Project 3 Study the number of unimodular triangulations of families of lattice polygons, as was done for lattice rectangles in [30]. This can involve finding upper and lower bounds that improve those in the literature. Study the proportion of these unimodular triangulations that are regular. For all these endeavors, polymake and TOPCOM are fantastically useful computational tools.

2.3   The Geometry of Tropical Plane Curves

Many theorems about classical plane curves have analogs within the tropical world. A prime example of this is Bézout's Theorem.

Theorem 2 (Bézout's Theorem) *Let C and D be two smooth algebraic plane curves of degrees d and e. If C and D have no common components, then $C \cap D$ has at most $d \times e$ points. If we are working in projective space over an algebraically closed field, and counting intersection points with multiplicity, then $C \cap D$ has exactly $d \times e$ points.*

As shown in [49], the same result holds for tropical plane curves, once we determine how to count intersection points with multiplicity, and how to deal with tropical curves that intersect "badly."

Definition 2 Suppose two tropical plane curves C_1 and C_2 intersect at an isolated point (a, b) that is not a vertex of either curve. Such a point is called a *transversal intersection*. Let $\langle u_1, v_1 \rangle$ and $\langle u_2, v_2 \rangle$ be integer vectors describing the slopes of the edges or rays of C and D containing (a, b), where $\gcd(u_1, v_1) = \gcd(u_2, v_2) = 1$, and let the weights of the edges or rays be m_1 and m_2. Then the *multiplicity* of (a, b) is

$$\mu(a, b) := m_1 \times m_2 \times \left| \det \left(\begin{smallmatrix} u_1 & v_1 \\ u_2 & v_2 \end{smallmatrix} \right) \right|. \tag{7}$$

Example 6 Consider the tropical polynomials

$$f = (-1 \odot x^2) \oplus (xy) \oplus (-1 \odot y^2) \oplus x \oplus y \oplus (-1) \tag{8}$$

and

$$g = \left(-\frac{1}{2} \odot x^2\right) \oplus (1 \odot xy) \oplus (-2 \odot y^2) \oplus x \oplus y \oplus 0. \tag{9}$$

They both have the triangle of degree 2 as their Newton polygon and have induced subdivisions as illustrated on the left in Fig. 10. As shown on the right, the tropical curves $\mathcal{T}(f)$ and $\mathcal{T}(g)$ intersect in three points. The multiplicities of these points can be computed as 1, 1, and 2.

Let us push this example a little further. If we think of $\mathcal{T}(f) \cup \mathcal{T}(g)$ as $\mathcal{T}(f \odot g)$, then we can consider the dual subdivision of Newt($f \odot g$), illustrated in Fig. 11. Every polygon in this subdivision is dual to a vertex of $\mathcal{T}(f \odot g)$, and each vertex in $\mathcal{T}(f \odot g)$ is either a vertex of $\mathcal{T}(f)$, a vertex of $\mathcal{T}(g)$, or an intersection point. Note that each polygon dual to an intersection point (a, b) has area equal to $\mu(a, b)$.

Fig. 10 The subdivisions induced by f and g, and the two tropical curves

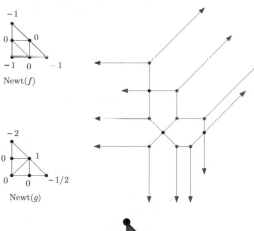

Newt(f)

Newt(g)

Fig. 11 The subdivisions induced by $f \odot g$, with blue triangles coming from vertices in $\mathcal{T}(f)$ and red triangles coming from $\mathcal{T}(g)$

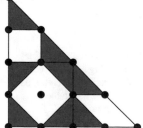

Exercise 7 Show that if f and g are tropical polynomials of degrees d and e, then $f \odot g$ is a tropical polynomial of degree $d \odot e$, and that $\mathscr{T}(f \odot g) = \mathscr{T}(f) \cup \mathscr{T}(g)$. Then show that the multiplicity of a transversal intersection point of f and g is equal to the area of the corresponding polygon in the subdivision of $\mathrm{Newt}(f \odot g)$ induced by $f \odot g$.

Theorem 3 (Tropical Bézout's Theorem, Transversal Case) *Let C and D be two tropical plane curves of degrees d and e with finitely many intersection points $(a_1, b_1), \cdots, (a_n, b_n)$, all of which are transversal. Then*

$$\sum_{i=1}^{n} \mu(a_i, b_i). \tag{10}$$

Note that we did not need to assume C and D were smooth. For an even more general result, we need to deal with the possibility that C and D have intersections that are not transversal. For two tropical curves C and D, we compute the *stable tropical intersection* as follows. Let $\mathbf{v} = \langle v_1, v_2 \rangle$ be a vector not parallel to any edge or ray of C and D, and for $\varepsilon \in \mathbb{R}^+$ let D_ε be a translation of D by $\varepsilon \mathbf{v}$. We then define

$$C \cap_{st} D = \lim_{\varepsilon \to 0} C \cap D_\varepsilon. \tag{11}$$

The *multiplicity* of a point in $C \cap_{st} D$ is the sum of the multiplicities of the corresponding points in a small enough perturbation $C \cap D_\varepsilon$.

Example 7 If $f(x, y) = x \oplus y \oplus 0$ and $g(x, y) = (1 \odot x) \oplus y \oplus 0$, then $C = \mathscr{T}(f)$ and $D = \mathscr{T}(g)$ are the tropical lines pictured in Fig. 12. Their set-theoretic intersection is a ray emanating from the point $(-1, 0)$. To find $C \cap_{st} D$, we move D slightly to D_ε, and then move it back to D. In the limit, we find a single stable intersection point at $(-1, 0)$.

Exercise 8 Show that $C \cap_{st} D$ is a well-defined set of finitely many points and is independent of the choice of v. Also show that the multiplicity of each point is well defined.

Fig. 12 Two tropical lines intersecting non-transversally, and a small perturbation used to compute the stable intersection

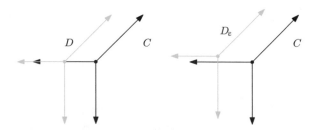

Theorem 4 (Tropical Bézout's Theorem, General Case) *Let C and D be two tropical plane curves of degrees d and e with $C \cap_{st} D = (a_1, b_1), \cdots, (a_n, b_n)$. Then*

$$\sum_{i=1}^{n} \mu(a_i, b_i). \tag{12}$$

Challenge Problem 5 Prove the transversal case of tropical Bézout's Theorem using an area-based argument involving the Newton polygon of $f \odot g$. Then use this result to prove the general case of tropical Bézout's Theorem.

Many classical results about algebraic plane curves involve when two curves are *tangent* to one another at some collection of points. Recently much work has been done to build up machinery to pose and study these sorts of results in the tropical world.

Definition 3 Let C and D be tropical curves. A *tangency* between C and D is a component of $C \cap D$ such that the stable intersection $C \cap_{st} D$ has more than one point in that component, counted with multiplicity. We say C and D are *tangent* at that component of $C \cap D$.

A tropical line that is tangent to a degree 4 curve at two distinct components are illustrated in Fig. 13. Such an intersection is called a *bitangent line*, which is also used to refer to an intersection component of multiplicity 4 or more.

Exercise 9 Find all the bitangent lines of the curve from Fig. 13. (Hint: there are infinitely many of them, but they still admit a nice classification.)

Counting bitangent lines is a very classical problem in algebraic geometry. In 1834, Plücker proved that a smooth algebraic plane curve of degree 4 has 28 bitangent lines [48]. A tropical analog of this fact was proved in [2].

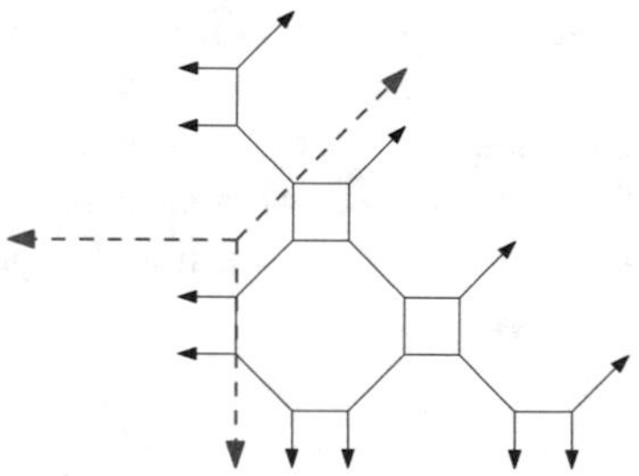

Fig. 13 A tropical line that is tangent to a tropical curve at two components

Theorem 5 (Theorem 3.9 in [2]) *Let C be a smooth tropical plane curve of degree 4. Then C has exactly seven classes[10] of bitangent lines.*

Later work was done to relate this theorem to Plücker's count, starting in [14] and culminating in [34], which showed how to recover the classical count of 28 bitangent lines from the tropical count, at least in sufficiently general cases.

Research Project 4 One great starting point for asking tropical questions is to study tropical versions of algebraic results. Study, prove, or disprove tropical analogs of these classical results. You may have to assume something about positions being sufficiently general.

- The De Bruijn–Erdös Theorem [18]: for any n points not all on a line determining t points, then $t \geq n$ and if $t = n$, any two lines have exactly one of the n points in common. (In this latter case, $n - 1$ of the points are collinear.)
- Steiner's conic problem [5]: given 5 curves of degree 2, how many curves of degree 2 are tangent to all of them? (Classically, the answer is 3264, although Steiner incorrectly computed it as 7776.)
- The Three Conics Theorem [20]: given three conics that pass through two given points, the three lines joining the other two intersections of each pair of conics all intersect at a point. Dually: given three conics that share two common tangents, the remaining pairs of common tangents intersect at three points that are collinear.
- The Four Conics Theorem [20]: Suppose we are given three conics, where two intersections of each pair lie on a fourth conic. Then the three lines joining the other two intersections of each pair of conics intersect in a point.

It is also worth determining when tropical geometry does *not* nicely mirror classical algebraic geometry. We say that an algebraic or a tropical curve C is *irreducible* if it cannot be written as $C_1 \cup C_2$, where $C_1 \subsetneq C$ and $C_2 \subsetneq C$ are curves as well. One nice property of algebraic curves (and more generally algebraic varieties) is that they admit a unique decomposition into irreducible components [17, Theorem 4.6.4], just as any integer $n \geq 2$ can be written as a product of primes uniquely (up to reordering). Tropical curves, however, do not.

---

[10]Loosely speaking, we say two bitangent lines intersecting at (P, Q) and (P', Q') with multiplicity 2 at each point are equivalent if (P, Q) and (P', Q') are equivalent in the language of *divisor theory* [22].

Fig. 14 A tropical curve that can be decomposed into irreducible tropical curves in two distinct ways

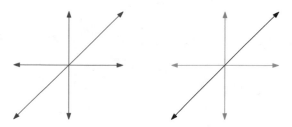

Example 8 Consider the set C in $\mathbb{R}^2$ consisting of the (usual) lines $x = 0$, $y = 0$, and $x = y$. We claim that C is a tropical curve; you will show this in Exercise 10. We can also write C as $\mathscr{T}(x \oplus y \oplus 0) \cup \mathscr{T}((xy) \oplus x \oplus y)$, or as $\mathscr{T}(x \oplus y) \cup \mathscr{T}(x \oplus 0) \cup \mathscr{T}(y \oplus 0)$, as illustrated in Fig. 14.

Exercise 10 Find a polynomial f such that $C = \mathscr{T}(f)$, where C is the set from Example 8. How does this polynomial relate to the polynomials defining the two decompositions of C as a union of tropical curves?

> **Research Project 5** Study how many decompositions a tropical curve can have as a union of tropical curves properly contained within it. You could stratify this study by the Newton polygon of the curve. (This is closely related to the research project on factoring tropical polynomials; see if you can see why, especially after you try Exercise 10!)

A new approach in tropical geometry that avoids non-uniqueness of decompositions is to develop *tropical schemes* [24, 38], just as algebraic geometers study *algebraic schemes* [27]. This model does not consider the tropical curves from the second decomposition in Example 8 to be tropical curves, and in fact gives us a unique decomposition in general.

2.4   Skeletons of Tropical Plane Curves

Choose a lattice polygon P with g interior lattice points, where g is at least 2. Write P_{int} for the convex hull of the g interior lattice points; this is either a line segment, or a polygon. Let $p(x, y)$ be a tropical polynomial with Newton polygon P. Rather than study the full tropical curve $\mathscr{T}(p)$, we can focus on a portion of it called its *skeleton*. To find the skeleton, we delete all rays from our tropical curve, and then successively remove any vertices incident to exactly one edge, along with such edges. This will lead to a collection of vertices and edges, where each vertex is incident to at least two edges. We "smooth over" the vertices incident to two edges,

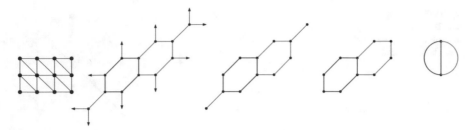

Fig. 15 A tropical curve (with its dual subdivision) undergoing the process of skeletonization. The edges of the tropical curve that end up contributing to the skeleton are color-coded based on which final edge they become a part of

removing such vertices and fusing the two edges into one. The resulting collection of edges and vertices is called the *skeleton* of the tropical curve. This process is illustrated in Fig. 15.

The structure that remains after "skeletonizing" a tropical curve is a *graph*.[11] A graph is simply a collection of vertices collected by edges; in our setting, two vertices may be connected to each other by multiple edges, and a vertex may be connected to itself by an edge, which we call a loop. This leads us to the following major question: Which graphs can appear as the skeleton of tropical plane curve? To simplify, let us assume that our tropical curves are smooth.

Definition 4 A graph that is the skeleton of some smooth tropical plane curve is called *tropically planar*, or *troplanar* for short. The *genus*[12] of the graph is the number of bounded regions in the plane formed by a drawing on the graph. By Euler's formula relating the number of vertices, edges, and faces of a planar graph, we could also define the genus as $E - V + 1$ for a graph with E edges and V vertices.

With these definitions, we can say that the graph on the right in Fig. 15 is troplanar, and has genus 2.

Exercise 11 Let G be a troplanar graph. Show that G is connected (all one piece), planar (able to be drawn in the plane without any edges crossing), and trivalent (meaning that every vertex has three edges coming from it, where a loop counts as two edges). Also show that the genus of the graph is equal to g, the number of interior lattice points of the Newton polygon of any smooth tropical curve that has G as its skeleton.

---

[11] In fact, there is a bit more structure: it is a *metric graph*, meaning the edges have lengths. We will come back to that later in this subsection.

[12] There is another, unrelated definition of *genus* in graph theory, dealing with the smallest number of holes a surface must have to allow a given graph to be embedded on it.

A daunting task is to try to determine which graphs are tropically planar. Even for fixed g, it is not immediately obvious that there is an algorithmic way to do this. There are several things working in our favor:

1. There are only finitely many polygons with $g \geq 1$ interior lattice points, up to equivalence.[13] As discussed in [7, Proposition 2.3], this follows from results in [51] and [32]. An algorithm for finding all such polygons for a given g is presented in [10].
2. If P and Q are lattice polygons with $P \subset Q$ and $P_{int} = Q_{int}$, all the troplanar graphs arising from P also arise from Q [7, Lemma 2.6].

Exercise 12 Prove item 2 above.

Item 1 means that we only need to consider a finite collection of possible Newton polygons for each genus g; item 2 decreases that number considerably. It means that we need to only consider *maximal* polygons, which are those that are not properly contained in any polygon with the same interior lattice points.

Even when we have restricted to maximal polygons, there are two different flavors of polygons: the *hyperelliptic* polygons, for which P_{int} as a line segment, and the *nonhyperelliptic* polygons, for which P_{int} is a two-dimensional polygon. See Fig. 16 for all the maximal polygons with 4 interior lattice points, up to equivalence. The leftmost three are nonhyperelliptic, and the other six are hyperelliptic.

How do we know there are not any other maximal polygons with 4 interior lattice points? For the hyperelliptic case, [31] classifies all maximal hyperelliptic curves: they are a family of trapezoids interpolating between a hyperelliptic rectangle and a

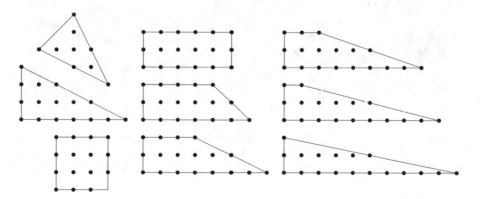

Fig. 16 The maximal polygons with 4 interior lattice points

---

[13]Here we say two lattice polygons are *equivalent* if one is the image of the other under a matrix transformation $\left(\begin{smallmatrix} a & b \\ c & d \end{smallmatrix}\right)$, where $ad - bc = \pm 1$.

hyperelliptic triangle (this result is also presented in [10]). For the nonhyperelliptic polygons, we have the following result.

Proposition 1 (Lemma 2.2.13 in [31]; Also Theorem 5 in [10]) *Let P be a maximal nonhyperelliptic polygon, with P_{int} its interior polygon. Then P is obtained from P_{int} by "pushing out" the edges of P_{int}. More formally, if $P_{int} = \bigcap_{i=s}^{s} H_i$, where H_i is the half-plane defined by the inequality $a_i x + b_i y \leq c_i$ (with a_i, b_i, c_i relatively prime integers), then $P_{int} = \bigcap_{i=1}^{s} H'_i$, where H'_i is the half-plane defined by the inequality $a_i x + b_i y \leq c_i + 1$.*

This means that in order to find all maximal nonhyperelliptic lattice polygons with g interior lattice points, one can first all lattice polygons with g lattice points total, and then determine which can be pushed out to form a lattice polygon.

Exercise 13 Using Proposition 1, verify that Fig. 16 does indeed contain all maximal nonhyperelliptic polygons with 4 interior lattice points. Then find all maximal nonhyperelliptic polygons with 5 interior lattice points.

Exercise 14 Determine which troplanar graphs of genus g come from hyperelliptic Newton polygons. (**Hint**: if $g = 3$, there are three such graphs, namely the middle three graphs from Fig. 17.)

Research Project 6 Study the properties of lattice polygons, stratified by the number of interior lattice points g. (A great starting point for exploring these topics are the papers [10] and [11].) For example: Given a maximal polygon P, let $n(P)$ be the number of subpolygons of P with the same set of interior lattice points. For which polygons is $n(P)$ equal to 1? What upper bounds can we find on $n(P)$, in terms of g? How big is $n(P)$ on average? (This gives us an idea of how much time we save by considering only maximal polygons when studying troplanar graphs.)

Example 9 Let us find all troplanar graphs of genus 3 (This will mirror arguments found in [2] and [7].) There are exactly five trivalent connected graphs of genus 3 [4], namely those appearing in Fig. 17. By Exercise 11, these are the only possible

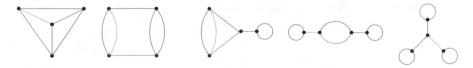

Fig. 17 The five candidate graphs of genus 3

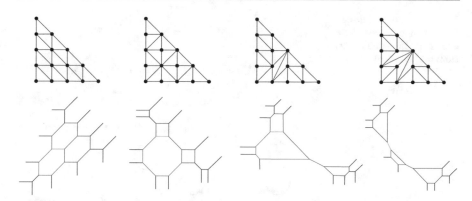

Fig. 18 Four triangulations, giving us four tropical curves whose skeletons are the first four graphs in Fig. 17

graphs that could be troplanar. We now must determine which of the five are actually achievable.

Let us determine which Newton polygons are possible. As mentioned previously, it suffices to take P maximal. We will focus on nonhyperelliptic polygons; the hyperelliptic ones are covered by Exercise 14. It turns out that the only nonhyperelliptic polygon with 3 interior lattice points, up to equivalence, is T_4, the triangle of degree 4. This is because the only lattice polygon (again, up to equivalence) with three lattice points is the triangle of degree 1, which pushes out to T_4. Figure 18 shows triangulations of T_4 that give tropical curves whose skeletons are the first four graphs from Fig. 17, so we know that those four graphs are all troplanar.

Let us now argue that the fifth graph, sometimes called the lollipop graph of genus 3, is not troplanar. Note that any bridge[14] in troplanar graph must be dual to a *split* in the subdivision of T_4, which is an edge goes from one boundary point to another, with some interior lattice points on each side and none in the edge's interior. So, any triangulation of the triangle of degree 4 that gives us the lollipop graph would have three splits. All possible splits in the triangle are illustrated in Fig. 19; however, no more than two of them can coexist in the same triangulation due to intersections, meaning we cannot obtain the lollipop graph. We conclude that there are four troplanar graphs of genus 3: the first four graphs in Fig. 17.

The fact that the lollipop graph did not appear also follows from a more general result about structures that cannot appear in troplanar graphs. We say a connected, trivalent graph is *sprawling* if removing a single vertex splits the graph into three pieces. Several examples of sprawling graphs appear in Fig. 20.

---

[14] A *bridge* in a connected graph is an edge that, if removed from the graph, would disconnect the graph.

Fig. 19 Twelve splits, any
three of which have at least
one intersection point away
from the boundary

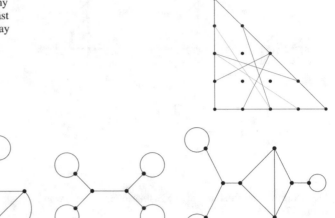

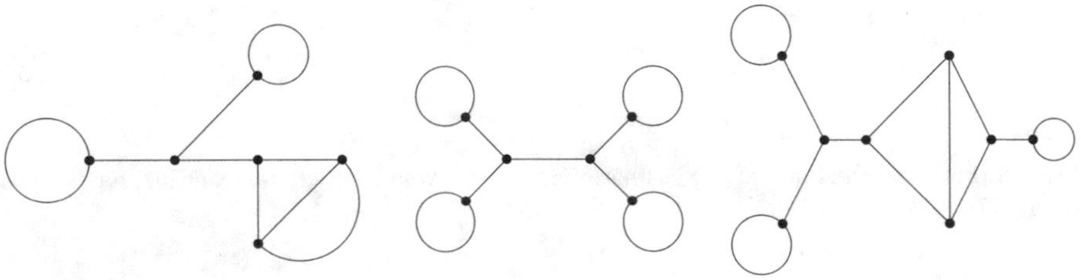

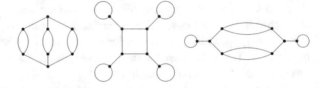

Fig. 20 Three sprawling graphs. Note that the vertex that disconnects the graph into three pieces
need not be unique

Fig. 21 Three graphs of
genus 5 that are *not* troplanar

Proposition 2 (Proposition 4.1 in [9]) *A sprawling graph cannot be troplanar.*

Although this result was originally proved in [9], the "sprawling" terminology
comes from [2], which offers an alternate proof.

Challenge Problem 6 Prove Proposition 2. (**Hint:** Consider the structure of the
dual triangulation of a smooth tropical curve with a sprawling skeleton.)

Challenge Problem 7 Show that the graphs in Fig. 21 are not troplanar.

> **Research Project 7** Find "forbidden structures" that never appear in tropla-
> nar graphs. (Proposition 2 gives an example of such a forbidden structure.
> Another is given in [42].)

Challenge Problem 8 There are 17 trivalent connected graphs of genus 4 [4].
Determine which of them are troplanar. Note that the only Newton polygons

you need to consider are those illustrated in Fig. 16. (If you have already done Exercise 14, you can ignore six of the polygons!)

In general, counting the number of tropically planar graphs of genus g can be accomplished as follows:

1. Find all maximal lattice polygons P with g interior lattice points, perhaps following [10].
2. Find all regular unimodular triangulations of each P from step 1, perhaps with `polymake` or `TOPCOM`.
3. Find the dual skeletons to the triangulations from step 2, and sort them into isomorphism classes.

This algorithm was implemented in [7] and was used to determine that the numbers of troplanar graphs of genus 2, 3, 4, and 5 are 2, 4, 13, and 37, respectively. This was pushed further as part of the Williams SMALL 2017 REU to genus 6 (151 troplanar graphs) and genus 7 (672 troplanar graphs).

Research Project 8 Find a more efficient way to determine the number of troplanar graphs of genus g than the algorithm outlined above.

Research Project 9 Study how the number of troplanar graphs of genus g grows with g. Can you find upper and lower bounds? Can you determine its asymptotic behavior? (Preliminary work in this direction was done in the Williams College SMALL REU in 2017.)

So far we have considered skeletons from a purely combinatorial perspective. Now we include the data of lengths on each edge of the graph, giving us a *metric graph*. A natural impulse is to sum up all the Euclidean lengths of the edges of the embedded tropical curve that make up a given edge of the skeleton and declare that to be its length. Unfortunately this definition of length is not invariant under the natural transformations that we apply to our Newton polygons. This leads us to use the following definition.

Definition 5 Let $P_1, P_2 \in \mathbb{R}^2$ be distinct points such that the line segment $\overline{P_1 P_2}$ has rational slope (or is vertical). Write the vector from P_1 to P_2 as $\lambda \times \langle a, b \rangle$, where $a, b \in \mathbb{Z}$ with $\gcd(a, b) = 1$ and $\lambda \in \mathbb{R}^+$. The *lattice length* of the line segment $\overline{P_1 P_2}$ is defined to be λ.

Fig. 22 A tropical curve
with lattice lengths labelled,
and the resulting lengths on
the skeleton

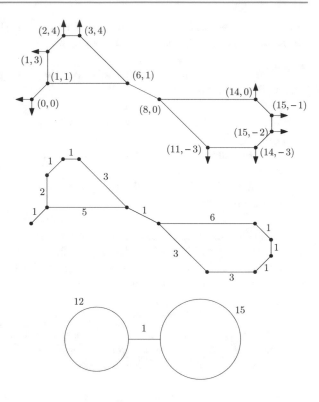

When considering a tropical plane curve, we measure the lengths of its finite
edges by lattice length. These lengths are then added up appropriately to assign
lengths to the edges of the skeleton.

Example 10 Consider the tropical plane curve illustrated on the top in Fig. 22.
Below it is the collection of all bounded edges in the curve, labelled with their
lattice lengths. As pictured, the skeleton is a graph consisting of two vertices joined
by an edge, with a loop attached to each vertex. The length of the middle edge
in the skeleton is 1; the lengths of the loops are $2 + 1 + 1 + 3 + 5 = 12$ and
$6 + 3 + 3 + 1 + 1 + 1 = 15$. (Note that one bounded edge from the tropical curve
does not contribute to the skeleton.)

When we say that a metric graph is troplanar, we mean that it is the skeleton of
a smooth tropical plane curve *giving those edge lengths*. So the metric graph at the
bottom of Fig. 22 is troplanar.

Challenge Problem 9 Let P be a 2×3 lattice rectangle. Find all troplanar metric
graphs that are the skeleton of a smooth tropical curve with that Newton polygon.
(**Hint**: in some sense you can get most, but not all, graphs of genus 2.)

Fig. 23 Two metric graphs that are not troplanar

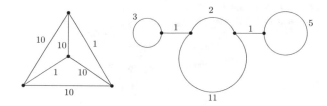

The algorithm presented in [2] did not simply find the combinatorial types of troplanar graphs; it computed, up to closure, all *metric* graphs of genus at most 5 that appeared as the skeleton of a smooth tropical plane curve. In their Theorem 5.1, they use this computation to characterize exactly which metric graphs of genus 3 are troplanar. Beyond the lollipop graph not appearing (regardless of the edge lengths), there are nontrivial edge length restrictions on the other four combinatorial types of graphs. Rather than presenting their full result here, we give a consequence of it.

Theorem 6 (Corollary 5.2 in [7]) *Approximately* 29.5% *of all metric graphs of genus* 3 *are troplanar.*

This probability is computed by considering the *moduli space of graphs of genus* 3 [6, 12]. This is a six-dimensional space, corresponding to the six edges a trivalent graph of genus 3 has. This space is not compact, since edge lengths can be arbitrarily long; so consider the subspace consisting of graphs with total length equal to 1; up to scaling, every metric graph can be represented in this way. Give each of the five combinatorial types of graphs (as illustrated in Fig. 17) an equal weight, and compute the volume of the space of troplanar graphs within this 5-dimensional space. This computation gives about 0.295 or 29.5%.

Challenge Problem 10 Show that neither of the metric graphs illustrated in Fig. 23 are troplanar. (This follows from the characterization given in [7, Theorem 5.1]; try to give your own argument.)

Research Project 10 Determine which metric graphs arise as the skeleton of a smooth tropical plane curve, perhaps under certain restrictions. For instance:

- Characterize exactly which metric graphs arise from hyperelliptic polygons, as explored in [42].
- Characterize which metric graphs arise from *honeycomb polygons*, a key tool in [7].

(continued)

Fig. 24 A nodal tropical
curve and its skeleton

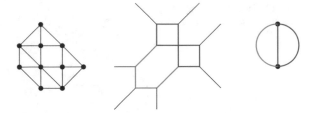

- Characterize which metric graphs are troplanar with as many degrees of freedom as possible on their edge lengths. In [7], this maximum number of degrees of freedom was shown to be $2g + 1$, at least for $g \geq 8$.

All of our questions have been posed for smooth tropical plane curves. Of course, we can also consider tropical curves with singularities. We say a tropical curve is *nodal* if, in the dual subdivision, all polygons besides the triangles of area $1/2$ are quadrilaterals of area 1. A vertex in a nodal tropical curve dual to such a quadrilateral is called a *node*.

Example 11 Figure 24 presents an example of a nodal tropical curve with its dual Newton subdivision. We can still consider a skeleton of the curve by interpreting each nodal crossing in the tropical curve as two edges in the graph that happen to look like they are crossing. The resulting skeleton is pictured on the right.

It was shown in [9] that *every* connected trivalent graph can be realized in a nodal tropical plane curve. Given a connected trivalent graph G, let $N(G)$ be the *tropical crossing number* of G, which is the smallest number of nodes required to achieve G as the skeleton of a nodal tropical curve. For instance, $N(G) = 0$ if and only if G is troplanar.

Research Project 11 Study the tropical crossing number. Can you determine its value explicitly for certain families of graphs? (Note that if this question is being posed for metric graphs, $N(G)$ does depend on the edge lengths.)

3   Tropical Geometry in Three Dimensions

Moving beyond the plane into three-dimensional space, we consider tropical polynomials in three variables x, y, and z. Such a polynomial can be written as

Fig. 25 The tropical plane
defined by $x \oplus y \oplus z \oplus 0$

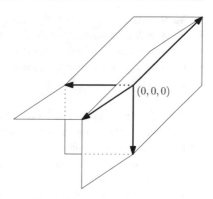

$$p(x, y, z) = \bigoplus_{(i,j,k) \in S} c_{ijk} \odot x^i \odot y^j \odot z^k, \qquad (13)$$

where S is the set of all exponent vectors that appear in $p(x, y, z)$. This polynomial defines a *tropical surface*, the set of all points in $\mathbb{R}^3$ where the maximum defined by the polynomial is achieved at least twice. Again, we denote this tropical surface $\mathcal{T}(p)$.

Example 12 Let $p(x, y, z) = x \oplus y \oplus z \oplus 0$. The tropical surface $\mathcal{T}(p)$ is illustrated in Fig. 25. It consists of the origin $(0, 0, 0)$; four rays, pointing in the directions $\langle -1, 0, 0 \rangle$, $\langle 0, -1, 0 \rangle$, $\langle 0, 0, -1 \rangle$, and $\langle 1, 1, 1 \rangle$; and six two-dimensional pieces, each obtained as the positive linear span of two of the rays. Such two-dimensional pieces of a tropical surface are called *two-dimensional cells*. Because of the form of $p(x, y, z)$, we call $\mathcal{T}(p)$ a *tropical plane*.

3.1   Tropical Surfaces and the Duality Theorem

The Duality Theorem still holds for tropical polynomials in three variables and the surfaces they define.[15] This time, instead of a Newton polygon we consider a *Newton polytope*, the convex hull of all exponent vectors appearing in the polynomial. (We will assume that the Newton polytope is three-dimensional to avoid certain degenerate cases.) To find an induced subdivision, we again associate heights to each lattice point of the Newton polytope; this time, however, we must compute our upper convex hull in four-dimensional space. We then have the following correspondence between parts of the tropical surface $S = \mathcal{T}(p)$ and the subdivision of Newt(p):

---

[15]Indeed, a Duality Theorem holds for all tropical varieties defined by a single equation in any number of variables; see [39, Proposition 3.1.6].

- Vertices in S correspond to 3-dimensional polytopes in the subdivision.
- Rays in S correspond to boundary two-dimensional faces.
- Edges in S correspond to interior two-dimensional faces.
- Unbounded two-dimensional cells in S correspond to boundary edges.
- Bounded two-dimensional cells in S correspond to interior edges.

As was the case for tropical plane curves, the relationships and geometry of all these pieces of the tropical surface are dictated by the subdivision. For instance, two vertices are joined by an edge if and only if the corresponding polytopes share a face; and that edge is perpendicular to the shared face.

We say that a subdivision of a polytope is a *unimodular tetrahedralization* if all polytopes in the subdivision are tetrahedra of volume $\frac{1}{6}$, which is the smallest possible volume. We say that a tropical surface $\mathscr{T}(p)$ is *smooth* if the induced subdivision of Newt(p) is a unimodular tetrahedralization. If Newt(p) is the tetrahedron with vertices at $(0, 0, 0)$, $(d, 0, 0)$, $(0, d, 0)$, and $(0, 0, d)$, we say that $p(x, y, z)$ *has degree d*.

Example 13 Let

$$f(x, y, z) = (xy \odot z) \oplus (-42 \odot xy) \oplus x \oplus y \oplus z \oplus (-42), \tag{14}$$

and let $P = \text{Newt}(f)$. The polytope P looks like a cube with two tetrahedra sliced off, as illustrated to the left in Fig. 26. Every term has coefficient 0, except for the $(0, 0, 0)$ and $(1, 1, 0)$ terms, which have a very negative coefficient. This means that in the subdivision, we will end up with two smaller tetrahedra with vertices at $(1, 0, 0)$, $(0, 1, 0)$, $(1, 1, 0)$, and $(1, 1, 1)$; and at $(0, 0, 0)$, $(0, 0, 1)$, $(1, 0, 0)$, and $(0, 1, 0)$; as well as a larger tetrahedron at $(0, 0, 1)$, $(1, 0, 0)$, $(0, 1, 0)$, and $(1, 1, 1)$.[16] This is illustrated in Fig. 26.

The tropical surface $\mathscr{T}(f)$ has three vertices, corresponding to the three tetrahedra. We can find their coordinates by computing the four-way ties.

- From $-42 = x = y = z$, we have a vertex at $(-42, -42, -42)$.
- From $x = y = z = x + y + z$, we have a vertex at $(0, 0, 0)$.
- From $x = y = -42 + x + y = x + y + z$, we have a vertex at $(42, 42, -42)$.

The vertex at $(0, 0, 0)$ connects to the other two vertices by a line segment. The vertex $(-42, -42, -42)$ will have three rays, pointing in the directions $\langle -1, 0, 0 \rangle, \langle 0, -1, 0 \rangle$, and $\langle 0, 0, -1 \rangle$. The vertex $(0, 0, 0)$ will have two rays, point-

---

[16]To prove this rigorously, we would need to show that the hyperplane in $\mathbb{R}^4$ containing the points $(1, 0, 0, 0)$, $(0, 1, 0, 0)$, $(1, 1, 0, 0)$, and $(1, 1, 1, -42)$ lies strictly above the points $(0, 0, 1, 0)$ and $(0, 0, 0, -42)$; as well as two other similar such statements, one for each of the other tetrahedra. (In fact, the hyperplane we get from the middle tetrahedron in (x, y, z, w)-space is just defined by $w = 0$, and certainly the other two lifted points $(0, 0, 0, -42)$ and $(1, 1, 0, -42)$ lie below this hyperplane.)

Fig. 26 The subdivided
Newton polytope from
Example 13

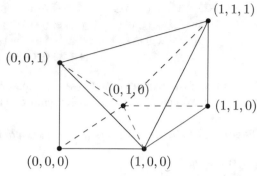

Fig. 27 The
one-dimensional pieces of the
surface from Example 13

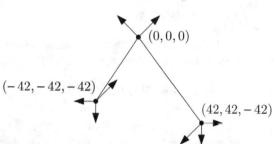

ing in the directions $\langle 1, -1, 1, \rangle$ and $\langle -1, 1, 1, \rangle$. Finally, the vertex $(42, 42, -42)$
will have three rays, pointing in the directions $\langle 1, 0, 0 \rangle, \langle 0, 1, 0 \rangle$, and $\langle 0, 0, 1 \rangle$. Ignoring the two-dimensional pieces, our tropical surface looks as pictured in Fig. 27.

We fill in two-dimensional pieces between adjacent rays and edges. This will
give a total of 12 unbounded two-dimensional pieces, corresponding to the 12 edges
in our tetrahedralization. All are unbounded, since all edges in the tetrahedralization
are exterior.

Note that this tropical surface is *not* smooth. Even though our induced subdivision is a tetrahedralization, it is not unimodular since the tetrahedra do not all have
volume $\frac{1}{6}$: the tetrahedron in the middle has volume $1/3$.

Challenge Problem 11 Show that the tropical polynomial of degree 2 defined by

$$f = (-3 \odot x^2) \oplus (-4 \odot xy) \oplus xz \oplus (-7 \odot y^2) \oplus (-2 \odot yz)$$
$$\oplus (-1 \odot z^2) \oplus x \oplus y \oplus (-2 \odot z) \oplus (-7) \tag{15}$$

is a smooth tropical surface. Determine how many vertices, edges, rays, bounded
two-dimensional cells, and unbounded two-dimensional cells there are. Do the same
for the tropical polynomial of degree 3 defined by

$$g = (-23 \odot x^3) \oplus (-15 \odot x^2 y) \oplus (-7 \odot x^2 z) \oplus (-15 \odot x y^2) \oplus xyz \oplus (-3 \odot xz^2)$$
$$\oplus (-25 \odot y^3) \oplus (-6 \odot y^2 z) \oplus (-10 \odot yz^2) \oplus (-20 \odot z^3) \oplus (-2 \odot x^2)$$
$$\oplus (-6 \odot xy) \oplus (-1 \odot xz) \oplus (-14 \odot y^2) \oplus yz \oplus (-9 \odot z^2) \oplus (-11 \odot x)$$
$$\oplus (-4 \odot y) \oplus (-9 \odot z) \oplus (-21).$$

(16)

You will almost certainly want to use a computer to help with this! After you try this Challenge Problem, you should check your counts against the following theorem.

Theorem 7 (Theorem 4.5.2 in [39]) *A smooth tropical surface of degree d has*

- d^3 *vertices,*
- $2d^2(d-1)$ *edges,*
- $4d^2$ *rays,*
- $d(d-1)(7d-11)/6$ *bounded two-dimensional cells,*
- $6d^2$ *unbounded two-dimensional cells.*

Its Euler characteristic[17] is $\frac{(d-1)(d-2)(d-3)}{6} + 1$.

> **Research Project 12** Study the geometry of smooth tropical surfaces. For instance:
>
> - A smooth surface of degree 3 has 10 bounded two-dimensional cells, each of which is a polygon, say with n_i sides for the ith polygon. What are the possible values for $n_1, \ldots, n_{10}$? How can these 10 polygons be arranged relative to each other?
> - A smooth surface of degree 4 has Euler characteristic 1, and so contains one polytope bounding a three-dimensional region. Can we characterize which polytopes are possible? (How many faces, how many edges, etc.)
> - Moving on to smooth surfaces of degree greater than 4, which have Euler characteristic greater than 1, there are multiple polytopes that are part of the surface. How can these polytopes be arranged? (This is the surface analog of asking what the skeleton of a smooth tropical plane curve can be.)

3.2   Tropical Curves in $\mathbb{R}^3$

In usual geometry, if we intersect a pair of two-dimensional surfaces in $\mathbb{R}^3$, we expect to get a one-dimensional curve. This also holds in tropical geometry, if we are willing to assume stable intersections to avoid the overlap of two-dimensional

---

[17]Intuitively, this is the number of bounded regions of $\mathbb{R}^3$ encapsulated by part of the surface.

pieces. There is still a Duality Theorem for tropical curves in $\mathbb{R}^3$ that arise as the intersection of two tropical surfaces, although it requires a bit more machinery.

Given two lattice polytopes $P, Q \subset \mathbb{R}^n$, place P and Q in $(n + 1)$-dimensional space by giving every point in P an extra coordinate of 0 and every coordinate of Q an extra coordinate of 1. The *Cayley polytope* of P and Q, written $\mathrm{Cay}(P, Q)$, is the convex hull in $\mathbb{R}^{n+1}$ of this arrangement.

Example 14 If $P = Q = \mathrm{conv}(\{(0, 0, 0), (1, 0, 0), (0, 1, 0), (0, 0, 1)\})$, then $\mathrm{Cay}(P, Q)$ is the convex hull of the eight points $(0, 0, 0, 0)$, $(1, 0, 0, 0)$, $(0, 1, 0, 0)$, $(0, 0, 1, 0)$, $(0, 0, 0, 1)$, $(1, 0, 0, 1)$, $(0, 1, 0, 1)$, and $(0, 0, 1, 1)$ in $\mathbb{R}^4$.

Suppose $p(x, y, z)$ and $q(x, y, z)$ are tropical polynomials in three variables, defining tropical surfaces S_1 and S_2, with intersection curve $C = S_1 \cap_{st} S_2$. Let $P = \mathrm{Newt}(p)$ and $Q = \mathrm{Newt}(q)$. As we did with Newton polygons and Newton polytopes of single polynomials, we can find an induced subdivision of $\mathrm{Cay}(P, Q)$. Each lattice point of $\mathrm{Cay}(P, Q)$ is either a lattice point of P with an extra coordinate of 0 or a lattice point of Q with an extra coordinate of 1, so we assign to each such lattice point a "height" based on the corresponding coefficient from the relevant polynomial. We can then compute the induced subdivision of $\mathrm{Cay}(P, Q)$ by looking at the upper convex hull in $\mathbb{R}^5$ of these lifted points. This subdivision then splits $\mathrm{Cay}(P, Q)$ into 4-dimensional polytopes. Some of these polytopes have one vertex from P and all others from Q, or vice versa; the other polytopes, with at least two vertices coming from each of P and Q, are called the *mixed cells* of the subdivision. The Duality Theorem for complete intersection curves, stated fully in [39, §4.6], then says that the vertices of C correspond to the mixed cells of this subdivision.

If all cells in the Cayley subdivision have the minimum possible volume (which turns out to be $1/24$), we call the tropical curve *smooth*. In this case it turns out that $P \cap_{st} Q = P \cap Q$. We can still talk about the skeletons of tropical curves in $\mathbb{R}^3$, retracting rays and leaves to obtain the desired graph. Again we still refer to the genus of the graph, although since it might not be a planar graph we need to define genus as $E - V + 1$.

Theorem 8 (Theorem 4.6.20 in [39]) *Let $f(x, y, z)$ and $g(x, y, z)$ be tropical polynomials with degrees d and e, respectively, such that $C = \mathcal{T}(f) \cap \mathcal{T}(g)$ is a smooth tropical curve. Then C has*

- *$d^2 e + d e^2$ vertices,*
- *$(3/2)d^2 e + (3/2)d e^2 - 2de$ edges,*
- *$4de$ rays, and*
- *genus equal to $(1/2)d^2 e + (1/2)d e^2 - 2de + 1$.*

Example 15 Let $p(x, y, z) = (-1 \odot x) \oplus (-1 \odot y) \oplus z \oplus 1$ and $q(x, y, z) = (-2 \odot x) \oplus (1 \odot y) \oplus (1 \odot z) \oplus (-1)$. Then $\mathrm{Newt}(p)$ and $\mathrm{Newt}(q)$ are P and Q

from Example 14. Using the `Macaulay2` package `Polyhedra`,[18] we compute
the subdivision of Cay(P, Q). It consists of four cells:

$$\Delta_1 = \text{conv}(\{(0, 0, 0, 0), (1, 0, 0, 0), (0, 1, 0, 0), (0, 0, 1, 0), (0, 1, 0, 1)\}) \qquad (17)$$

$$\Delta_2 = \text{conv}(\{(0, 0, 0, 0), (1, 0, 0, 0), (0, 0, 1, 0), (0, 1, 0, 1), (0, 0, 1, 1)\}) \qquad (18)$$

$$\Delta_3 = \text{conv}(\{(0, 0, 0, 0), (1, 0, 0, 0), (1, 0, 0, 1), (0, 1, 0, 1), (0, 0, 1, 1)\}) \qquad (19)$$

$$\Delta_4 = \text{conv}(\{(0, 0, 0, 0), (0, 0, 0, 1), (1, 0, 0, 1), (0, 1, 0, 1), (0, 0, 1, 1)))\})$$
$$(20)$$

Each cell has volume 1/24, so the tropical intersection curve is smooth; as the
intersection of two tropical planes, we call it a *tropical line in* $\mathbb{R}^3$. Of the four cells,
only Δ_2 and Δ_3 are mixed cells. This means the line $P \cap Q$ has two vertices. The
vertex (a, b, c) coming from Δ_2 arises from a three-way tie between the $(0, 0, 0)$,
$(1, 0, 0)$, $(0, 0, 1)$ terms of p and a two-way tie between the $(0, 1, 0)$ and $(0, 0, 1)$
terms of q. Written in conventional notation, we have $1 = -1 + a = c$, so $a = 2$
and $c = 1$. We also have $1 + b = 1 + c$, so $b = c = 1$. Thus there is a vertex at
$(a, b, c) = (2, 1, 1)$. In the next exercise, you will find the other vertex, as well as
the rest of the line.

Exercise 15 Draw the tropical line from the previous example. Be sure to check
your answer against Theorem 8 with $d = e = 1$.

Challenge Problem 12 Show that the tropical surfaces from Challenge Problem 11
intersect in a smooth tropical curve. Show that the skeleton of the curve is the
complete bipartite graph $K_{3,3}$.

Research Project 13 Which graphs of genus 4 arise in smooth tropical
curves that are the intersection of a tropical surface of degree 2 and a tropical
surface of degree 3? For instance, are any of these graphs sprawling?
  More generally: which graphs of genus $(1/2)d^2e + (1/2)de^2 - 2de + 1$
arise as the skeleton of a smooth tropical curve that is the intersection of a
surface of degree d with a surface of degree e?
  (You can approach these questions considering the graphs either combinatorially, or as metric graphs.)

---

[18]The Polyhedra package defaults to the min convention rather than the max. This means we have
to negate all the coefficients before we find the decomposition.

> **Research Project 14** Let Q_1 and Q_2 be two smooth tropical surfaces of degree 2. Study the possibilities of the intersection $Q_1 \cap_{st} Q_2$, possibly through a similar lens as [21]. (If the intersection is a smooth curve, then it has genus 1 by Theorem 8, and we understand its combinatorial properties very well. What other intersections are possible?)

One noteworthy difference between classical geometry and tropical geometry is that in tropical geometry, not all planes look the same. In the previous section, we studied tropical curves as a subset of the usual plane $\mathbb{R}^2$. But this plane is combinatorially different from, say, the tropical plane from Example 12. A natural question is then whether or not there are "tropical plane curves" besides those we studied in Sect. 2; that is, whether certain tropical skeletons appear on tropical planes in $\mathbb{R}^3$ that did not arise from tropical curves in $\mathbb{R}^2$. (We could ask the same for 2-dimensional tropical planes in $\mathbb{R}^4$, or $\mathbb{R}^5$, or in general $\mathbb{R}^n$.)

Recent work shows that the answer is yes! Recall that only 29.1% of all graphs of genus 3 appear in tropical curves in $\mathbb{R}^2$. It is shown in [26] that *every* metric graph of genus 3, besides a family of measure zero, appears as a tropical curve in a tropical plane in $\mathbb{R}^3$, $\mathbb{R}^4$, or $\mathbb{R}^5$. For example, they show that the lollipop graph appears as a tropical curve on a tropical plane in $\mathbb{R}^5$. It is not known if their result is sharp; for instance, it is an open question if there are any graphs of genus 3 that do not appear on a tropical plane in $\mathbb{R}^3$.

> **Research Project 15** Can the lollipop graph be realized on a tropical plane in $\mathbb{R}^3$ or $\mathbb{R}^4$? More generally, which graphs can be realized on a tropical plane in $\mathbb{R}^n$, for different values of n?

4   Tropicalization

In this section we present the connections between *algebraic geometry*, which studies solutions to usual polynomial equations, and tropical geometry, which studies solutions to tropical polynomial equations. See [39] for a more complete treatment of this connection, and [17] for an undergraduate introduction to algebraic geometry.

Let k be a field, and let $k[x_1, \ldots, x_n]$ be the polynomial ring in n variables over k. For an ideal $I \subset k[x_1, \ldots, x_n]$, the *affine variety* defined by I is

$$\mathbf{V}(I) = \{ (a_1, \ldots, a_n) \mid f(a_1, \ldots, a_n) = 0 \text{ for all } f \in I \} \subset k^n. \tag{21}$$

Given $f_1, \ldots, f_s \in k[x_1, \ldots, x_n]$, we can also define

$$\mathbf{V}(f_1, \ldots, f_s) = \{(a_1, \ldots, a_n) \mid f_i(a_1, \ldots, a_n) = 0 \text{ for all } i\} \subset k^n. \qquad (22)$$

If $I = \langle f_1, \cdots, f_s \rangle$, then $\mathbf{V}(I) = \mathbf{V}(f_1, \ldots, f_s)$. By Hilbert's Basis Theorem [28][19] every ideal in $k[x_1, \ldots x_n]$ has a finite set of generators, so these two characterizations of affine varieties are equivalent.

Sometimes it is useful to work within the ambient space of the *algebraic torus* $(k^*)^n$, where $k^* = k \setminus \{0\}$. To do this we can let our ideal I be a subset of $k[x_1^{\pm 1}, \ldots, x_n^{\pm 1}]$, so that $\mathbf{V}(I) \subset (k^*)^n$.

4.1   Fields with Valuation

We will work with fields with an additional structure called a *valuation*. A valuation on a field k is a function val : $k \rightarrow (\mathbb{R} \cup \{\infty\})$ such that

- val$(a) = \infty$ if and only if $a = 0$.
- val$(ab) = $ val$(a) + $ val(b).
- val$(a + b) \geq \min\{$val$(a),$ val$(b)\}$ with equality if val$(a) \neq$ val(b).

Every field has an example of a valuation called the *trivial valuation*, defined by val$(0) = \infty$ and val$(a) = 0$ for all $a \neq 0$. Let us find some nontrivial valuations.

Exercise 16 Let $\mathbb{Q}$ be the field of rational numbers, and let p be a prime number. Define the *p-adic valuation* on $\mathbb{Q}$ by

$$\mathrm{val}_p\left(p^k \frac{a}{b}\right) = k, \qquad (23)$$

where a and b are integers that are not divisible by p. Show that this is a valuation on $\mathbb{Q}$.

> **Research Project 16** Study the sequences obtained by applying p-adic valuations to sequences of integers. For instance, applying the 2-adic valuation to the sequence of Fibonacci numbers
>
> $$1, 1, 2, 3, 5, 8, 13, 21, 34, 55, \ldots \qquad (24)$$

(continued)

---

[19]For a presentation in English, see [17, §2.5].

gives the sequence

$$0, 0, 1, 0, 0, 3, 0, 0, 1, 0, \ldots \tag{25}$$

We can think of this as *tropicalizing* sequences of integers. See [1, 36, 40] for work done in this direction.

Exercise 17 Let K be a field and let $K((t))$ be the field of *Laurent series* over K, the nonzero elements of which are power series in t with integer exponents that are bounded below:

$$a_m t^m + a_{m+1} t^{m+1} + a_{m+2} t^{m+2} + \cdots, \tag{26}$$

where $m \in \mathbb{Z}$, $a_i \in K$ for all i, and $a_m \neq 0$. We define a valuation on $K((t))$ by reading off the exponent of the smallest nonzero term:

$$\mathrm{val}\left(a_m t^m + a_{m+1} t^{m+1} + a_{m+2} t^{m+2} + \cdots\right) = m. \tag{27}$$

Show that this is indeed a valuation on $K((t))$.

Challenge Problem 13 It turns out that the field $K((t))$ is not algebraically closed, even if K is. For an example of an algebraically closed field with a nontrivial valuation, we turn to the *field of Puiseux series over K*, written $K\{\{t\}\}$. A nonzero element of this field is of the form

$$a_m t^{m/n} + a_{m+1} t^{(m+1)/n} + a_{m+2} t^{(m+2)/n} + \cdots, \tag{28}$$

where $m \in \mathbb{Z}$, $n \in \mathbb{Z}^+$, $a_i \in k$ for all i, and $a_k \neq 0$. Note that the value of n can vary between different elements of $K\{\{t\}\}$, so we could equivalently define a single Puiseux series as a power series in t with rational exponents, where there is a lower bound on the denominator of the exponents. Again, we can define a valuation by reading off the lowest exponent:

$$\mathrm{val}\left(a_m t^{m/n} + a_{m+1} t^{(m+1)/n} + a_{m+2} t^{(m+2)/n} + \cdots\right) = m/n. \tag{29}$$

Show that if K is algebraically closed and char$(K) = 0$, then $K\{\{t\}\}$ is algebraically closed.

Valuations have a similar flavor to tropical arithmetic, at least if we use the min convention instead of the max convention: they introduce an infinity element ∞, they turn multiplication into addition, and they turn addition into a minimum (except possibly when the valuations tie). They also justify the notation of "vanishing" as being connected to a minimum or maximum being achieved at least twice, as you will show in the following exercise.

Exercise 18 Let val be a valuation on a field k, and let $a_1, \ldots, a_n \in k$ with $n \geq 2$. Show that if $a_1 + a_2 + \cdots + a_n = 0$, then the minimum value among $\mathrm{val}(a_1), \ldots, \mathrm{val}(a_n)$ occurs at least twice.

4.2   Two Ways to Tropicalize

To stay consistent with the rest of this chapter, we will continue working in the max convention.[20] We now explore two ways of taking a variety $V(I) \subset (k^*)^n$ and moving it into $\mathbb{R}^n$. One way is to take coordinate-wise valuation of points in $V(I)$ and append a minus sign onto each coordinate. That is, we consider the set image of $V(I)$ under the map

$$- \mathrm{val} : (k^*)^n \to \mathbb{R}^n, \tag{30}$$

$$- \mathrm{val}(a_1, \cdots, a_n) := (-\mathrm{val}(a_1), \cdots, -\mathrm{val}(a_1)). \tag{31}$$

The other way is to consider polynomials $f \in I$, and to turn them into tropical polynomials. Given $f \in I$ with $f = \sum_\alpha c_\alpha x_1^{\alpha_1} \cdots x_n^{\alpha^n}$, consider the tropical polynomial

$$\mathrm{trop}(f) := \bigoplus_\alpha (-\mathrm{val}(c_\alpha)) \odot x_1^{\alpha_1} \odot \cdots \odot x_n^{\alpha_n}. \tag{32}$$

Since $V(I) = \bigcap_{f \in I} V(f)$, we consider $\bigcap_{f \in I} \mathcal{T}(\mathrm{trop}(f))$ as a tropical version of $V(I)$. We call this intersection the *tropicalization* of $V(I)$.

Exercise 19 Let $k = \mathbb{C}\{\{t\}\}$, and define $f \in k[x, y]$ by

$$\begin{aligned} f(x, y) = \quad & \left(\tfrac{\sqrt{-1}}{\pi} t^3 - 3t^{10/3} + \cdots \right) x^2 + 1000xy \\ & + (1 - t^{1/2} + t^{5/8} + \cdots)x + y + (\sqrt{5}t - t^{100}). \end{aligned} \tag{33}$$

Find the tropicalization of $V(f)$.

Example 16 Let $k = \mathbb{C}\{\{t\}\}$ where $\mathbb{C}$ is the field of complex numbers, and consider the set $V(I) \subset (k^*)^2$ where I is generated by the single polynomial $x + ty + 2 \in k[x, y]$. A point $(a, b) \in V(I)$ is sent to $(-\mathrm{val}(a), -\mathrm{val}(b))$ by the map $-\mathrm{val}$. Note that if $(a, b) \in V(I)$, then $a = -tb - 2$. This means that either $\mathrm{val}(a) =$

---

[20]Because we are working in the max convention, there are many instances when we have to consider -1 times a valuation. In the min convention, we can just consider valuations.

$\min\{\text{val}(-tb), -2\} = \min\{\text{val}(b) + 1, 0\}$, or $\text{val}(a) \geq \min\{\text{val}(b) + 1, 0\}$ with $\text{val}(b) + 1 = 0$. Equivalently, either $-\text{val}(a) = \max\{-\text{val}(b) - 1, 0\}$ or $-\text{val}(a) \geq \max\{-\text{val}(b) - 1, 0\}$ with $-\text{val}(b) - 1 = 0$. So, all points (A, B) in $-\text{val}(\mathbf{V}(I))$ fall into one of three classes:

- $A = B - 1 \leq 0$
- $A = 0 \leq B - 1$
- $B - 1 = 0 \leq A$

So, the minimum between A, $B - 1$, and 0 is achieved at least twice. In other words, $-\text{val}(\mathbf{V}(I)) \subset \mathcal{T}(x \oplus (-1 \odot y) \oplus 0)$. We do not have equality, since all points in $-\text{val}(\mathbf{V}(I))$ have rational coordinates; we leave it as an exercise to show that $-\text{val}(\mathbf{V}(I)) = \mathcal{T}(x \oplus (-1 \odot y) \oplus 0) \cap \mathbb{Q}^2$

Note that $\text{trop}(x + ty + 2) = x \oplus (-1 \odot y) \oplus 0$. All polynomials in I are multiples of $x + ty + 2$, which means that $\bigcap_{f \in I} \mathcal{T}(\text{trop}(f)) = \mathcal{T}(x \oplus (-1 \odot y) \oplus 0)$. So, the tropicalization of $\mathbf{V}(I)$ is the tropical line defined by $x \oplus (-1 \odot y) \oplus 0$.

These two constructions gave us similar, but not identical, subsets of $\mathbb{R}^2$: we had containment of $-\text{val}(\mathbf{V}(I))$ in the tropicalization of $\mathbf{V}(I)$, though these sets were not equal.

Exercise 20 Show that we always have $-\text{val}(\mathbf{V}(I)) \subset \bigcap_{f \in I} \mathcal{T}(\text{trop}(f))$.

It turns out that, as long as we are working over an algebraically closed field, these two sets are equal up to taking a closure in the usual Euclidean topology of $\mathbb{R}^n$.

Theorem 9 (The Fundamental Theorem of Tropical Geometry) *Let k be an algebraically closed field with a nontrivial valuation val, and let I be an ideal of $k[x_1^{\pm}, \cdots, x_n^{\pm 1}]$. Then*

$$\overline{-val(\mathbf{V}(I))} = \bigcap_{f \in I} \mathcal{T}(\text{trop}(f)). \tag{34}$$

This fact is a key result of tropical geometry, originally proved by Kapranov in an unpublished manuscript when I is generated by a single polynomial. A proof of the more general result appears in [39, Theorem 3.2.3].

Given $X = \mathbf{V}(I) \subset (k^*)^n$, let $\text{Trop}(X)$ denote the set $\overline{-\text{val}(\mathbf{V}(I))}$. Understanding the relationship between X and $\text{Trop}(X)$ is one of the core themes in tropical geometry.

4.3   Tropical Intersections

Let X and Y be varieties in $(k^*)^n$. Let us consider how $\mathrm{Trop}(X \cap Y)$ and $\mathrm{Trop}(X) \cap \mathrm{Trop}(Y)$ relate to one another.

Exercise 21 Show that we always have $\mathrm{Trop}(X \cap Y) \subset \mathrm{Trop}(X) \cap \mathrm{Trop}(Y)$. (This is mostly an exercise in set theory.)

The question then becomes whether we have an equality of these sets. If we do, then every tropical intersection point in $\mathrm{Trop}(X) \cap \mathrm{Trop}(Y)$ "lifts" to an intersection point in $X \cap Y$. One core result from [44] is that if $\mathrm{Trop}(X)$ and $\mathrm{Trop}(Y)$ intersect in components of the expected dimensions, then indeed the points do lift; if $n = 2$ and $\mathrm{Trop}(X)$ and $\mathrm{Trop}(Y)$ are tropical plane curves, this means they intersect in isolated points. Not only that, these points lift with the expected multiplicity! If $\mathrm{Trop}(X)$ and $\mathrm{Trop}(Y)$ intersect in higher dimensional components, the story is more complicated.

Example 17 Let $k = \mathbb{C}\{\{t\}\}$, and let $f, g \in k[x, y]$ be defined by $f(x, y) = ax + by + c$ and $g(x, y) = dx + ey + f$, where $\mathrm{val}(a) = \mathrm{val}(b) = \mathrm{val}(c) = \mathrm{val}(d) = \mathrm{val}(e) = \mathrm{val}(a) = 0$. Let $X = \mathbf{V}(f)$ and $Y = \mathbf{V}(g)$ be the two lines defined by these equations. Then $\mathrm{Trop}(X) = \mathrm{Trop}(Y) = \mathscr{T}(x \oplus y \oplus 0)$, the tropical line in Fig. 3. This means $\mathrm{Trop}(X) \cap \mathrm{Trop}(Y) = \mathscr{T}(x \oplus y \oplus 0)$. Unless X and Y are the same line, at most one of these infinitely many tropical intersection points can lift to an intersection point of X and Y. Let us determine which point might lift.

Assume that $X \cap Y$ consists of one point. We can solve the equations $ax + by + c = dx + ey + f = 0$ to find the intersection point as $\left(\frac{ce - bf}{bd - ae}, \frac{af - cd}{bd - ae} \right)$. So we know that

$$\mathrm{Trop}(X \cap Y) = \mathrm{Trop}\left(\left\{ \left(\frac{ce - bf}{bd - ae}, \frac{af - cd}{bd - ae} \right) \right\}\right) = \left\{ \left(-\mathrm{val}\left(\frac{ce - bf}{bd - ae} \right), -\mathrm{val}\left(\frac{af - cd}{bd - ae} \right) \right) \right\}.$$
(35)

If there is no cancellation in $ce - bf$, $bd - ae$, $af - cd$, and $bd - ae$, then $\mathrm{Trop}(X \cap Y)$ is $\{(0, 0)\}$, which is the stable tropical intersection $\mathrm{Trop}(X) \cap_{st} \mathrm{Trop}(Y)$. However, there are cases that give different values for $\mathrm{Trop}(X \cap Y)$. Let r be a positive rational number, and note that:

- If $f = x + 2y + (1 + t^r)$ and $g = x + y + 1$, then the intersection point $X \cap Y$ is $(-1 + t^r, -t^r)$, which is sent to $(0, -r)$.
- If $f = 2x + y + (1 + t^r)$ and $g = x + y + 1$, then the intersection point $X \cap Y$ is $(-t^r, -1 + t^r,)$, which is sent to $(-r, 0)$.
- If $f = (2 + t^r)x + 2y + 1$ and $g = x + y + 1$, then the intersection point $X \cap Y$ is $\left(\frac{1}{t^r}, \frac{1 + t^r}{t^r} \right) = \left(t^{-r}, t^{-r}(1 + t^r) \right)$, which is sent to (r, r).

This means if all we know about X and Y is that $\mathrm{Trop}(X) \cap \mathrm{Trop}(Y) = \mathscr{T}(x \oplus y \oplus 0)$, then *any* point in $\mathscr{T}(x \oplus y \oplus 0) \cap \mathbb{Q}^2$ could be the image of the intersection point of X and Y.

Challenge Problem 14 Let $a, b, c, d, e, f \in k = \mathbb{C}\{\{t\}\}$, where $\mathrm{val}(a) = \mathrm{val}(b) = \mathrm{val}(c) = \mathrm{val}(d) = \mathrm{val}(e) = 0$ and $\mathrm{val}(f) = 1$. Consider the two polynomials $f, g \in k[x, y]$ defined by

$$f(x, y) = ax + by + c, \tag{36}$$

$$g(x, y) = dxy + ex + fy. \tag{37}$$

Let $X = \mathbf{V}(f)$, and $Y = \mathbf{V}(g)$. What are the possible configurations of $\mathrm{Trop}(X \cap Y)$ inside $\mathrm{Trop}(X) \cap \mathrm{Trop}(Y)$?

Research Project 17 Study the possibilities for $\mathrm{Trop}(X \cap Y)$ inside $\mathrm{Trop}(X) \cap \mathrm{Trop}(Y)$, for plane curves or in higher dimensions. Some resources to check are [35, 43–45].

References

1. Amdeberhan, T., Medina, L. A., Moll, V. H.: Asymptotic valuations of sequences satisfying first order recurrences. Proc. Amer. Math. Soc. **137**, no. 3 (2009)
2. Baker, M., Len, Y., Morrison, R., Pflueger, N., Ren, Q.: Bitangents of tropical plane quartic curves. Math. Z. **282**, no. 3-4 (2016)
3. Baker, M., Norine, S.: Riemann-Roch and Abel-Jacobi theory on a finite graph. Adv. Math 215 no. 2 (2007)
4. Balaban, A.T.: Enumeration of cyclic graphs. Chemical Applications of Graph Theory (A.T. Balaban, ed.) 63?105, Academic Press (1976)
5. Bashelor, A., Ksir, A., Traves, W.: Enumerative algebraic geometry of conics. Amer. Math. Monthly **115**, no. 8 (2008)
6. Brannetti, S., Melo, M., Viviani, F.: On the tropical Torelli map. Adv. Math., 226(3) (2011)
7. Brodsky, S., Joswig, M., Morrison, R. Sturmfels, B.: Moduli of tropical plane curves. Res. Math. Sci. **2**, Art. 4 (2015)
8. Butkovič, P.: Max-linear systems: theory and algorithms. Springer Monographs in Mathematics. Springer-Verlag London, Ltd., London (2010).
9. Cartwright, D., Dudzik, A., Manjunath, M., Yao, Y.: Embeddings and immersions of tropical curves. Collect. Math. **67**, no. 1 (2016)
10. Castryck, W.: Moving out the edges of a lattice polygon. Discrete and Computational Geometry **47**, no. 3 (2012)
11. Castryck, W., Voight, J.: On nondegeneracy of curves. Algebra and Number Theory **3** (2009)
12. Chan, M.: Combinatorics of the tropical Torelli map. Algebra Number Theory, 6(6) (2012)
13. Chan, M.: Tropical hyperelliptic curves. J. Algebraic Combin. **37**, no. 2 (2013)
14. Chan, M., Jiradilok, P.: Theta characteristics of tropical K_4-curves. Combinatorial algebraic geometry, 65–86, Fields Inst. Commun., 80, Fields Inst. Res. Math. Sci., Toronto, ON (2017)
15. Cools, F., Draisma, J.: On metric graphs with prescribed gonality. J. Combin. Theory Ser. A **156** (2018)
16. Corry, S., Perkinson, D: Divisors and sandpiles. An introduction to chip-firing. American Mathematical Society, Providence, RI (2018)

17. Cox, D. A., Little, J., O'Shea, D.: Ideals, varieties, and algorithms. An introduction to computational algebraic geometry and commutative algebra. Fourth edition. Undergraduate Texts in Mathematics. Springer, Cham, xvi+646 pp. (2015)
18. de Bruijn, N. G., Erdös, P: On a combinatorial problem. Nederl. Akad. Wetensch., Proc. **51** (1948)
19. De Loera, J.A., Rambau, J., Santos, F.: Triangulations. Structures for algorithms and applications. Algorithms and Computation in Mathematics, 25. Springer-Verlag, Berlin (2010)
20. Evelyn, C. J. A., Money-Coutts, G. B.,Tyrrell, J. A.: The seven circles theorem and other new theorems. Stacey International, London (1974)
21. Farouki, R.T., Neff, C., O'Conner, M.A.: Automatic parsing of degenerate quadric-surface intersections. ACM Transactions on Graphics (TOG) **8**, No. 3 (1989)
22. Gathmann, A., Kerber, M.: A Riemann-Roch theorem in tropical geometry. Math. Z. **259**, no. 1 (2008)
23. Gawrilow, E., Joswig, M.: polymake: a framework for analyzing convex polytopes. Polytopes-combinatorics and computation (Oberwolfach, 1997), 43–73, DMV Sem., 29, Birkhäuser, Basel (2000)
24. Giansiracusa, J., Giansiracusa, N.: Equations of tropical varieties. Duke Math. J. **165**, no. 18, 3379–3433 (2016)
25. Grayson, D., Stillman, M. E.: Macaulay2, a software system for research in algebraic geometry. Available at http://www.math.uiuc.edu/Macaulay2/
26. Hahn, M.A., Markwig, H., Ren, Y., Tyomkin, I.: Tropicalized quartics and canonical embeddings for tropical curves of genus 3. arXiv preprint arXiv:1802.02440 (2018)
27. Hartshorne, R.: Algebraic geometry. Graduate Texts in Mathematics, No. 52. Springer-Verlag, New York-Heidelberg xvi+496 pp. (1977)
28. Hilbert, D.: Ueber die Theorie der algebraischen Formen. (German) Math. Ann. **36**, no. 4 (1890)
29. Jensen, A.: Gfan, a software system for Gröbner fans and tropical varieties. Available at http://home.imf.au.dk/jensen/software/gfan/gfan.html.
30. Kaibel, V., Ziegler, G.M.: Counting lattice triangulations. Surveys in combinatorics, 2003 (Bangor), 277–307, London Math. Soc. Lecture Note Ser., 307, Cambridge Univ. Press, Cambridge, (2003)
31. Koelman, R.: The number of moduli of families of curves on toric surfaces. Ph.D. thesis, Katholieke Universiteit Nijmegen (1991)
32. Lagarias, J.C., Ziegler, G.M.: Bounds for lattice polytopes containing a fixed number of interior points in a sublattice. Canadian J. Math. **43** (1991)
33. Le Gall, François: Powers of tensors and fast matrix multiplication. Proceedings of the 39th International Symposium on Symbolic and Algebraic Computation (ISSAC 2014)
34. Len, Y., Markwig, H..: Lifting tropical bitangents. arXiv preprint arXiv:1708.04480 (2018)
35. Len, Y., Satriano., M.: Lifting tropical self intersections. arXiv preprint arXiv:1806.01334 (2018)
36. Lengyel, T.: On the divisibility by 2 of the Stirling numbers of the second kind. Fibonacci Quart. **32**, no. 3 (1994)
37. Lin, B., Tran, N. M.: Linear and rational factorization of tropical polynomials. arXiv preprint arXiv:1707.03332 (2017)
38. Maclagan, D., Rincón, F.: Tropical ideals. Compos. Math. **154**, no. 3 (2018)
39. Maclagan, D., Sturmfels, B.: Introduction to tropical geometry. Graduate Studies in Mathematics, 161. American Mathematical Society, Providence, RI (2015)
40. Medina, L. A., Rowland, E.: p-regularity of the p-adic valuation of the Fibonacci sequence. Fibonacci Quart. **53**, no. 3 (2015)
41. Mikhalkin, G., Zharkov, I.: Tropical curves, their Jacobians and theta functions. Curves and abelian varieties, 203–230, Contemp. Math., 465, Amer. Math. Soc., Providence, RI (2008)
42. Morrison, R.: Tropical hyperelliptic curves in the plane. arXiv preprint arXiv:1708.00571 (2017)
43. Morrison, R: Tropical images of intersection points. Collect. Math. **66**, no. 2 (2015)

44. Osserman, B., Payne, S.: Lifting tropical intersections. Doc. Math. **18** (2013)
45. Osserman, B., Rabinoff, J.: Lifting nonproper tropical intersections. Tropical and non-Archimedean geometry, 15–44, Contemp. Math., 605, Centre Rech. Math. Proc., Amer. Math. Soc., Providence, RI, (2013)
46. Pick, G. A.: Geometrisches zur Zahlenlehre, Sitzenber. Lotos (Prague) **19** (1899)
47. Pin, J.: Tropical semirings. Idempotency (Bristol, 1994), 50–69, Publ. Newton Inst., 11, Cambridge Univ. Press, Cambridge (1998)
48. Plücker, J.: Solution d'une question fondamentale concernant la théorie générale des courbes. (French) J. Reine Angew. Math. **12** (1834)
49. Richter-Gebert, J., Sturmfels, B., Theobald, T.: First steps in tropical geometry. Idempotent mathematics and mathematical physics, 289–317, Contemp. Math., **377**, Amer. Math. Soc., Providence, RI (2005)
50. Rambau, J.: TOPCOM: Triangulations of Point Configurations and Oriented Matroids, Mathematical Software - ICMS 2002 (Cohen, Arjeh M. and Gao, Xiao-Shan and Takayama, Nobuki, eds.), World Scientific, pp. 330–340 (2002)
51. Scott, P.R.: On convex lattice polygons. Bull. Austral. Math. Soc. **15** (1976)
52. Simon, I.: Recognizable sets with multiplicities in the tropical semiring. Mathematical foundations of computer science, 1988 (Carlsbad, 1988), 107–120, Lecture Notes in Comput. Sci., 324, Springer, Berlin, (1988)
53. Strassen, V.: Gaussian Elimination is not Optimal. Numer. Math. **13** (1969)

Chip-Firing Games and Critical Groups

Darren Glass and Nathan Kaplan

Abstract

In this note we introduce a finite abelian group that can be associated with any finite connected graph. This group can be defined in an elementary combinatorial way in terms of chip-firing operations, and has been an object of interest in combinatorics, algebraic geometry, statistical physics, and several other areas of mathematics. We will begin with basic definitions and examples and develop a number of properties that can be derived by looking at this group from different angles. Throughout, we will give exercises, some of which are straightforward and some of which are open questions. We will also highlight some of the many contributions to this area made by undergraduate students.

Suggested Prerequisites The basic definitions and themes of this note should be accessible to any student with some knowledge of linear algebra and group theory. As we go along, deeper understanding of graph theory, abstract algebra, and algebraic geometry will be of use in some sections.

D. Glass (✉)
Gettysburg College, Gettysburg, PA, USA
e-mail: dglass@gettysburg.edu

N. Kaplan
University of California, Irvine, CA, USA
e-mail: nckaplan@math.uci.edu

© Springer Nature Switzerland AG 2020   107
P. E. Harris et al. (eds.), *A Project-Based Guide to Undergraduate Research in Mathematics*, Foundations for Undergraduate Research in Mathematics,
https://doi.org/10.1007/978-3-030-37853-0_4

1   Critical Groups

The primary object of interest in this chapter will be a finite abelian group that is associated with a graph. This group has been studied from a variety of different perspectives, and as such it goes by several different names, including the *sandpile group*, the *component group*, the *critical group*, or the *Jacobian* of a graph. We will give definitions and some results about critical groups of graphs and pose some questions that we think would be interesting for an undergraduate to tackle. For additional background and motivation for this topic as well as a more in-depth treatment, we recommend the books by Klivans [48] and Corry and Perkinson [32].

We will highlight several significant contributions to the study of critical groups made by undergraduates—papers with at least one undergraduate author are highlighted in red in the bibliography—and we will discuss some open problems that would make excellent topics for future undergraduate research.

1.1   Definitions and Examples

Part of what makes the study of critical groups such a good topic for undergraduate research is that the definitions are very concrete and one can get started computing examples right away.

Let G be a connected, undirected graph with vertex set $V(G)$ of finite size n and edge set $E(G)$. Choose an ordering of $V(G)$: $v_1, \ldots, v_n$. We define the *adjacency matrix* of the graph G to be the $n \times n$ matrix A where the entry $a_{i,j}$ in the ith row and j^{th} column of A is the number of edges between v_i and v_j. We also define the matrix D to be the diagonal matrix where the entry $d_{i,i}$ is equal to the degree of v_i. Finally, we let $L(G)$ be the matrix $D - A$; this matrix is referred to as the *Laplacian matrix*, or *combinatorial Laplacian*, of the graph G. We often write L for this matrix when the graph is clear from context.

Note We defined the adjacency matrix A of G by saying that $a_{i,j}$ is the *number of edges* between v_i and v_j, implying that this number can be greater than 1. For most of this paper we focus on the case of *simple* graphs (at most one edge between any pair of vertices), with no *self-loops* (edges from v_i to v_i), that are *connected* (for any pair of vertices v_i, v_j there is a path from v_i to v_j in G), and where edges are *undirected*. In this case we will denote an edge between v_i and v_j as $\overline{v_i v_j}$. Much of the theory of critical groups carries over to more general settings, but we find that it is most helpful to first focus on this simplest case.

Example 1 We will consider the graph below:

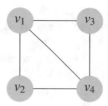

One can see that the adjacency matrix, degree matrix, and Laplacian of this graph are given by

$$A = \begin{pmatrix} 0 & 1 & 1 & 1 \\ 1 & 0 & 0 & 1 \\ 1 & 0 & 0 & 1 \\ 1 & 1 & 1 & 0 \end{pmatrix}, \quad D = \begin{pmatrix} 3 & 0 & 0 & 0 \\ 0 & 2 & 0 & 0 \\ 0 & 0 & 2 & 0 \\ 0 & 0 & 0 & 3 \end{pmatrix}, \quad L = \begin{pmatrix} 3 & -1 & -1 & -1 \\ -1 & 2 & 0 & -1 \\ -1 & 0 & 2 & -1 \\ -1 & -1 & -1 & 3 \end{pmatrix}.$$

It follows from the definition of the Laplacian matrix of a graph that the entries in any row or in any column sum to 0. This implies that the vector consisting of all ones, $\mathbf{1}$, is in the null space of the matrix. In fact, we have the following result:

Theorem 1 *For any finite connected graph G, the null space of the Laplacian matrix of G is generated by the vector $\mathbf{1}$.*

Proof Since $\mathbf{1}$ is in the null space, all multiples of it are as well. Let $\mathbf{x} = (x_1, \ldots, x_n)$ be a vector in the null space of L, so that $L\mathbf{x} = \mathbf{0}$, the all zero vector. Note that this implies that $\mathbf{x}^T L\mathbf{x} = 0$. One can check that

$$\mathbf{x}^T L\mathbf{x} = \sum_{v_i v_j \in E(G)} (x_i - x_j)^2.$$

Each of these terms is nonnegative so the entries of $\mathbf{x}$ corresponding to any pair of neighboring vertices must be equal. Because G is connected we must have that for any vector in the null space all of the entries in $\mathbf{x}$ are equal, concluding the proof.

More generally, we can determine the number of connected components of G in terms of its Laplacian.

Proposition 1 *For any finite graph G, the dimension of the null space of the Laplacian matrix of G is the number of connected components of G.*

Exercise 1 If G is a graph with c connected components, describe c linearly independent vectors in the null space of L. Mimic the proof of Theorem 1 to show that the dimension of the null space is, in fact, c.

This result is the first of many results relating the Laplacian matrix of a graph to other seemingly combinatorial properties of the graph. The eigenvalues of the Laplacian turn out to be particularly interesting, and the area of *spectral graph theory* is largely dedicated to studying this relationship. We refer the interested reader to the survey article [67] or the book [22].

In order to discuss our main object of interest, we note that any $n \times n$ integer matrix A can be thought of as a linear map $A \colon \mathbb{Z}^n \to \mathbb{Z}^n$. The *cokernel* of A, denoted $\mathrm{cok}(A)$, is $\mathbb{Z}^n / \mathrm{Im}(A)$. Theorem 1 implies that if L is the Laplacian of a connected graph G then $\dim(\mathrm{Im}(L)) = n - 1$, so $\mathrm{cok}(L) \cong \mathbb{Z} \oplus K$ for some finite abelian group K. This group K is the *critical group* of the graph G. We will denote it by either K or $K(G)$ depending on whether the graph is understood by context.

The main goal of this article is to outline problems about critical groups. What interesting information does $K(G)$ tell us about G? In Sect. 1.5 we will see that the order of $K(G)$ tells us about the subgraphs of G, in particular, that $|K(G)|$ is the number of spanning trees of G. In the next section we will introduce divisors on G and see that the structure of the finite abelian group $K(G)$ tells us something about how these divisors on G behave under chip-firing operations.

1.2   Divisors on a Graph and the Chip-Firing Game

We started by giving an algebraic description of the critical group as the torsion part of the cokernel of the Laplacian matrix of G, but one can also approach it from a more combinatorial point of view via the *chip-firing game*, which was originally introduced by Biggs in [14]. In order to define this game, we set some notation. A *divisor* on a graph G is a function $\delta \colon V(G) \to \mathbb{Z}$, which we think of as assigning an integer number of chips to each vertex of G. We can think of a divisor as an element of $\mathbb{Z}^{|V(G)|}$. The *degree* of a divisor is defined by $\deg(\delta) = \sum_v \delta(v)$. We define an addition of divisors by $(\delta_1 + \delta_2)(v) = \delta_1(v) + \delta_2(v)$. In this way, we see that the set of all divisors on G, denoted $\mathrm{Div}(G)$, is isomorphic to a free abelian group with $|V(G)|$ generators. We let $\mathrm{Div}^0(G)$ denote the subgroup of all degree 0 divisors on G. One can see that $\mathrm{Div}^0(G)$ is isomorphic to a free abelian group with $|V(G)| - 1$ generators.

Exercise 2 Describe a set Δ of $|V(G)| - 1$ divisors on G so that $\mathrm{Div}^0(G)$ is isomorphic to the free abelian group on Δ.

We next define two types of transitions between divisors, which are called *chip-firing moves*. In the first, we choose a vertex and *borrow* a chip from each of its neighbors. The second is an inverse to the first, where we choose a vertex and *fire* it, sending a chip to each one of its neighbors. We will treat these two as inverses in an algebraic sense, so, for example, when we say "perform -2 borrowings at v" one should think of it as the same as "perform 2 firings at v." Note that each one of these chip-firing moves preserves the degree of a divisor. Two divisors D_1 and D_2 are *equivalent* if we can get from D_1 to D_2 by a sequence of chip-firing moves (Fig. 1).

Fig. 1 A divisor on the cycle graph C_3, followed by the divisor obtained by first "firing" at the lower-left vertex and then "borrowing" at the upper-left vertex

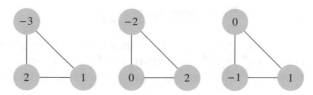

The set of divisors that are equivalent to the all zero divisor is exactly $\text{Im}(L(G))$. Starting with a divisor δ, which we think of a column vector in $\mathbb{Z}^{|V(G)|}$, firing v_i corresponds to subtracting the ith column of $L(G)$ from this vector. Similarly, borrowing at v_i corresponds to adding i^{th} column of $L(G)$. This gives a second interpretation of the critical group.

Proposition 2 *Let G be a finite connected graph. The critical group $K(G)$ is isomorphic to $\text{Div}^0(G)/\sim$, the set of all degree 0 divisors of G modulo chip-firing equivalence.*

Example 2 Let G be the cycle on three vertices. Consider any divisor δ of degree zero on G. Let $\hat{\delta}$ be the divisor attained after performing $\delta(v_3)$ borrowing operations at v_1, so in particular $\hat{\delta}(v_3) = 0$. Because the degree of $\hat{\delta}$ is zero we must have that $\hat{\delta}(v_2) = -\hat{\delta}(v_1)$ so in particular $\hat{\delta}$ is a multiple of the divisor $\delta_{1,2}$ which is defined by setting $\delta_{1,2}(v_1) = 1$, $\delta_{1,2}(v_2) = -1$, and $\delta_{1,2}(v_3) = 0$. This implies that every element of $\text{Div}^0(G)$ is equivalent to a multiple of $\delta_{1,2}$. Therefore, $K(G)$ is cyclic. One can also show that $3\delta_{1,2}$ is chip-firing equivalent to the zero divisor, but that $\delta_{1,2}$ and $2\delta_{1,2}$ are not. We conclude that $K(C_3) \cong \mathbb{Z}/3\mathbb{Z}$.

Remark 1 These definitions are in parallel with a family of ideas in algebraic geometry, and many recent results in the field have come from trying to better understand this analogy. In particular, given a curve C defined as the solution set to a polynomial equation $f(x, y) = 0$, algebraic geometers define a *divisor* on the curve to be a formal finite linear combination $\sum a_i P_i$ of points on the curve. The degree of the divisor is defined to be the sum $\sum a_i$, and the set of divisors of degree zero is denoted by $\text{Div}^0(C)$. The *Jacobian* of the curve is then defined to be $\text{Div}^0(C)/\sim$, where two divisors δ_1 and δ_2 are said to be equivalent if $\delta_1 - \delta_2$ is the divisor corresponding to a rational function on C. For more details about Jacobians in algebraic geometry, we recommend [43].

Exercise 3 Show that if δ is a divisor of degree zero on the graph from Example 1, then δ is equivalent after some number of firing/borrowing operations to a divisor $\hat{\delta}$ so that $\hat{\delta}(v_3) = \hat{\delta}(v_4) = 0$. This result implies that every divisor of degree zero is equivalent to a multiple of the divisor $\delta_{1,2}$ which is defined by setting $\delta_{1,2}(v_1) = 1$, $\delta_{1,2}(v_2) = -1$, and $\delta_{1,2}(v_3) = \delta_{1,2}(v_4) = 0$.

Next, show that the order of $\delta_{1,2}$ in $K(G)$ is 8, proving that the critical group of this graph is $\mathbb{Z}/8\mathbb{Z}$.

1.3   Smith Normal Forms

We have defined the critical group of a connected graph G as the torsion part of the cokernel of the Laplacian matrix of G, but it is not so clear how to determine the structure of this finite abelian group. Linear algebra provides a nice solution.

Proposition 3 *Let L be a $n \times n$ integer matrix of rank r. There exist matrices U and V with integer entries so that $\det(U) = \pm \det(V) = \pm 1$ and $S = ULV$ is a diagonal matrix where $s_{r+1,r+1} = s_{r+2,r+2} = \cdots = s_{n,n} = 0$ and $s_{i,i} \mid s_{i+1,i+1}$ for all $1 \le i < r$. The matrix S is called the Smith Normal Form of L.*
   Moreover,

$$\mathrm{cok}(L) \cong \mathrm{cok}(S) \cong (\mathbb{Z}/s_{1,1}\mathbb{Z}) \oplus (\mathbb{Z}/s_{2,2}\mathbb{Z}) \oplus \cdots \oplus (\mathbb{Z}/s_{r,r}\mathbb{Z}) \oplus \mathbb{Z}^{n-r}.$$

In particular, one can read off the critical group of G directly from the Smith normal form of $L(G)$. The hard part here is showing the existence of the invertible matrices U and V. For a proof see [32, Theorem 2.33]. Once one knows that U and V satisfying these properties exist, the fact that the cokernels are isomorphic follows from the commutative diagram below. Note that the fact that U and V have determinant ± 1 means that they define isomorphisms $\mathbb{Z}^n \to \mathbb{Z}^n$.

$$
\begin{array}{ccccccccc}
1 & \longrightarrow & \mathbb{Z}^{n-r} & \longrightarrow & \mathbb{Z}^n & \xrightarrow{\ S\ } & \mathbb{Z}^n & \longrightarrow & \mathrm{cok}(S) & \longrightarrow & 1 \\
 & & \downarrow{\scriptstyle \wr} & & \downarrow{\scriptstyle U} & & \uparrow{\scriptstyle V} & & & & \\
1 & \longrightarrow & \mathbb{Z}^{n-r} & \longrightarrow & \mathbb{Z}^n & \xrightarrow{\ L\ } & \mathbb{Z}^n & \longrightarrow & \mathrm{cok}(L) & \longrightarrow & 1.
\end{array}
$$

Finally, it is straightforward to determine the cokernel of a diagonal matrix, so the last claim follows.

Example 3 Consider the graph G below:

We can see that

$$
L(G) = \begin{pmatrix}
2 & -1 & 0 & -1 & 0 & 0 \\
-1 & 4 & -1 & -1 & -1 & 0 \\
0 & -1 & 2 & 0 & -1 & 0 \\
-1 & -1 & 0 & 4 & -1 & -1 \\
0 & -1 & -1 & -1 & 4 & -1 \\
0 & 0 & 0 & -1 & -1 & 2
\end{pmatrix},
$$

and can write

$$ULV = \begin{pmatrix} 0 & -1 & 0 & 0 & 0 & 0 \\ 0 & 0 & -1 & 0 & 0 & 0 \\ 0 & 0 & 1 & 0 & -1 & 0 \\ 0 & 0 & 1 & 0 & -1 & 1 \\ 1 & 2 & 3 & 0 & 4 & -7 \\ 1 & 1 & 1 & 1 & 1 & 1 \end{pmatrix} L \begin{pmatrix} 1 & 4 & -1 & 10 & 10 & 1 \\ 0 & 1 & 0 & 2 & 3 & 1 \\ 0 & 0 & 0 & 1 & 2 & 1 \\ 0 & 0 & 1 & -3 & -1 & 1 \\ 0 & 0 & 0 & 0 & 1 & 1 \\ 0 & 0 & 0 & 0 & 0 & 1 \end{pmatrix} = \begin{pmatrix} 1 & 0 & 0 & 0 & 0 & 0 \\ 0 & 1 & 0 & 0 & 0 & 0 \\ 0 & 0 & 1 & 0 & 0 & 0 \\ 0 & 0 & 0 & 3 & 0 & 0 \\ 0 & 0 & 0 & 0 & 18 & 0 \\ 0 & 0 & 0 & 0 & 0 & 0 \end{pmatrix} = S.$$

In particular, U and V both have determinant -1, so S is the Smith normal form of L. This implies that the critical group of the graph is $\mathbb{Z}/3\mathbb{Z} \oplus \mathbb{Z}/18\mathbb{Z}$.

How do we actually compute the Smith normal form of a matrix? One useful fact (see, for example, [68, Theorem 2.4]) is the following:

Theorem 2 *Let L be an $n \times n$ integer matrix of rank r whose Smith normal form has nonzero diagonal entries $s_1, \ldots, s_r$ where $s_i \mid s_{i+1}$ for all $1 \le i < r$. For each $i \le r$, we have that $s_1 s_2 \cdots s_i$ is equal to the greatest common divisor of all $i \times i$ minors of L.*

Example 4 Consider the complete graph K_n on n vertices. One sees that

$$L(K_n) = \begin{pmatrix} n-1 & -1 & -1 & \cdots & -1 \\ -1 & n-1 & -1 & \cdots & -1 \\ -1 & -1 & n-1 & \cdots & -1 \\ \vdots & \vdots & \vdots & \ddots & \vdots \\ -1 & -1 & -1 & \cdots & n-1 \end{pmatrix}.$$

The greatest common divisor of the entries of this matrix is 1, so $s_1 = 1$. The 2×2 submatrices of this matrix are all of the following form:

$$\begin{pmatrix} -1 & -1 \\ -1 & -1 \end{pmatrix}, \begin{pmatrix} n-1 & -1 \\ -1 & -1 \end{pmatrix}, \begin{pmatrix} -1 & -1 \\ n-1 & -1 \end{pmatrix},$$

$$\begin{pmatrix} -1 & -1 \\ -1 & n-1 \end{pmatrix}, \begin{pmatrix} -1 & n-1 \\ -1 & -1 \end{pmatrix}, \begin{pmatrix} n-1 & -1 \\ -1 & n-1 \end{pmatrix}.$$

In particular, the 2×2 minors are all in the set $\{0, \pm n, n^2 - 2n\}$, and the greatest common divisor of these values is n. This implies $s_2 = n$, which in turn tells us that $n \mid s_i$ for all $2 \le i \le n - 1$. The determinant of the $(n-1) \times (n-1)$ matrix that we get by deleting the last row and column of $L(K_n)$ is n^{n-2}. We conclude that $s_i = n$ for each $2 \le i \le n - 1$. This implies that the Smith normal form of the Laplacian is

$$S = \begin{pmatrix} 1 & 0 & 0 & \cdots & 0 & 0 \\ 0 & n & 0 & \cdots & 0 & 0 \\ 0 & 0 & n & \cdots & 0 & 0 \\ \vdots & \vdots & \vdots & \ddots & \vdots & \vdots \\ 0 & 0 & 0 & \cdots & n & 0 \\ 0 & 0 & 0 & \cdots & 0 & 0 \end{pmatrix}$$

and therefore the critical group of the complete graph is $(\mathbb{Z}/n\mathbb{Z})^{n-2}$.

Exercise 4 Verify that the determinant of the $(n-1) \times (n-1)$ matrix that we get by deleting the last row and column of $L(K_n)$ is n^{n-2}.

Theorem 2 gives an explicit (if not very effective) way to compute the Smith normal form, and thus the critical group, of any graph by computing many determinants of submatrices and their greatest common divisors. However, it can also be used in other ways to tell us about the structure of the critical group. For example, using the notation from Proposition 3, if G is a connected graph with n vertices, then the product $s_1 \cdots s_{n-2}$ is the greatest common divisor of the $(n-2) \times (n-2)$ minors of $L(G)$. So if any one of these minors is equal to 1, then $s_1 \cdots s_{n-2} = 1$ and $|K(G)| = s_{n-1}$. This gives the following result:

Corollary 1 *Let G be a connected graph on n vertices. If there exists an $(n-2) \times (n-2)$ minor of L equal to 1, then the critical group of G is cyclic.*

We have defined the critical group of a connected graph as the torsion part of the cokernel of the Laplacian matrix, but it is often convenient to think of the critical group as the cokernel of an invertible matrix. Let the *reduced Laplacian* of a connected graph G be the matrix $L_0(G)$ (or just L_0 when the graph is clear from context) that we get from deleting the final row and column of $L(G)$. Because all of the rows and columns of L sum to 0, the torsion part of $\mathrm{cok}(L)$ is equal to $\mathrm{cok}(L_0)$. In fact, it is a special property of Laplacian matrices that one can remove any row and column from L and the cokernels of the matrices will be isomorphic. See [32, Section 2.2.1] or [13, Chapter 6] for more detail. The following result then follows from Theorem 2.

Corollary 2 *Let G be a graph on n vertices. For any i, j satisfying $1 \leq i, j \leq n$, let $L^{i,j}$ be the $(n-1) \times (n-1)$ matrix that we get by deleting the ith row and j^{th} column of $L(G)$. Then $K(G) \cong \mathrm{cok}(L^{i,j})$. In particular, the order of $K(G)$ is equal to the determinant of the reduced Laplacian $L^{i,j}$.*

As mentioned earlier, the algorithm suggested by Theorem 2 is not very efficient. There are much more efficient algorithms for computing Smith normal forms that proceed similarly to how one row reduces matrices into reduced echelon form in a linear algebra class. In particular, one can put any $n \times n$ integer matrix into a unique matrix in Smith normal form by a sequence of the following operations:

1. Multiply rows or columns by -1,
2. Swap two rows,
3. Swap two columns,
4. Add any integer multiple of one row to another row, or
5. Add any integer multiple of one column to another column.

Just as when putting matrices into reduced echelon form there are many choices one makes along the way which may speed up or slow down the process. For details of how to optimize this procedure, we refer the reader to [39] and [69]. There are efficient implementations of these algorithms in most computer algebra systems including Sage, Maple, and Mathematica.

Example 5 Consider again the graph from Example 1. Let us use row and column reduction in order to find the Smith normal form of the Laplacian of this graph.

$$
\begin{pmatrix} 3 & -1 & -1 & -1 \\ -1 & 2 & 0 & -1 \\ -1 & 0 & 2 & -1 \\ -1 & -1 & -1 & 3 \end{pmatrix} \xrightarrow[r_4-r_3]{r_1 \leftrightarrow r_2} \begin{pmatrix} -1 & 2 & 0 & -1 \\ 3 & -1 & -1 & -1 \\ -1 & 0 & 2 & -1 \\ 0 & -1 & -3 & 4 \end{pmatrix} \xrightarrow[r_3-r_1]{r_2+3r_1} \begin{pmatrix} -1 & 2 & 0 & -1 \\ 0 & 5 & -1 & -4 \\ 0 & -2 & 2 & 0 \\ 0 & -1 & -3 & 4 \end{pmatrix}
$$

$$
\xrightarrow[-r_1]{r_2+2r_3} \begin{pmatrix} 1 & -2 & 0 & -1 \\ 0 & 1 & 3 & -4 \\ 0 & -2 & 2 & 0 \\ 0 & -1 & -3 & 4 \end{pmatrix} \xrightarrow[r_4+r_2]{r_3+2r_2} \begin{pmatrix} 1 & -2 & 0 & 1 \\ 0 & 1 & 3 & -4 \\ 0 & 0 & 8 & -8 \\ 0 & 0 & 0 & 0 \end{pmatrix}
$$

$$
\xrightarrow[c_4-c_1]{c_2+2c_1} \begin{pmatrix} 1 & 0 & 0 & 0 \\ 0 & 1 & 3 & -4 \\ 0 & 0 & 8 & -8 \\ 0 & 0 & 0 & 0 \end{pmatrix} \xrightarrow[c_4+4c_2+c_3]{c_3-3c_2} \begin{pmatrix} 1 & 0 & 0 & 0 \\ 0 & 1 & 0 & 0 \\ 0 & 0 & 8 & 0 \\ 0 & 0 & 0 & 0 \end{pmatrix}.
$$

It is often interesting to look at specific families of graphs and ask how to compute their critical groups. As an example, the *complete bipartite graph* $K_{m,n}$ has vertex set $\{x_1, \ldots, x_m, y_1, \ldots, y_n\}$ and edge set consisting of the edges between each x_i and y_j and no others.

Exercise 5 Find the critical group of $K_{3,3}$ by computing the Smith normal form of its Laplacian matrix. How would your results generalize to other complete bipartite graphs $K_{m,n}$?

We note that a formula for the critical groups of all complete multipartite graphs is given in [45].

Fig. 2 The circulant graphs
$C_6(1, 2)$ and $C_{12}(2, 3)$

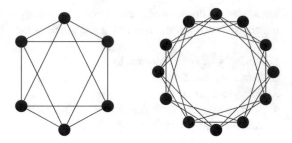

Many authors have worked on problems about computing critical groups for other special families of graphs. For example, the critical groups of *wheel graphs* are described in [14, §9], *rook graphs* are considered in [35], and *Paley graphs* in [20]. Much of this work has been done by undergraduate students, and there are many families of graphs that one could still explore!

We close this section by discussing one family of graphs where there are still many open questions about the critical groups, the circulant graphs. To be explicit, the *circulant graph* $C_n(a_1, \ldots, a_k)$ is formed by placing n points on a circle and drawing the edges from each vertex to the vertices that are $a_1, a_2, \ldots, a_k$ positions further in the clockwise direction. Two examples are given in Fig. 2. Some graphs of this form have been analyzed in several papers, [40, 44, 59, 60], where results like the following are shown:

Theorem 3 *Let F_n be the nth Fibonacci number and let $d = \gcd(n, F_n)$. Then the critical group of $C_n(1, 2)$ is isomorphic to $\mathbb{Z}/d\mathbb{Z} \oplus \mathbb{Z}/F_n\mathbb{Z} \oplus \mathbb{Z}/(nF_n/d)\mathbb{Z}$.*

Exercise 6 Write down the Laplacian matrix for the graph $C_6(1, 2)$. Verify that the critical group of this graph is $\mathbb{Z}/6\mathbb{Z} \oplus (\mathbb{Z}/13\mathbb{Z})^2$.

An unpublished note [26] argues that in general the critical group of the circulant graph $C_n(a, b)$ can be generated by at most $2b - 1$ elements. The authors also describe explicit calculations giving a library of the critical groups of all circulant graphs with at most 27 vertices.

Research Project 1 Compute the critical groups of $C_n(1, 3)$ for $n \leq 10$, either by hand or using a computer algebra system. Try to find patterns. Compare your results to [59, Theorem 2].

What kinds of patterns can you find for other families of circulant graphs?

1.4   Elements of the Critical Group

In the previous section we saw how to determine the critical group of a graph by computing the Smith normal form of the Laplacian of G. We also have seen how equivalence classes of divisors on a graph give elements of the critical group.

Question 1 How do we write down representatives for the elements of $K(G)$? In particular, we know that $K(G)$ is isomorphic to the group of all classes of degree 0 divisors on G under chip-firing equivalence. How do we make a "good choice" of one divisor from each class? How do we determine if two divisors are in the same class?

There are several different approaches to choosing a representative from each class, and we will give one here. Let $\delta \in \mathrm{Div}(G)$ and $v \in V(G)$. We say that v is *in debt* if $\delta(v) < 0$. We fix a vertex $q \in V(G)$ and define a divisor $\delta \in \mathrm{Div}(G)$ to be *q-reduced* if $\delta(v) \geq 0$ for all $v \neq q$ and, moreover, for every nonempty set of vertices $A \subseteq V(G) \setminus \{q\}$, if one starts with the divisor δ and simultaneously fires every vertex in A, then some vertex in A goes into debt.

Example 6 Once again we consider the graph from Example 1. We denote the upper-left vertex as $q = v_1$, the lower-left as v_2, the upper-right as v_3, and the lower-right as v_4. In order for a divisor δ to be q-reduced, one first notes that $\delta(v) < \deg(v)$ for all $v \neq q$ to account for the situation when A is a single vertex. On the other hand, if we fire all three of the vertices in $A = \{v_2, v_3, v_4\}$, then δ decreases by one at each of these vertices, so if firing at each vertex of A causes one of the vertices to go into debt we know that the value of δ is zero for at least one of them (Fig. 3).

Firing both vertices in $A = \{v_2, v_3\}$ decreases the value of the divisor at each of these vertices by two, which will already make both of the values negative by our above reasoning. If $A = \{v_2, v_4\}$, then firing both vertices in A decreases $\delta(v_2)$ by one and $\delta(v_4)$ by two. In particular, if $\delta(v_4) = 2$, then $\delta(v_2) = 0$ and if $\delta(v_2) = 1$, then $\delta(v_4) = 0$ or 1. Considering $A = \{v_3, v_4\}$ gives the analogous results for v_3.

Combining these facts, one can see that there are eight q-reduced divisors of degree zero on this graph, given by the 4-tuples $(\delta(v_1), \delta(v_2), \delta(v_3), \delta(v_4))$:

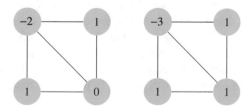

Fig. 3 Two divisors on the graph from Example 1. The first is q-reduced, as one can see by firing each of the seven nonempty subsets of $\{v_1, v_2, v_3\}$. The second is not q-reduced, as one can see by noting that firing all of the vertices in $\{v_1, v_2, v_3\}$ will not put any of these vertices into debt

$$\{(0, 0, 0, 0), (-1, 1, 0, 0), (-1, 0, 1, 0), (-1, 0, 0, 1),$$

$$(-2, 1, 1, 0), (-2, 1, 0, 1), (-2, 0, 1, 1), (-2, 0, 2, 0)\}.$$

Exercise 7 Show that if we had chosen q to be the vertex v_2 instead of v_1 that there would still be eight q-reduced divisors of degree zero.

As was suggested by the previous example, the number of q-reduced divisors does not depend on the choice of vertex q, even though the specific set of divisors certainly does. In fact, a much stronger result is true:

Theorem 4 ([7, Prop 3.1]) *Let G be a finite connected graph and $q \in V(G)$. Then every divisor class in $K(G)$ contains a unique q-reduced divisor.*

Checking whether or not a divisor is q-reduced directly from the above definition is difficult for large graphs as there are exponentially many subsets A one needs to check. However, there is a fast algorithm due to Dhar known as the *Burning Algorithm* that verifies whether a divisor is q-reduced by checking only a linear number of firing sets. We will not give the details of this algorithm but refer the interested reader to [48, Section 2.6.7]. It is worth noting that q-reduced divisors were independently developed under the name of G-parking functions in order to generalize what are now called classical parking functions; for more details about this story, we refer the reader to [48, Section 3.6].

1.5   Spanning Trees and the Matrix Tree Theorem

A *spanning tree* of a connected graph G is a subgraph $\mathcal{T}$ consisting of all of the vertices of G and a subset of the edges of G so that the graph $\mathcal{T}$ is connected and contains no cycles. It follows from elementary results in graph theory that if G (and hence $\mathcal{T}$) has n vertices then $\mathcal{T}$ will have $n - 1$ edges.

Example 7 Consider the cycle on n vertices, C_n. We get a spanning tree by deleting any single edge. Thus, C_n has n spanning trees.

The graph from Example 1 consists of 4 vertices and 5 edges, so any spanning tree will be obtained by deleting two of the edges from the graph. However, in this case we cannot just delete any two edges; for example, deleting the edges $\overline{v_1 v_2}$ and $\overline{v_2 v_4}$ will leave us with a graph that is both disconnected and contains a cycle (see Fig. 4). In particular, if we delete the edge $\overline{v_1 v_4}$, then we can delete any of the remaining edges as our second edge. Otherwise, we must delete exactly one edge from $\{\overline{v_1 v_2}, \overline{v_2 v_3}\}$ and one from $\{\overline{v_1 v_3}, \overline{v_3 v_4}\}$. In particular, there are eight spanning trees of this graph.

In general, it might appear to be a difficult question to ask for the number of spanning trees a given graph, but there is a nice answer given in terms of the Laplacian of the graph. The result is often attributed to Kirchhoff based on work

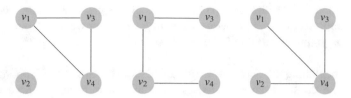

Fig. 4 Three subgraphs of the graph from Example 1 which each have three edges. The first one is not a spanning tree but the other two are

he did as an undergraduate in the 1840s. Many different proofs have been given over the years. For a discussion of the history of this theorem as well as a proof and some related results, see [47] and [13, Chapter 6]. Recall that the reduced Laplacian $L^{i,j}(G)$ of a graph G is the matrix we get by deleting the ith row and j^{th} column from $L(G)$.

Theorem 5 (Matrix Tree Theorem) *The number of spanning trees of G is equal to* $|\det(L^{i,j}(G))|$ *for any i, j.*

Combining this theorem with the discussion in Sect. 1.3 gives us the following result which we will make use of repeatedly:

Corollary 3 *The order of the critical group $K(G)$ is the number of spanning trees of G.*

In fact, Cori and Le Borgne give an explicit bijection between spanning trees of a graph and reduced divisors in the critical group in [27]. In [9], Baker and Shokrieh reformulate the question in terms of minimizing energy potential to generalize these results further. We will not discuss these refinements here.

Corollary 3 immediately tells us that any tree has trivial critical group, a fact that we will give a different proof of in Corollary 5. It also tells us that the critical group of a cycle on n vertices has order n and that the critical group of the graph in Example 1 has order 8, although it does not help us pin down the group exactly. We will return to critical groups of cycles in the next section.

Exercise 8 Consider the "house graph" pictured here:

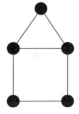

Show that there are 11 different spanning trees of this graph, and conclude that the critical group must be $\mathbb{Z}/11\mathbb{Z}$. More generally, what can we say about the critical group of the graph consisting of two cycles sharing a common edge?

At the beginning of Sect. 1.4 we noted that there are several approaches to choosing one divisor from each divisor class and then discussed the example of q-reduced divisors. Another interesting choice comes from the theory of *break divisors*, which are defined in terms of the spanning trees of G. An, Baker, Kuperberg, and Shokrieh use these divisors to give a decomposition of $\mathrm{Pic}^g(G)$, the set of all divisors of degree d on G modulo chip-firing equivalence [3]. This leads to a "geometric proof" of Theorem 5.

1.6   How Does the Critical Group Change Under Graph Operations?

To this point, we have used techniques from linear algebra to compute critical groups. One can also often use combinatorial properties of graphs to help with these computations. In this section, we will consider several such approaches.

The Dual of a Planar Graph A graph G is *planar* if it can be drawn on a sheet of paper without any edges crossing. The *dual graph* $\hat{G}$ is defined as follows. Choose a drawing of G. The vertices of $\hat{G}$ are in bijection with the planar regions of the drawing. There is an edge connecting two vertices of $\hat{G}$ precisely when the corresponding regions of the drawing of G share an edge. Two examples are given in Fig. 5. We note that the dual of a planar simple graph may have multiple edges between two vertices.

This definition of the dual depends on a choice of embedding into the plane. In particular there are graphs where different embeddings into the plane lead to non-isomorphic dual graphs. That said, we have the following result of Berman [12, Proposition 4.1] that was rediscovered by Cori and Rossin [28, Theorem 2], and by Bacher, de la Harpe, and Nagnibeda [6, Proposition 8].

Theorem 6 *If G is a planar graph and $\hat{G}$ is its dual graph, then $K(G) \cong K(\hat{G})$.*

Fig. 5 Two planar graphs and their duals. The vertices of the original graphs are given in gray and the edges are solid. The vertices of the dual graphs are given in black and the edges are dashed

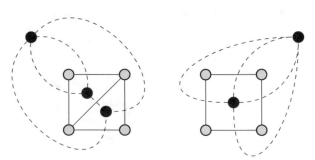

Corollary 4 *The critical group of the cycle graph C_n is $\mathbb{Z}/n\mathbb{Z}$.*

Proof The dual graph to C_n consists of two vertices (one representing the inside of the cycle and one representing the outside) with n edges between them, as illustrated in Fig. 5. Therefore,

$$L(\hat{C}_n) = \begin{pmatrix} n & -n \\ -n & n \end{pmatrix}.$$

We easily deduce that $K(\hat{C}_n) \cong \mathbb{Z}/n\mathbb{Z}$. The result follows from Theorem 6.

In this argument we took the dual graph of a cycle and got a graph that had n distinct edges between our pair of vertices. As we noted earlier, standard facts about critical groups work in this more general *multigraph* setting– it is a good exercise to check that you believe us!

There is a construction similar to the dual graph known as the *line graph* of a graph G. In particular, the line graph of G is the graph G_L whose vertices are in bijection with the edges of G and two vertices in G_L have an edge between them if and only if the corresponding edges share a vertex. For information on critical groups of line graphs see [11].

The Wedge of Two Graphs Let G_1 and G_2 be two finite graphs with designated vertices $v_1 \in G_1$ and $v_2 \in G_2$. The *wedge* of G_1 and G_2 is the graph G consisting of the two graphs G_1 and G_2 with the vertices v_1 and v_2 identified.

Example 8 Let G be the wedge of two triangles, as shown below.

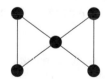

One can check that

$$L(G) = \begin{pmatrix} 2 & -1 & -1 & 0 & 0 \\ -1 & 2 & -1 & 0 & 0 \\ -1 & -1 & 4 & -1 & -1 \\ 0 & 0 & -1 & 2 & -1 \\ 0 & 0 & -1 & -1 & 2 \end{pmatrix}.$$

Deleting the third row and third column of $L(G)$, gives block matrix consisting of two copies of the 2×2 matrix $\begin{pmatrix} 2 & -1 \\ -1 & 2 \end{pmatrix}$. It is straightforward to see that each of these blocks is the reduced Laplacian of a single triangle graph, and therefore the reduced Laplacian of the original graph can be reduced through row and column operations to

$$\begin{pmatrix} 1 & 0 & 0 & 0 \\ 0 & 3 & 0 & 0 \\ 0 & 0 & 1 & 0 \\ 0 & 0 & 0 & 3 \end{pmatrix}.$$

By Corollary 2, the critical group of G is $\mathbb{Z}/3\mathbb{Z} \oplus \mathbb{Z}/3\mathbb{Z}$.

This example generalizes, as shown in the following theorem:

Theorem 7 *Let G_1 and G_2 be two finite graphs and let G be the wedge of G_1 and G_2. Then $K(G) \cong K(G_1) \oplus K(G_2)$.*

Exercise 9 Give a proof of Theorem 7 in the spirit of the previous example. In particular, if G is the wedge of G_1 and G_2, determine the relationship between $L(G)$, $L(G_1)$, and $L(G_2)$ and use this to compute the cokernel of $L(G)$ in terms of $\mathrm{cok}(L(G_1))$ and $\mathrm{cok}(L(G_2))$.

The following result follows immediately from Corollary 3, but we will give an additional proof illustrating the ideas of this section.

Corollary 5 *Let G be any tree. Then the critical group $K(G)$ is trivial.*

Proof If H is the graph consisting of two vertices and a single edge, then $L(H) = \left(\begin{smallmatrix} 1 & -1 \\ -1 & 1 \end{smallmatrix} \right)$. In particular it is clear that $K(H)$ is trivial. Any tree can be constructed as the successive wedges of graphs isomorphic to H and therefore the critical group of a tree is itself trivial.

Adding/Subtracting an Edge The fundamental theorem of finite abelian groups tells us that any finite abelian group H can be written uniquely as a direct sum

$$H \cong \mathbb{Z}/n_1\mathbb{Z} \oplus \mathbb{Z}/n_2\mathbb{Z} \oplus \cdots \oplus \mathbb{Z}/n_r\mathbb{Z},$$

where $n_i \mid n_{i+1}$ for all i and $n_r > 1$. The n_i are the *invariant factors* of H, and the integer r is the *rank* of H, the minimum size of a generating set of H. Let G be a finite connected graph and G' be a graph on the same set of vertices where we have added one additional edge. Lorenzini shows that the rank of $K(G)$ and the rank of $K(G')$ differ by at most 1 [54, Lemma 5.3].

Lorenzini uses this result to give an upper bound for the rank of the critical group of a connected graph G. Since $K(G)$ is isomorphic to the cokernel of an $(n-1) \times (n-1)$ matrix, it is clear that the rank of $K(G)$ is at most $n-1$. This bound is in general not good, and in fact we will see evidence in Sect. 1.9 that most graphs have cyclic critical groups. Recall that the *genus* of a graph is the number of independent cycles that the graph contains; in particular, it can be computed as $g(G) = |E(G)| - |V(G)| + 1$.

Theorem 8 ([54, Proposition 5.2]) *Let G be a connected graph and let $h(G)$ denote the rank of $K(G)$. Then $h(G) \leq g(G)$.*

One can see that this bound is sharp by considering the graph formed as the wedge of k copies of the triangle C_3. This graph has genus k and critical group $(\mathbb{Z}/3\mathbb{Z})^k$. In general, finding a minimal set of generators is an open problem. We will return to this question in Sect. 1.8.

Subdividing an Edge Let G be a graph with $v_1, v_2 \in V(G)$ and $\overline{v_1 v_2} \in E(G)$. Let G' be the graph whose vertex set is the same as G except with the edge $\overline{v_1 v_2}$ replaced with a path of k edges. We see that $V(G')$ consists of $V(G)$ together with $k - 1$ new vertices along this path.

Subdividing a single edge of a graph can have all kinds of different effects on the critical group; If you subdivide an edge on a path, then it does not change the critical group, as it will still be trivial, but if you subdivide an edge on the cycle C_n, replacing it with a path of length 2, it changes the critical group from $\mathbb{Z}/n\mathbb{Z}$ to $\mathbb{Z}/(n + 1)\mathbb{Z}$. Subdividing an edge can change not only the order of $K(G)$, but can also change whether or not this group is cyclic, as illustrated in Fig. 6.

The following result from [21] shows that after a suitable choice of subdivisions one can always make the critical group cyclic.

Theorem 9 *Let G be a graph of genus $g \geq 1$. Then there is a choice of at most $g - 1$ subdivisions after which the critical group becomes cyclic.*

Exercise 10 Show that Theorem 9 is true in the case where G is the wedge of two cycle graphs C_m and C_n. In particular, this graph has genus two so you should show that either $K(G)$ is already cyclic or $K(G)$ can be made cyclic after a single subdivision. Can you generalize this argument to the wedge of three or more cycles?

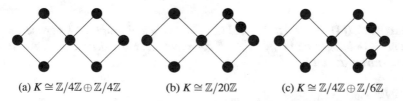

(a) $K \cong \mathbb{Z}/4\mathbb{Z} \oplus \mathbb{Z}/4\mathbb{Z}$ (b) $K \cong \mathbb{Z}/20\mathbb{Z}$ (c) $K \cong \mathbb{Z}/4\mathbb{Z} \oplus \mathbb{Z}/6\mathbb{Z}$

Fig. 6 Pictured above is (**a**) a graph with critical group $\mathbb{Z}/4\mathbb{Z} \oplus \mathbb{Z}/4\mathbb{Z}$, (**b**) a graph with critical group $\mathbb{Z}/4\mathbb{Z} \oplus \mathbb{Z}/5\mathbb{Z} \cong \mathbb{Z}/20\mathbb{Z}$ obtained by subdividing the previous graph, and (**c**) a graph with the noncyclic critical group $\mathbb{Z}/4\mathbb{Z} \oplus \mathbb{Z}/6\mathbb{Z}$ obtained by another subdivision

In a different vein, one can explicitly describe what happens after simultaneously subdividing all edges. We begin with an example:

Example 9 Let G be the graph consisting of the wedge of the cycles C_3 and C_4. We have already seen that the critical group of G is $K(G) \cong \mathbb{Z}/3\mathbb{Z} \oplus \mathbb{Z}/4\mathbb{Z}$. Note that if we subdivide each edge of G into k edges then the new graph G_k will be the wedge of the cycles C_{3k} and C_{4k} and therefore has critical group $K(G_k) \cong \mathbb{Z}/3k\mathbb{Z} \oplus \mathbb{Z}/4k\mathbb{Z}$.

It turns out that the previous example generalizes in a natural way. Recall that Theorem 8 tells us that if g is the genus of a graph G, then the critical group of G can be written as $\mathbb{Z}/m_1\mathbb{Z} \oplus \ldots \oplus \mathbb{Z}/m_g\mathbb{Z}$, where it may be the case that some of the $m_i = 1$. We can use this decomposition to get the following result:

Theorem 10 ([56, Proposition 2]) *Let $G_{sub(k)}$ be the graph obtained by subdividing each edge of G into k edges. Then, writing*

$$K(G) \cong \mathbb{Z}/m_1\mathbb{Z} \oplus \ldots \oplus \mathbb{Z}/m_g\mathbb{Z}$$

as above we see that

$$K(G_{sub(k)}) \cong \mathbb{Z}/km_1\mathbb{Z} \oplus \ldots \oplus \mathbb{Z}/km_g\mathbb{Z}.$$

Exercise 11 Let G be the graph from Example 1 and let $G_{\text{sub}(2)}$ be the graph obtained by subdividing each edge of G into two edges. Compute the critical group of $G_{\text{sub}(2)}$ both by using Theorem 10 and by using results about the Laplacian matrix of $G_{\text{sub}(2)}$.

The Cone Over a Graph The *join* of two graphs G and H consists of disjoint copies of G and H together with edges $\overline{uv}$ for all pairs $u \in V(G)$ and $v \in V(H)$. The nth *cone over* G, denoted G_n, is the join of G and the complete graph K_n. Several authors have studied how the critical group of G_n is related to the critical group of G [1,19]. The following result of Goel and Perkinson builds on these earlier efforts.

Theorem 11 ([41, Theorem 1]) *Let G be a connected graph on k vertices, $n \geq 2$ be a positive integer, and G_n be the nth cone over G. Let $\mathbf{1}$ denote the $k \times k$ matrix whose entries are all 1.*

1. We have

$$K(G_n) \cong (\mathbb{Z}/(n+k)\mathbb{Z})^{n-2} \oplus \text{cok}\,(nI_k + L(G) + \mathbf{1}).$$

2. The group $\text{cok}\,(nI_k + L(G) + \mathbf{1})$ has a subgroup isomorphic to $\mathbb{Z}/(n+k)\mathbb{Z}$.

3. We have

$$|K(G_n)| = \frac{|p_{L(G)}(-n)|}{n}(n+k)^{n-1},$$

where $p_{L(G)}$ is the characteristic polynomial of $L(G)$.

The last of these statements is Corollary B in [19].

Example 10 Let G be the path graph on two vertices. One can see that $L(G) = \left(\begin{smallmatrix} 1 & -1 \\ -1 & 1 \end{smallmatrix}\right)$, so that $p_{L(G)}(t) = t^2 - 2t$. The third statement of this theorem therefore implies that $|K(G_n)| = (n+2)^n$ for all choices of n. This does not tell us the specific group structure, although in this case we can see from the first statement that

$$K(G_n) \cong (\mathbb{Z}/(n+2)\mathbb{Z})^{n-2} \oplus \text{cok} \left(\begin{smallmatrix} n+2 & 0 \\ 0 & n+2 \end{smallmatrix}\right) \cong (\mathbb{Z}/(n+2)\mathbb{Z})^n.$$

When the graph is more complicated, Theorem 11 is more useful in determining the order of the critical group of the cone of a graph than in determining its group structure, something which [19, Question 1.2] asks about in a slightly different form. Goel and Perkinson show that this involves understanding when $\mathbb{Z}/(n+k)\mathbb{Z}$ is a direct summand of $\text{cok}(nI_k + L(G) + 1)$. This question is analyzed for the path on 4 vertices in [41, Example 5].

Research Project 2 How much more can one say about the structure of $K(G_n)$ for a general graph G and positive integer n, where G_n is the nth cone over G?

Functions Between Graphs There are various results that look at the functorial properties of the critical groups of graphs. One particularly nice example is given by *Harmonic morphisms* between graphs, which Baker and Norine use to prove a graph-theoretic analogue of the Riemann–Hurwitz formula from algebraic geometry [8]. These morphisms induce different kinds of functorial maps between divisors on graphs and between their critical groups. Reiner and Tseng examine the situation where one has a map between two graphs $\phi : G \to H$ that satisfies certain technical conditions and show that this induces a surjection of the critical groups $K(G) \twoheadrightarrow K(H)$ whose kernel can be understood [64]. Other papers look at graphs that admit automorphisms and what one can say about either $|K(G)|$ or the structure of $K(G)$ in relation to its quotients. For examples related to reflective symmetry see [23] and for dihedral group actions see [40].

1.7   Which Finite Abelian Groups Occur as the Critical Group of a Graph?

Up to this point, we have primarily been concerned with the situation where we are given a graph G and try to determine $K(G)$. One could also ask how to construct graphs that have a given critical group. Combining Theorems 4 and 7 implies that we can construct a graph with critical group

$$\mathbb{Z}/m_1\mathbb{Z} \oplus \ldots \oplus \mathbb{Z}/m_d\mathbb{Z}$$

by taking the wedge of cycles $C_{m_1}, C_{m_2}, \ldots, C_{m_d}$.

> **Research Project 3** Let H be a finite abelian group. We know that there is *some* graph G with $K(G) \cong H$. This G is clearly far from unique. What is the graph G with the smallest number of vertices and given critical group?

This is related to a problem of Rosa, which asks for the smallest number of vertices of a graph with a given number of spanning trees. Even this simpler sounding problem is not well understood. See [65] for partial results.

There is a technical detail related to our discussion so far. If any of the m_i are equal to 2, this construction taking a wedge of cycles C_{m_i} does not result in a simple graph. In fact, it is not difficult to show that there is no simple graph G with $K(G) \cong \mathbb{Z}/2\mathbb{Z}$. Suppose G were such a graph and let $\mathcal{T}$ be one of its spanning trees. There must be some $e \in E(G)$ so that $\mathcal{T} \cup \{e\}$ contains a cycle. Since G is a simple graph, this cycle has at least three edges. Removing any edge in this cycle gives a spanning tree of G. Therefore, G has at least three spanning trees, so $|K(G)| \geq 3$. In [38], the authors significantly strengthen these ideas, and prove that there are no simple connected graphs with any of the following critical groups:

$$\mathbb{Z}/2\mathbb{Z} \oplus \mathbb{Z}/4\mathbb{Z}, \ (\mathbb{Z}/2\mathbb{Z})^2 \oplus \mathbb{Z}/4\mathbb{Z}, \ \mathbb{Z}/2\mathbb{Z} \oplus (\mathbb{Z}/4\mathbb{Z})^2, \ \text{or} \ (\mathbb{Z}/2\mathbb{Z})^k \text{ for any } k \geq 1.$$

Moreover, they show the following:

Theorem 12 *Let H be any finite abelian group. There exists some positive integer k_H so that there are no connected simple graphs with critical group $H \oplus (\mathbb{Z}/2\mathbb{Z})^k$ for any $k \geq k_H$.*

> **Research Project 4** Let $H \cong \mathbb{Z}/8\mathbb{Z}$. For what values of k is there a connected simple graph with critical group $\mathbb{Z}/8\mathbb{Z} \oplus (\mathbb{Z}/2\mathbb{Z})^k$?

(continued)

More generally, for other finite abelian groups H, what can we say about the value of k_H? One approach to constructing such graphs might be to find graphs of a given genus and critical group and then subdividing each edge into two edges and using Theorem 10.

So far in this section we have asked only about the existence of a simple graph with a given critical group. We can ask stronger questions about the existence of graphs with additional properties and given critical group. For example, a graph G has *connectivity at least κ* if G remains connected even if one deletes any set of $\kappa - 1$ vertices and all edges incident to a vertex in this set. In particular, a graph is said to be *biconnected* if it remains connected after deleting any single vertex and all edges incident to it. The authors of [38] show that if a graph is biconnected and has maximum vertex degree δ, then the critical group must contain some element whose order is at least δ. This result is one of the ingredients in proving that there are no simple graphs with critical group $(\mathbb{Z}/2\mathbb{Z})^k$. These observations lead them to make the following conjecture.

Research Project 5 Is it true that for any positive integer n, there exists k_n such that if $k > k_n$, there is no biconnected graph G with critical group $(\mathbb{Z}/n\mathbb{Z})^k$?

1.8   Generators of Critical Groups

In Sect. 1.9, we will study properties of critical groups of random graphs and see that we often expect these critical groups to be cyclic. The simplest possible nonzero divisor on G is of the form δ_{xy} where $x, y \in V(G)$, $\delta_{xy}(x) = 1$, $\delta_{xy}(y) = -1$ and $\delta_{xy}(v) = 0$ at all other vertices.

Question 2 Let G be a connected finite graph with $K(G)$ cyclic. When does $K(G)$ have a generator of the form δ_{xy}?

In [10], the authors give a number of examples of graphs with cyclic critical groups and generators of this form, and also give examples of graphs with $K(G)$ cyclic that do not have a generator of this form. They propose a general criterion for when a graph G has such a generator. This conjecture was proven in [17].

Theorem 13 *Let x and y be vertices on a finite connected graph G and let G' be the graph obtained by adding $\overline{xy}$ if $\overline{xy} \notin E(G)$ and deleting $\overline{xy}$ if $\overline{xy} \in E(G)$. Let*

δ_{xy} *be defined as above and let* $S \subseteq K(G)$ *be the subgroup of the critical group of* G *generated by* δ_{xy}. *Then we have the following relationships:*

- $[K(G) : S]$ *divides* $\gcd(|K(G)|, |K(G')|)$
- $\gcd(|K(G)|, |K(G')|)$ *divides* $[K(G) : S]^2$.

In particular, δ_{xy} *is a generator of* $K(G)$ *if and only if* $\gcd(|K(G)|, |K(G')|) = 1$.

Research Project 6 Theorem 13 gives a way of testing whether a given pair of vertices x, y gives a divisor δ_{xy} that generates $K(G)$. Is there a simple way to test whether there exists a pair of vertices x, y such that δ_{xy} generates $K(G)$?

For example, the wedge of a triangle, square, and pentagon has critical group $\mathbb{Z}/60\mathbb{Z}$, but there is no pair of vertices x, y such that δ_{xy} generates $K(G)$.

Research Project 7 What happens when the critical group of G is not cyclic? For example, is there a way of testing whether two divisors $\delta_{x_1 y_1}$ and $\delta_{x_2 y_2}$ generate $K(G)$?

1.9   Critical Groups of Random Graphs

In Sect. 1.7, we saw that every finite abelian group occurs as the critical group of a graph if we allow multiple edges between vertices, and that every finite abelian group of odd order occurs as the critical group of a simple graph. Instead of asking whether a group occurs as $K(G)$ for at least one graph G, we could ask about which kinds of groups occur often as the critical group of a graph. Throughout this section we restrict our attention to simple graphs.

Question 3 What can we say about critical groups in families of "random graphs"?

Here is one way to make this question precise. There are $\binom{n}{2}$ possible edges between vertices $v_1, \ldots v_n$, so there are $2^{\binom{n}{2}}$ labeled simple graphs on this vertex set. As a warmup, we can ask the following.

Question 4 How many of these $2^{\binom{n}{2}}$ graphs are connected and have trivial critical group?

Corollary 3 implies that a connected graph has trivial critical group if and only if it is a tree. It follows from Example 4 that the number of labeled trees of n vertices is n^{n-2}. So the proportion of graphs on n vertices that are connected and have trivial critical group is $n^{n-2}/2^{\binom{n}{2}}$, which goes to zero as n goes to infinity. This tells us that the size of $K(G)$ is not often equal to 1, but does not tell us how large we should expect it to be.

In order to determine the average size of the critical group of a graph on n vertices, we introduce some ideas from probabilistic combinatorics. There are n^{n-2} trees on n vertices, and each tree has exactly $n-1$ edges. Fix a choice of a spanning tree $\mathcal{T}$ on n vertices. The number of graphs on n vertices containing $\mathcal{T}$ as a subgraph will be $2^{\binom{n}{2}-(n-1)}$ since, for each edge not in $\mathcal{T}$, we can choose whether it is present in our graph. This implies that the probability that $\mathcal{T}$ is contained in a random graph is $1/2^{n-1}$. It then follows from linearity of expectation that the expected number of spanning trees of a graph on n vertices is $n^{n-2}/2^{n-1}$. It is easy to check this formula in small cases.

Example 11 There are 8 graphs with vertex set $\{v_1, v_2, v_3\}$, and 4 of these are connected: the complete graph K_3, which has 3 spanning trees, and $3^{3-2} = 3$ trees, which have 1 spanning tree each. We conclude that the average number of spanning trees of a graph on 3 vertices is $3/4$.

A graph G on n vertices is not connected if and only if it does not contain any of the n^{n-2} spanning trees of the complete graph with vertex set $V(G)$.

Exercise 12 Show that as n goes to infinity, the proportion of graphs on n vertices that are connected goes to 1.

Here is one approach: A graph G with n vertices is connected if every one of the $\binom{n}{2}$ pairs of vertices $v_i, v_j \in V(G)$ share a common neighbor. What is the probability that v_k is a common neighbor of both v_i and v_j? What is the probability that v_i and v_j do not share a common neighbor?

For the rest of this section, when we ask about the proportion of graphs G on n vertices for which $K(G)$ satisfies some property, what we really mean is the proportion of graphs G that are connected and such that $K(G)$ has this property. By Exercise 12, as n goes to infinity the proportion of connected graphs goes to 1, so we do not need to keep writing this extra assumption.

Since $n^{n-2}/2^{n-1}$ goes to infinity with n, we see that the average size of $K(G)$ gets large as $|V(G)|$ gets large. In fact, something stronger is true:

Proposition 4 *Let X be a positive integer. The proportion of graphs G on n vertices for which $|K(G)| \leq X$ goes to 0 as n goes to infinity.*

Note that if G has at most X spanning trees, then we can make G disconnected by removing at most X edges, so X has *edge connectivity* at most X. We leave the proof of this proposition as an exercise, but refer the interested reader to [37, Chapter 4] for results on connectivity of random graphs.

A consequence of Proposition 4 is that for any particular finite abelian group H, the probability that $K(G) \cong H$ goes to 0 as $|V(G)|$ goes to infinity. Instead of asking for $K(G)$ to be isomorphic to a particular group, we can ask for the probability that this group has some chosen property.

Question 5 What proportion of the $2^{\binom{n}{2}}$ graphs on n vertices have $K(G)$ cyclic?

This question has been the subject of much recent research including work of Wagner [70], Lorenzini [57], and Wood [72]. One nice thing about this type of question is that it is not so difficult to do large experiments using a computer algebra system, for example Sage, and to get a sense for what to expect. Building on work of [25], the authors of [24] make the following conjecture.

Conjecture 1 We have

$$\lim_{n \to \infty} \frac{\#\{\text{Connected graphs } G \text{ with } |V(G)| = n \text{ and } K(G) \text{ cyclic}\}}{2^{\binom{n}{2}}}$$

$$= \zeta(3)^{-1} \zeta(5)^{-1} \zeta(7)^{-1} \zeta(9)^{-1} \zeta(11)^{-1} \cdots \approx .7935212.$$

In this conjecture, $\zeta(s) = \sum_{n=1}^{\infty} n^{-s}$ denotes the Riemann zeta function. Wood has proven that this conjectured value is an upper bound for the probability that the critical group of a random graph is cyclic [72, Corollary 9.5]. Showing that equality holds appears to be quite difficult.

It is also interesting to ask questions about other properties of the order of the critical group, such as the following:

Question 6 What proportion of the $2^{\binom{n}{2}}$ graphs on n vertices have $|K(G)|$ odd?

That is, we would like to understand the following limit:

$$\lim_{n \to \infty} \frac{\#\{\text{Connected graphs } G \text{ with } |V(G)| = n \text{ and } |K(G)| \text{ odd}\}}{2^{\binom{n}{2}}}. \tag{1}$$

One of the main ideas that goes into the study of these questions is that a finite abelian group H decomposes as a direct sum of its Sylow p-subgroups. Recall that the *Sylow p-subgroup* of a finite abelian group H is the subgroup of all of its elements of p-power order. We denote this subgroup by H_p. We can interpret many questions about $K(G)$ in terms of the Sylow p-subgroups $K(G)_p$. For example, a connected graph G has a cyclic critical group if and only if $K(G)_p$ is cyclic for each prime p. Similarly, G has an odd number of spanning trees if and only if $K(G)_2$ is trivial. This suggests that a good starting place is to try to understand how the Sylow p-subgroups of critical groups of random graphs behave. The following result of Wood answers this question.

Theorem 14 ([72, Theorem 1.1]) *Let p be a prime and H a finite abelian p-group. Then*

$$\lim_{n \to \infty} \frac{\#\{Connected\ graphs\ G\ with\ |V(G)| = n\ \ and\ \ K(G)_p \cong H\}}{2^{\binom{n}{2}}}$$

$$= \frac{\#\{symmetric,\ bilinear,\ perfect\ pairings\ \phi \colon H \times H \to \mathbb{C}^*\}}{|H||\mathrm{Aut}(H)|} \prod_{k \geq 0}(1 - p^{-2k-1}).$$

We will discuss pairings on finite abelian p-groups and this theorem in more detail in Sect. 1.10. In the meantime, taking $p = 2$ and H equal to the trivial group, we see that the probability that a random graph has an odd number of spanning trees is $\prod_{k \geq 0}^{\infty}(1 - 2^{-2i-1}) \approx 0.4194$, answering Question 6.

Critical Groups of Random Graphs and Cokernels of Random Integer Matrices Questions about critical groups of random graphs are closely connected to questions about random symmetric integer matrices. When R is equal to either $\mathbb{Z}$ or $\mathbb{Z}/p\mathbb{Z}$, we let $\mathrm{Sym}_n(R)$ denote the set of $n \times n$ symmetric matrices with entries in R. To see the connection between random graphs and matrices, we note that half of the $2^{\binom{n}{2}}$ graphs G with $V(G) = \{v_1, \ldots, v_n\}$ have $\overline{v_i v_j} \in E(G)$. So choosing one of these $2^{\binom{n}{2}}$ graphs uniformly at random is the same as flipping a coin for each of the $\binom{n}{2}$ potential edges of the graph to decide whether to include it. This implies that choosing a random graph on n vertices and computing its critical group is the same as the following process:

1. Choose a random matrix $A \in \mathrm{Sym}_n(\mathbb{Z})$ with all diagonal entries equal to 0 by taking each pair $1 \leq i < j \leq n$ and setting $a_{i,j} = 0$ with probability $1/2$ and $a_{i,j} = 1$ with probability $1/2$.
2. Compute the diagonal matrix D with (i, i)-entry equal to the negative of the sum of the entries in the ith row of A. Let L_0 be the $(n - 1) \times (n - 1)$ matrix that we get by deleting the last row and column of $D - A$.
3. Take the cokernel of L_0.

Many questions about properties of random graphs can be phrased as questions about this family of random integer matrices. For example, we have seen that a graph G is connected if and only if $L_0(G)$ has rank $n - 1$, so the proportion of graphs with n vertices that are connected is the same as the probability that a random matrix L_0 chosen by the procedure above has rank $n - 1$.

We will use the fact that $K(G)_p$ only depends on the entries of $L_0(G)$ modulo powers of p.

Exercise 13 Let G be a connected graph.

(a) Prove that $K(G)_p$ is trivial if and only if $p \nmid \det(L_0(G))$.
(b) Conclude that $K(G)_p$ is trivial if and only if we reduce the entries of $L_0(G)$ modulo p and get a matrix with entries in $\mathbb{Z}/p\mathbb{Z}$ of rank $n - 1$.

How often should we expect $K(G)_p$ to be trivial? Exercise 13 suggests that a good first step is to compute the proportion of all matrices in $\mathrm{Sym}_{n-1}(\mathbb{Z}/p\mathbb{Z})$ that have rank $n - 1$.

Theorem 15 ([58, Theorem 2]) *The number of invertible matrices in* $\mathrm{Sym}_{n-1}(\mathbb{Z}/p\mathbb{Z})$ *is*

$$
p^{\binom{n}{2}} \prod_{j=1}^{\lceil \frac{n-1}{2} \rceil} (1 - p^{1-2j}).
$$

We leave the proof as a nice exercise in linear algebra over finite fields.

As we take n to infinity, Theorem 15 implies that the proportion of invertible matrices in $\mathrm{Sym}_{n-1}(\mathbb{Z}/p\mathbb{Z})$ approaches $\prod_{k \geq 0}^{\infty}(1 - p^{-2i-1})$. This is the same probability that we get by taking the trivial group in Theorem 14, the probability that the number of spanning trees of a large random graph is not divisible by p. Wood's theorem demonstrates a deep type of *universality for cokernels of random matrices*. Even though the reduced Laplacian of a random graph *does not* give a uniformly random element of $\mathrm{Sym}_{n-1}(\mathbb{Z}/p\mathbb{Z})$, as n goes to infinity the probability that the reduced Laplacian modulo p is an invertible matrix is the same as the proportion of matrices in $\mathrm{Sym}_{n-1}(\mathbb{Z}/p\mathbb{Z})$ that are invertible.

In order to understand the Sylow p-subgroup of $\mathrm{cok}(L_0(G))_p$, we must consider not only the entries of $L_0(G)$ modulo p, but also modulo higher powers of p. There is a nice algebraic setting for these questions. Instead of thinking about $L_0(G)$ as a matrix with integer entries, we think of it as a matrix with entries in the *p-adic integers*, which we denote by $\mathbb{Z}_p$. A p-adic integer consists of an element of $\mathbb{Z}/p^k\mathbb{Z}$ for each k that is compatible with the canonical surjections $\mathbb{Z}/p^k\mathbb{Z} \twoheadrightarrow \mathbb{Z}/p^{k-1}\mathbb{Z}$. For any prime p, $\mathbb{Z} \subset \mathbb{Z}_p$ since the integer n corresponds to choosing the residue class $n \pmod{p^k}$ for each k. There is a nice description of how to choose a random matrix with p-adic entries that comes from the existence of Haar measure for $\mathbb{Z}_p$. We do not give details here. For an accessible introduction to p-adic numbers, we recommend Gouvea's book [42].

Clancy, Leake, and Payne performed large computational experiments about critical groups of random graphs and made conjectures based on their data [25]. Motivated by these conjectures, these authors together with Kaplan and Wood determine the distribution of cokernels of random elements of $\mathrm{Sym}_n(\mathbb{Z}_p)$ as n goes to infinity [24]. Theorem 14 is a consequence of a much stronger result of Wood about cokernels of families of random p-adic matrices [72]. Wood proves that for a large class of distributions on the entries of such a matrix the distribution of the cokernels does not change. This class is large enough to include reduced Laplacians of random graphs, so even though these matrices are very far from being uniformly random modulo powers of p, the distribution of their cokernels matches the distribution in the uniformly random setting.

Choosing a Random Graph So far in this section we have chosen a random graph by choosing one of the $2^{\binom{n}{2}}$ graphs on n vertices uniformly at random. It is common in the study of random graphs to allow the probability of choosing a particular graph to be weighted by its number of edges. Let $0 < q < 1$. An *Erdős–Rényi random graph* on n vertices, $G(n, q)$, is a graph on n vertices $v_1, \ldots, v_n$ where we independently include the edge $\overline{v_i v_j}$ with probability q. That is, $G(n, q)$ is a probability space on graphs with n vertices in which a graph with m edges is chosen with probability

$$q^m (1 - q)^{\binom{n}{2} - m}.$$

We see that our earlier model of choosing a random graph corresponds to $G(n, 1/2)$, in which each graph is chosen with equal probability.

The conjectures in [24, 25] and the results of [72] apply in this more general Erdős–Rényi random graph setting. That is, if we choose an Erdős–Rényi random graph G on n vertices with edge probability equal to some fixed constant q (for example, $1/2$, or $2/3$, or 10^{-100}), as n goes to infinity the probability that $K(G)_p$ is isomorphic to a particular finite abelian p-group H is given by the right-hand side of Theorem 14, no matter what value of q we choose. Again, this is a consequence of Wood's universality results for cokernels of random matrices [72].

An active area of current research involves allowing the edge probability q to change with n. Linearity of expectation implies that the expected number of edges of a random graph $G(n, q)$ is $\binom{n}{2} q$. Therefore, if we allow q to go to 0 as n goes to infinity, but not too fast, this random graph will still have an increasing number of edges.

Exercise 14 Show that the probability that an Erdős–Rényi random graph $G(n, n^{-1/2})$ is connected goes to 1 as n goes to infinity, even though $n^{-1/2}$ goes to 0.

This exercise is more challenging than Exercise 12. We again refer the interested reader to [37, Chapter 4].

It is likely that a version of Theorem 14 holds when q is allowed to go to 0 or 1 as n goes to infinity, as long as it does not approach 0 or 1 too fast. Determining the threshold where the behavior of the critical group changes is an interesting, and likely very challenging, open problem. For work in this direction see the recent paper of Nguyen and Wood [63].

Question 7 What can we say about Sylow p-subgroups of critical groups in other families of random graphs?

We give two concrete examples to show what Question 7 is all about. A graph G is *bipartite* if we can divide its vertex set $V(G)$ into disjoint sets V_1 and V_2 so that every edge in G connects a vertex in V_1 to a vertex in V_2. We can choose a random bipartite graph with vertex set $V(G) = V_1 \cup V_2$ as follows. Fix $0 < q < 1$. Independently include each of the $|V_1||V_2|$ possible edges between a vertex in V_1 and a vertex in V_2 with probability q.

> **Research Project 8** Consider a random bipartite graph with edge probability q and $|V_1| = |V_2| = n$. As n goes to infinity, how are the Sylow p-subgroups of the critical groups of these graphs distributed?

Koplewitz shows that if the sizes of the vertex sets V_1 and V_2, are too "unbalanced," that is $|V_1|/|V_2| < 1/p$, then the resulting distribution of Sylow p-subgroups of the critical groups of these random bipartite graphs does not match the distribution given in Theorem 14 [51].

To give a second example, a graph G is *d-regular* if every $v \in V(G)$ has degree d. Fix a positive integer $d \geq 3$. Choose a d-regular graph on n vertices uniformly at random. Mészáros has recently shown that as n goes to infinity, the distribution of Sylow p-subgroups of critical groups of random d-regular graphs is the same as the one given by Theorem 14, except when $p = 2$ and d is even, in which case we get a different distribution [61].

These are just two examples of a large family of problems to investigate.

> **Research Project 9** Choose your favorite graph property P. Is it true that the distribution of Sylow p-subgroups of large random graphs with property P matches the distribution of Sylow p-subgroups of all random graphs? For example, what is the distribution of Sylow p-subgroups of large random planar graphs? What about random triangle-free graphs?

1.10   The Monodromy Pairing on Divisors

The expression on the right side of Theorem 14 contains a term that involves the number of symmetric, bilinear, perfect pairings on a finite abelian group H. This is because the critical group of a graph comes with extra algebraic structure. More precisely, our goal is to explain a result of Bosch and Lorenzini [16] that the critical group of a connected graph comes equipped with a symmetric, bilinear, perfect pairing. In order to explain this result, we introduce some additional material about divisors on graphs closely following Shokrieh's presentation in [66].

We first show that the group of degree zero divisors on G comes with a pairing, that is, a function $\langle \cdot, \cdot \rangle \colon \mathrm{Div}^0(G) \times \mathrm{Div}^0(G) \to \mathbb{Q}$, and then, that this pairing descends to a pairing defined on $K(G)$. Much of the following terminology for divisors on graphs is motivated by the analogy with divisors on algebraic curves that we first mentioned in Remark 1.

Recall that a divisor on a graph G is a function $\delta \colon V(G) \to \mathbb{Z}$. Let $\mathcal{M}(G)$ denote the abelian group consisting of integer-valued functions defined on $V(G)$, that is, $\mathcal{M}(G) = \mathrm{Hom}(V(G), \mathbb{Z})$. Let $f \in \mathcal{M}(G)$. For $v \in V(G)$, we define

$$\mathrm{ord}_v(f) = \sum_{\substack{w \in V(G) \\ \overline{vw} \in E(G)}} (f(v) - f(w)).$$

The divisor of the function f, denoted $\mathrm{div}(f)$, is defined by setting $(\mathrm{div}(f))(v) = \mathrm{ord}_v(f)$ for any $v \in V(G)$. Every $\mathrm{div}(f)$ has degree 0, but not every degree 0 divisor is the divisor of a function f. We say that a divisor is *principal* if it is equal to $\mathrm{div}(f)$ for some $f \in \mathcal{M}(G)$ and denote the group of principal divisors on G by $\mathrm{Prin}(G)$.

Example 12 Consider the graph consisting of a cycle on three vertices $\{u, v, w\}$. For any function $f \in \mathcal{M}(G)$ we see that $\mathrm{ord}_u(f) = 2f(u) - f(v) - f(w)$, $\mathrm{ord}_v(f) = 2f(v) - f(u) - f(w)$, and $\mathrm{ord}_w(f) = 2f(w) - f(v) - f(u)$. It is clear that these three numbers sum to zero for any choice of f. On the other hand, if we set δ to be the divisor of degree zero with $\delta(u) = 0$, $\delta(v) = 1$, $\delta(w) = -1$ then in order for δ to be principal, there would have to be an integer-valued function so that

$$2f(u) - f(v) - f(w) = 0$$
$$2f(v) - f(u) - f(w) = 1$$
$$2f(w) - f(v) - f(u) = -1.$$

It is a simple exercise in linear algebra to see that this cannot happen.

Exercise 15 For the cycle from the previous example, describe which divisors of degree zero are principal and which are not.

Exercise 16 More generally, let G be any connected graph. If we identify $\mathrm{Div}(G)$ with column vectors of length $|V(G)|$ that have integer entries, we have seen that a divisor D is chip-firing equivalent to the all zero divisor if and only if it is in the image of $L(G)$. Show that D is chip-firing equivalent to the all zero divisor if and only if it is principal. Use this characterization to see that $K(G) \cong \mathrm{Div}^0(G)/\mathrm{Prin}(G)$.

We now describe the *monodromy pairing* on divisors on the critical group of a connected graph G, which is a graph-theoretic analogue of a notion called the *Weil pairing* on the Jacobian of an algebraic curve. Let $D_1, D_2 \in \mathrm{Div}^0(G)$ and let m_1, m_2 be integers such that $m_1 D_1$ and $m_2 D_2$ are principal. (Such integers must exist because $K(G)$ is finite.) In particular, there will be functions $f_1, f_2 \in \mathcal{M}(G)$ such that $m_1 D_1 = \mathrm{div}(f_1)$ and $m_2 D_2 = \mathrm{div}(f_2)$.

Exercise 17 Show that

$$\frac{1}{m_2} \sum_{v \in V(G)} D_1(v) f_2(v) = \frac{1}{m_1} \sum_{v \in V(G)} D_2(v) f_1(v).$$

We define a pairing $\langle \cdot, \cdot \rangle \colon \mathrm{Div}^0(G) \times \mathrm{Div}^0(G) \to \mathbb{Q}$ by

$$\langle D_1, D_2 \rangle = \frac{1}{m_2} \sum_{v \in V(G)} D_1(v) f_2(v).$$

By the previous exercise, $\langle D_1, D_2 \rangle = \langle D_2, D_1 \rangle$ for all $D_1, D_2 \in \mathrm{Div}(G)$, that is, this pairing is *symmetric*. It is also not difficult to check that it is *bilinear*, meaning that $\langle a D_1 + b D_2, D_3 \rangle = a \langle D_1, D_3 \rangle + b \langle D_2, D_3 \rangle$ for all divisors D_1, D_2, D_3 and all rational numbers a, b.

A symmetric bilinear pairing on a finite abelian group H is *non-degenerate* if the group homomorphism defined by $h \to \langle h, \cdot \rangle$ is injective. If it is an isomorphism, the pairing is called *perfect*. We write $\overline{D}$ for an element of $K(G)$ if $\overline{D}$ is the divisor class of D in $K(G)$. The following theorem of Bosch and Lorenzini states that the pairing on $\mathrm{Div}^0(G)$ descends to a well-defined perfect pairing on $K(G)$ [16]. For consistency with our notation in this section, we give the statement of this result from [66, Theorem 3.4].

Theorem 16 *The pairing* $\langle \cdot, \cdot \rangle \colon K(G) \times K(G) \to \mathbb{Q}/\mathbb{Z}$ *defined by*

$$\langle \overline{D_1}, \overline{D_2} \rangle = \frac{1}{m_2} \sum_{v \in V(G)} D_1(v) f_2(v) \ (\mathrm{mod} \ \mathbb{Z}),$$

where $m_2 D_2 = \mathrm{div}(f_2)$ *is a well-defined, perfect pairing on* $K(G)$.

This pairing is called the *monodromy pairing*. Shokrieh gives a concrete proof of Theorem 16 in [66, Appendix A].

The same underlying finite abelian group may have different perfect pairings defined on it. Let G be a finite abelian group and $\langle \cdot, \cdot \rangle_1$ and $\langle \cdot, \cdot \rangle_2$ be two pairings defined on G. We say that these pairings are isomorphic if there exists $\varphi \in \mathrm{Aut}(G)$ such that for all $x, y \in G$, $\langle x, y \rangle_1 = \langle \varphi(x), \varphi(y) \rangle_2$. The following exercise contains some of the basics of the classification of pairings on finite abelian groups. For much more on this topic see [62, 71].

Exercise 18 Let p be an odd prime and r be a positive integer.

(a) Show that every non-degenerate pairing $\langle \cdot, \cdot \rangle \colon \mathbb{Z}/p^r\mathbb{Z} \times \mathbb{Z}/p^r\mathbb{Z} \to \mathbb{Q}/\mathbb{Z}$ is of the form

$$\langle x, y \rangle_a = \frac{axy}{p^r}$$

for some integer a not divisible by p.

(b) Show that $\langle x, y \rangle_a$ is isomorphic to $\langle x, y \rangle_b$ if and only if the Legendre symbols $\left(\frac{a}{p}\right)$ and $\left(\frac{b}{p}\right)$ are equal.

(c) Show that every finite abelian p-group with a perfect pairing decomposes as an orthogonal direct sum of cyclic groups with pairings.

Like many things in algebra, the prime $p = 2$ behaves in a special way. The classification of perfect pairings on finite abelian 2-groups is significantly more complicated than in the case where p is odd. See [38, Section 2.4] for a discussion of these issues. For any finite abelian group H, this material can be used to compute the term #{symmetric, bilinear, perfect pairings $\phi \colon H \times H \to \mathbb{C}^*$} from Theorem 14; see equation (2) of [72, p. 916].

We can now revisit the material from each of the previous two sections and ask not only about finite abelian groups that occur as the critical group of a graph, but also about finite abelian groups with a chosen perfect pairing. In [38], the authors use a construction based on *subdivided banana graphs* to show that odd order groups with pairings occur as critical groups.

Theorem 17 ([38, Theorem 2]) *Assume the* generalized Riemann hypothesis. *Let Γ be a finite abelian group of odd order with a perfect pairing on Γ. Then there exists a graph G such that $K(G) \cong \Gamma$ as groups with pairing.*

It may seem surprising that the generalized Riemann hypothesis (GRH), one of the major unsolved problems in number theory, would play a role in a problem about critical groups of graphs. The connection comes via the existence of small quadratic non-residues that satisfy additional properties. In [38], the authors explain how a positive answer to the following conjecture would remove this dependence on GRH.

Conjecture 2 Let p be a prime. There exists a prime $q < 2\sqrt{p}$ with $q \equiv 3 \pmod 4$ such that q is a quadratic non-residue modulo p.

Theorem 14 gives the probability that the Sylow p-subgroup of the critical group of an Erdős–Rényi random graph $G(n, q)$ is isomorphic to a particular finite abelian p-group. Clancy, Leake, and Payne give the analogous conjecture for a finite abelian p-group together with a perfect pairing [25].

Conjecture 3 Fix $0 < q < 1$. Let Γ be a finite abelian p-group and $\langle \cdot, \cdot \rangle$ be a perfect pairing on Γ. Then, as n goes to infinity, the probability that the Sylow p-subgroup of the critical group of the Erdős–Rényi random graph $G(n, q)$ is isomorphic to Γ with its associated monodromy pairing isomorphic to $\langle \cdot, \cdot \rangle$ is

$$\frac{\prod_{i=1}^{\infty}(1 - p^{1-2i})}{|\Gamma| \cdot |\text{Aut}(\Gamma, \langle \cdot, \cdot \rangle)|},$$

where $\text{Aut}(\Gamma, \langle \cdot, \cdot \rangle)$ is the set of automorphisms of Γ that preserve the pairing $\langle \cdot, \cdot \rangle$.

We defined the critical group of a connected graph G as the cokernel of its reduced Laplacian L_0, so we should also be able to understand the pairing on $K(G)$ in terms of this matrix. In fact, this pairing is an instance of the pairing taking values in $\mathbb{Q}/\mathbb{Z}$ defined on the cokernel of any nonsingular symmetric integer matrix A induced by

$$\langle x, y \rangle = y^T A^{-1} x.$$

See [16, Section 1] and [24] for a discussion of the pairing on the cokernel of a symmetric matrix. In particular, Theorem 2 of [24] shows that Conjecture 3 is consistent with Sylow p-subgroups of critical groups of random graphs being distributed like Sylow p-subgroups of cokernels of random elements of $\text{Sym}_n(\mathbb{Z}_p)$ with their associated pairings. Conjecture 3 is likely to be very difficult since it implies Theorem 14, the proof of which was a significant achievement that required the introduction of several new ideas into the study of critical groups.

1.11   Ranks of Divisors and Gonality of Graphs

We next introduce additional material about divisors on graphs that is motivated by connections to Brill–Noether theory, an important topic in algebraic geometry. A divisor δ on G is *effective* if $\delta(v) \geq 0$ for all v. This property is not invariant under chip-firing. We have seen examples of divisors that are not effective but are chip-firing equivalent to divisors that are effective; for another example, see Fig. 7.

A divisor δ has *positive rank* if for any $v \in V(G)$ the divisor δ' we get by setting $\delta'(v) = \delta(v) - 1$ and $\delta'(u) = \delta(u)$ for all other vertices u is chip-firing equivalent

Fig. 7 Two divisors on the graph from Example 1 that are chip-firing equivalent. The first is not effective, but the second is

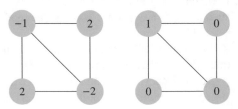

to an effective divisor. The *gonality* of G, denoted $\mathrm{gon}(G)$, is the smallest degree of an effective divisor with positive rank.

Example 13 Consider the following graph:

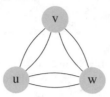

If δ is an effective divisor of degree one, then we may assume without loss of generality that $\delta(u) = 1$ and $\delta(v) = \delta(w) = 0$. One can show that the divisor δ' given by $\delta'(u) = 1$, $\delta'(v) = -1$, $\delta'(w) = 0$ is not equivalent to any effective divisor, which implies that δ does not have positive rank. We will leave it as an exercise to show that no effective divisor of degree two has positive rank, either. On the other hand, the divisor with $\delta(u) = \delta(v) = \delta(w) = 1$ is a degree 3 divisor of positive rank, showing that the gonality of this graph is 3.

Several authors have studied ranks of divisors and the gonality of graphs. For example, de Bruyn and Gijswijt connect the gonality of a graph to the notion of *treewidth*, an important concept in graph theory [33]. The authors of [34] study the gonality of Erdős–Rényi random graphs and prove the following theorem.

Theorem 18 ([34, Theorem 1.1]) *Let $p(n) = c(n)/n$, and suppose that $\log(n) \ll c(n) \ll n$. Then the expected value of the gonality of an Erdős–Rényi random graph $G(n, p(n))$ is asymptotic to n.*

Related work of Amini and Kool in the setting of divisors of metric graphs leads to the similar results, but with bounds that are not as tight [2].

Theorem 18 gives the expected value of the gonality of one model of a random graph, but there are many other questions to consider. Amini and Kool show in [2] that random d-regular graphs on n vertices have gonality bounded above and below by constant multiples of n. Connections to tropical geometry led the authors of [34] to ask about the gonality of random 3-regular graphs. Dutta and Jensen prove a lower bound for the gonality of a regular graph G in terms of the *Cheeger constant* of G, one of the most studied measures of graph expansion [36]. They also give a lower bound for gonality of a general graph G in terms of its *algebraic connectivity*, the second smallest eigenvalue of $L(G)$. As a consequence they prove the following.

Theorem 19 ([36, Theorem 1.3]) *Let G be a random 3-regular graph on n vertices. Then*

$$\mathrm{gon}(G) \geq 0.0072n$$

asymptotically almost surely.

Research Project 10 Can we improve the results about the expected gonality of a random k-regular graph? What can we say about the expected gonality of other families of random graphs?

There are several additional interesting directions in the Brill–Noether theory of graphs and metric graphs that have been the subject of successful research projects with undergraduate coauthors. See, for example, [29, 46, 52, 53].

1.12   Chip-Firing on Directed Graphs

Throughout this section, we have assumed that the graphs we consider are undirected. However, one can define a similar situation on directed graphs by considering the *directed Laplacian matrix* $\hat{L} = D - A$, where D is a diagonal matrix with (i, i)-entry equal to the outdegree of v_i, and the entries of the adjacency matrix A correspond to the number of edges from v_i to v_j. The critical group of this directed graph is the torsion part of the cokernel of $\hat{L}$.

Example 14 Let us consider the following version of the graph from our running example where we consider some of the edges as being unidirectional:

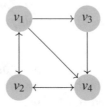

The adjacency matrix, degree matrix, and directed Laplacian of this graph are given by

$$A = \begin{pmatrix} 0 & 1 & 1 & 1 \\ 1 & 0 & 0 & 1 \\ 0 & 0 & 0 & 1 \\ 0 & 1 & 0 & 0 \end{pmatrix}, \quad D = \begin{pmatrix} 3 & 0 & 0 & 0 \\ 0 & 2 & 0 & 0 \\ 0 & 0 & 1 & 0 \\ 0 & 0 & 0 & 1 \end{pmatrix}, \quad \hat{L} = \begin{pmatrix} 3 & -1 & -1 & -1 \\ -1 & 2 & 0 & -1 \\ 0 & 0 & 1 & -1 \\ 0 & -1 & 0 & 1 \end{pmatrix}.$$

One can compute from the Smith normal form of $\hat{L}$ that $\mathrm{cok}(\hat{L}) \cong \mathbb{Z}$, so the associated critical group is trivial.

The notion of critical groups of directed graphs was first introduced in [15] and further developed in an unpublished note by Wagner [70]. However, there are still many questions to be considered.

> **Research Project 11** Consider a finite connected undirected graph G. For each edge of G make a choice of how to orient it. What can we say about the critical groups that occur as we vary over all possible choices? For starters, consider the graph from the previous example.

We can ask many of the questions considered in previous sections in this directed graph setting. For example, for information on critical groups of Erdős–Rényi random directed graphs see work of Koplewitz [50] and Wood [73].

2   Arithmetical Structures

In this section we consider a generalization of the Laplacian matrix and critical group of a graph that leads to interesting new enumerative problems. The Laplacian of G is defined by $L(G) = D - A$ where A is the adjacency matrix of G and D is the diagonal matrix whose entries consist of the degrees of the vertices of the graph. One generalization of this idea is to allow the entries on the diagonal of D to be other positive integers. This leads to the notion of arithmetical structures, the topic of this section.

2.1   Definitions and Examples

Let G be a finite connected graph with adjacency matrix A. We define an *arithmetical structure* on G by a vector $\mathbf{d} \in \mathbb{Z}_{\geq 0}^n$ so that there exists a vector $\mathbf{r} \in \mathbb{Z}_{>0}^n$ with $(D - A)\mathbf{r} = \mathbf{0}$, where D is the diagonal matrix with the entries of $\mathbf{d}$ along the diagonal. We will sometimes write $D = \text{diag}(\mathbf{d})$.

Exercise 19 In Sect. 1, we saw that for a connected graph G with $|V(G)| = n$, the Laplacian matrix $L(G) = D - A$ has rank $n - 1$. Show that for any arithmetical structure on G, the matrix $\text{diag}(\mathbf{d}) - A$ has rank $n - 1$.

This exercise shows that the null space of $\text{diag}(\mathbf{d}) - A$ is 1-dimensional, so there is a unique vector in it up to scalar multiplication. Unless stated otherwise, we will use $\mathbf{r}$ to denote the vector in $\text{Null}(D - A)$ whose entries are all relatively prime positive integers. This choice of $\mathbf{r}$ uniquely specifies an arithmetical structure on G. As such, we often refer to the pair $(\mathbf{r}, \mathbf{d})$ as an arithmetical structure, even though each one is uniquely determined by the other. We denote the matrix $\text{diag}(\mathbf{d}) - A$

by $L(G, \mathbf{r})$. In Sect. 1 we studied one arithmetical structure at length, $(\mathbf{1}, \mathbf{d})$, where $\mathbf{d}$ is the vector consisting of the degrees of the vertices of G. This is the *Laplacian arithmetical structure* on G. In this case, $L(G, \mathbf{1}) = L(G)$.

The $\mathbf{r}$-vector of an arithmetical structure has another interpretation based on elementary number theory. In particular, one can think of an arithmetical structure as a labeling of the vertices of G with relatively prime positive integers so that the label of any given vertex is a divisor of the (weighted, if necessary) sum of its neighbors.

Example 15 Consider again the situation from Example 1:

$$G = \begin{array}{c}\includegraphics\end{array}, \qquad A = \begin{pmatrix} 0 & 1 & 1 & 1 \\ 1 & 0 & 0 & 1 \\ 1 & 0 & 0 & 1 \\ 1 & 1 & 1 & 0 \end{pmatrix}.$$

Let $\mathbf{d} = \begin{pmatrix} 5 & 6 & 3 & 1 \end{pmatrix}^T$. The null space of the matrix

$$L(G, \mathbf{r}) = D - A = \begin{pmatrix} 5 & -1 & -1 & -1 \\ -1 & 6 & 0 & -1 \\ -1 & 0 & 3 & -1 \\ -1 & -1 & -1 & 1 \end{pmatrix}$$

is spanned by the vector $\mathbf{r} = \begin{pmatrix} 3 & 2 & 4 & 9 \end{pmatrix}^T$, so $(\mathbf{r}, \mathbf{d})$ is an arithmetical structure on G. If we label the graph as below, then the label of each vertex is a divisor of the sum of the labels of its neighbors.

Exercise 20 Find more arithmetical structures on the graph from this example. As a hint, there are a total of 63 structures, and the largest entry of any $\mathbf{r}$ that occurs is 18.

Just as we defined the critical group of a graph G to be the torsion part of the cokernel of $L(G)$, we can define the critical group associated with any arithmetical structure $(\mathbf{r}, \mathbf{d})$ to be the torsion part of the cokernel of $L(G, \mathbf{r})$. We denote this critical group by $\mathcal{K}(G; \mathbf{r})$. We described how to compute $\mathrm{cok}(L(G))$ by finding its Smith normal form and can proceed similarly in the more general setting with the matrix $L(G, \mathbf{r})$. If we do this for the matrix from Example 15, we see that the associated critical group is trivial. In Sect. 2.3 we will analyze the structure of this group in more depth.

The concept of arithmetical structures on graphs was originally developed by Lorenzini in [54] as a way of trying to understand the Néron models of certain algebraic curves where components might appear with multiplicity greater than one. Explaining these applications is beyond the scope of this note, but we refer the interested reader to [55]. We also refer the reader to [5, Section 4] where Asadi and Backman show that chip-firing on arithmetical graphs can be interpreted as a special case of the chip-firing for directed multigraphs that we introduced in Sect. 1.12, but do not pursue this perspective further here.

2.2 Counting Arithmetical Structures

In [54] Lorenzini proves that any finite connected graph has a finite number of arithmetical structures. However, the proof is nonconstructive and in general does not give an upper bound for the number of these arithmetical structures. In recent years, several authors have become interested in trying to count the number of arithmetical structures on certain types of graphs.

One general approach to counting arithmetical structures comes from the following observation. We first introduce some notation. Let G be a graph and $(\mathbf{r}, \mathbf{d})$ be an arithmetical structure on G. For $v \in V(G)$ we write $\mathbf{r}_v$ for the value of $\mathbf{r}$ corresponding to v and $\mathbf{d}_v$ for the value of $\mathbf{d}$ corresponding to v.

Theorem 20 *Let G be a graph and let $(\mathbf{r}, \mathbf{d})$ be an arithmetical structure on G. Assume that v is a vertex of degree 2 with neighbors u and w so that $\mathbf{r}_v > \mathbf{r}_u$ and $\mathbf{r}_v > \mathbf{r}_w$. Then $\mathbf{r}_v = \mathbf{r}_u + \mathbf{r}_w$.*

Moreover, if one defines the graph G' to be the graph whose vertex set is $V(G') = V(G) \setminus \{v\}$, and whose edge set is $E(G') = E(G) \cup \{\overline{uw}\} \setminus \{\overline{uv}, \overline{vw}\}$, then one gets a new arithmetical structure on G' by defining $\mathbf{r}'$ to have the same values as $\mathbf{r}$ on all remaining vertices.

Exercise 21 Verify that this theorem holds for the structures that you found in Exercise 20.

Proof The proof of the first claim follows from the fact that if we have such an arithmetical structure we know that $\mathbf{r}_v \mid (\mathbf{r}_u + \mathbf{r}_w)$. If we know that $\mathbf{r}_v > \mathbf{r}_u$ and $\mathbf{r}_v > \mathbf{r}_w$, then $\mathbf{r}_u + \mathbf{r}_w < 2\mathbf{r}_v$, which implies that $\mathbf{r}_u + \mathbf{r}_w = \mathbf{r}_v$.

The proof of the second claim is straightforward and can be best understood by considering a picture such as the one in Fig. 8, and making the observation that if $\mathbf{r}_u \mid \left((\mathbf{r}_u + \mathbf{r}_w) + \sum \mathbf{r}_i \right)$, then $\mathbf{r}_u \mid (\mathbf{r}_w + \sum \mathbf{r}_i)$.

We refer to the operation of removing a vertex of degree 2 corresponding to a local maximum of $\mathbf{r}$, such as the one described in the previous theorem, as *smoothing at vertex v*. One can also define a smoothing operation at a vertex of degree 1; in particular, if v is a vertex of degree 1 that is adjacent to the vertex u and if $\mathbf{r}_v = \mathbf{r}_u$, then one gets a new arithmetical structure on a smaller graph by

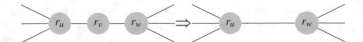

Fig. 8 Pictures showing the "smoothing" operation at a vertex of degree two

Fig. 9 Pictures showing the "smoothing" operation at a vertex of degree 1

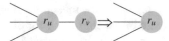

removing the vertex v, as illustrated in Fig. 9. An arithmetical structure $(\mathbf{r}, \mathbf{d})$ on G is *smooth* if there are no vertices of G at which we can apply a smoothing operation.

These smoothing operations are reversible, and in particular the number of ways that one can take an arithmetical structure and *subdivide* it can be described in terms of certain ballot numbers. (For details, see [18] and [4]). The approach taken in those papers is to count the number of smooth structures on smaller graphs and then count the number of ways they can be subdivided into general arithmetical structures on G. In particular, one can show theorems of the following type:

Theorem 21 *We can count the number of smooth structures on certain graphs in the following way:*

1. *The only smooth structure on a path is given by the Laplacian arithmetical structure on a single vertex. The total number of structures on a path of length n is given by the $(n-1)^{st}$ Catalan number, $C_{n-1} = \frac{1}{n}\binom{2(n-1)}{n-1}$.*
2. *The only smooth structure on a cycle of length n is given by the Laplacian arithmetical structure. The total number of structures on a cycle on n vertices is given by the binomial coefficient $\binom{2n-1}{n-1}$.*
3. *Let $n \geq 4$ and P'_n be the path graph on n vertices where the first edge is doubled. The number of smooth structures on P'_n is 4, and the total number of structures is $4C_{n-1} - 2C_{n-2}$.*

In general it appears to be quite difficult to count precisely the number of smooth arithmetical structures on a graph. For example, even for a *bident* graph, a path plus one additional vertex connected only to the second vertex on the path, it is only known that the number of smooth arithmetical structures is bounded between two cubic polynomials in the number of vertices [4].

Research Project 12 Consider the graph $\widetilde{C_4}$ obtained by taking the cycle C_4 and adding a second edge between two consecutive vertices.

1. How many smooth arithmetical structures are there on $\widetilde{C_4}$?

(continued)

2. How many total arithmetical structures are there on $\widetilde{C}_4$?
3. What if we instead consider bigger cycles or add more edges?

In the definition of smoothing at a vertex v of degree 2 or 1, we have $\mathbf{d}_v = 1$. One might wonder whether this idea could be generalized to vertices v of larger degree at which $\mathbf{d}_v = 1$. These smoothing operations are special cases of the *clique-star transform* defined in [30]. This operation replaces a subgraph that is isomorphic to a star graph, the complete bipartite graph $K_{1,n}$, by the complete graph on n vertices.

As an example, let us consider arithmetical structures on the complete graph K_n. Every such arithmetical structure is uniquely determined by a vector of relatively prime positive integers $\mathbf{r} = (r_1, \ldots, r_n)$ where each r_i divides the sum $\sum_{i=1}^n r_i$. The star graph $K_{1,n}$ consists of a vertex v_0 connected to n other vertices, each of which has degree 1. If an arithmetical structure on this graph has $\mathbf{d}_{v_0} = 1$, then $r_0 = \sum_{i=1}^n r_i$ (Fig. 10). Therefore, such arithmetical structures on $K_{1,n}$ are in bijection with the set of all arithmetical structures on K_n. It is interesting to further consider the remaining structures on $K_{1,n}$ that have $\mathbf{d}_{v_0} > 1$.

To further consider the set of arithmetical structures on K_n, we note that the definition of an arithmetical structure implies that for each i:

$$d_i r_i = \sum_{j \neq i} r_j$$

$$(d_i + 1) r_i = \sum_j r_j$$

Fig. 10 The complete graph K_6 and the star graph $K_{1,6}$

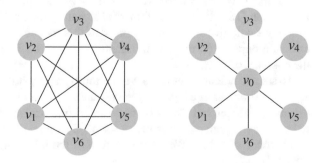

$$\frac{1}{d_i + 1} = \frac{r_i}{\sum_j r_j}.$$

In particular, if we sum over all i, we get that

$$\sum_{i=1}^{n} \frac{1}{d_i + 1} = 1.$$

The arithmetical structures of K_n are therefore in bijection with ways of writing 1 as a sum of reciprocals of n positive integers. Finding the number of ways of doing this is a difficult problem in additive number theory.

Exercise 22 Classify all sets of positive integers $\{a_1, a_2, a_3, a_4\}$ so that $\sum \frac{1}{a_i} = 1$. For each one find the corresponding arithmetical structure on K_4.

In general, there is no known formula for this number, but we do have a lower bound that is doubly exponential in n [49]. Corrales and Valencia get similar results for all structures on star graphs [31]. We close this section with a conjecture from [30], which is based on the observation that vertices with higher degree seem to lead to more arithmetical structures.

Research Project 13 Show that for any simple connected graph G with n vertices the number of arithmetical structures on G is at least the number on the path P_n and at most the number on the complete graph K_n.

2.3   Critical Groups of Arithmetical Structures

We have already seen that it is difficult to enumerate all arithmetical structures on a given graph. However, it might be easier to say something about the critical groups that occur associated with this set of arithmetical structures. For example, it is shown in [18] that every arithmetical structure on a path leads to a trivial critical group; we will give an alternative proof of this fact below. Recall that we define the critical group $\mathcal{K}(G; \mathbf{r})$ of an arithmetical structure $(\mathbf{d}, \mathbf{r})$ to be the torsion part of the cokernel of $L(G, \mathbf{r}) = \text{diag}(\mathbf{d}) - A$.

To set our notation, let G be a finite multigraph with $V(G) = \{v_1, \ldots, v_n\}$. Let $x_{i,j}$ be the number of edges between v_i and v_j. Since G is a multigraph we note that $x_{i,j}$ may be larger than 1. Let $(\mathbf{r}, \mathbf{d})$ be an arithmetical structure on G. We define $G_{\mathbf{r}}$ to be the graph with the same vertex set, $V(G)$, and with $x_{i,j} r_i r_j$ edges between any two vertices v_i and v_j. We leave the proof of the following lemma as an exercise in linear algebra:

Lemma 1 *We have* $L(G_{\mathbf{r}}, 1) = RL(G, \mathbf{r})R$, *where* $R = \mathrm{diag}(\mathbf{r})$.

Let $L(G, \mathbf{r})^{i,j}$ be the matrix we get from $L(G, \mathbf{r})$ by deleting its ith row and j^{th} column. Similar to the situation in Corollary 2, the determinant of $L(G, \mathbf{r})^{i,j}$ is given by $r_i r_j |\mathscr{K}(G; \mathbf{r})|$. From this, one can compute

$$
\begin{aligned}
|\mathscr{K}(G_{\mathbf{r}}; 1)| &= \det(L(G_{\mathbf{r}}, 1)^{1,1}) \\
&= \det((RL(G, \mathbf{r})R)^{1,1}) \\
&= \det(R^{1,1}) \det(L(G, \mathbf{r})^{1,1}) \det(R^{1,1}) \\
&= (r_2 \ldots r_n)^2 r_1^2 |\mathscr{K}(G; \mathbf{r})|.
\end{aligned}
$$

On the other hand, we know from Corollary 3 that $|\mathscr{K}(G_{\mathbf{r}}; 1)|$ is the number of spanning trees of $G_{\mathbf{r}}$. So, we can determine the order of $\mathscr{K}(G; \mathbf{r})$ by counting spanning trees of the graph $G_{\mathbf{r}}$.

Let us first consider the special case where the skeleton of G is a tree. Let $V(G) = \{v_1, \ldots, v_n\}$. The *skeleton* of a multigraph G is the graph $\overline{G}$ that has the same vertex set as G and has $\min(1, x_{i,j})$ edges between any pair of vertices v_i, v_j. Intuitively, this is what happens when you remove all "repeated" edges. If $\overline{G}$ is a tree, then it is easy to see that the number of spanning trees of G is $\prod_{x_{i,j} \neq 0} x_{i,j}$.

Moreover, it is clear that $\overline{G_{\mathbf{r}}}$ is also a tree and therefore that the number of spanning trees of $G_{\mathbf{r}}$ is given by

$$
|\mathscr{K}(G_{\mathbf{r}}; 1)| = \prod_{x_{i,j} \neq 0} x_{i,j} r_i r_j = \prod_{x_{i,j} \neq 0} x_{i,j} \prod_{i=1}^{n} r_i^{\deg(v_i)}.
$$

In particular, this proves the following result of Lorenzini [54, Corollary 2.3].

Corollary 6 *Let G be a graph with $V(G) = \{v_1, \ldots, v_n\}$ so that $\overline{G}$ is a tree and let $(\mathbf{d}, \mathbf{r})$ be an arithmetical structure on G. Then*

$$
|\mathscr{K}(G; \mathbf{r})| = \prod_{x_{i,j} \neq 0} x_{i,j} \prod_{i=1}^{n} r_i^{\deg(v_i)-2}.
$$

More generally, one can count spanning trees of $G_{\mathbf{r}}$ by noting that a spanning tree of G that includes an edge $\overline{v_i v_j}$ will lead to $r_i r_j$ spanning trees in $G_{\mathbf{r}}$, as we can choose any of the related edges. In particular, a spanning tree $\mathscr{T}$ of G leads to $\prod_i r_i^{\deg_{\mathscr{T}}(v_i)}$ spanning trees of $G_{\mathbf{r}}$, where $\deg_{\mathscr{T}}(v_i)$ denotes the degree of the vertex v_i in the tree $\mathscr{T}$. This discussion proves the following theorem:

Theorem 22 *Let G be a graph with $V(G) = \{v_1, \ldots, v_n\}$ and let $\mathbf{r}$ give an arithmetical structure on G. Then we have*

$$|\mathscr{K}(G; \mathbf{r})| = \sum_{\mathscr{T} \subseteq G} \left(\prod_{i=1}^{n} r_i^{\deg_{\mathscr{T}}(v_i)-2} \right),$$

where the sum ranges over all spanning trees of the graph G.

Example 16 If G is a path on n vertices, then there is a single spanning tree given by G itself. It follows from [18, Lemma 1] that any arithmetical structure on a path has $r_1 = r_n = 1$. Therefore one computes that

$$|\mathscr{K}(G; \mathbf{r})| = \sum_{\mathscr{T} \subseteq G} \left(\prod_{i=1}^{n} r_i^{\deg_{\mathscr{T}}(v_i)-2} \right)$$

$$= \prod_{i=1}^{n} r_i^{\deg_G(v_i)-2} = \frac{1}{r_1 r_n} = 1.$$

This gives an alternative proof to the first claim in [18, Theorem 7].

Example 17 Let G be a cycle on n vertices. A spanning tree of G corresponds to removing a single edge. In particular, Theorem 22 implies that

$$|\mathscr{K}(G; \mathbf{r})| = \sum_{i=1}^{n} \frac{1}{r_i r_{i+1}}.$$

If $\mathbf{r} \neq \mathbf{1}$ then the arithmetical structure has some vertex v_i with $r_i = r_{i-1} + r_{i+1}$, so we can smooth the structure at this vertex. In particular, we note that

$$\frac{1}{r_{i-1} r_i} + \frac{1}{r_i r_{i+1}} = \frac{1}{r_{i-1}(r_{i-1}+r_{i+1})} + \frac{1}{r_{i+1}(r_{i-1}+r_{i+1})} = \frac{1}{r_{i-1} r_{i+1}}.$$

This shows us that smoothing the structure at this vertex will not change the order of the critical group. Any arithmetical structure $(\mathbf{r}, \mathbf{d})$ on C_n can be smoothed to the Laplacian arithmetical structure on some C_k with $k \leq n$. For this value of k we see that

$$|\mathscr{K}(C_n; \mathbf{r})| = |\mathscr{K}(C_k; \mathbf{1})| = k.$$

Understanding the structure of the group $\mathscr{K}(G; \mathbf{r})$ rather than just its order requires a more careful analysis. The following theorem is a restatement of Proposition 1.12 in [54].

Theorem 23 *We have the following two short exact sequences:*

$$1 \to \bigoplus \mathbb{Z}/r_i\mathbb{Z} \to E \to \mathscr{K}(G; \mathbf{r}) \to 1$$

$$1 \to E \to \mathscr{K}(G_{\mathbf{r}}; \mathbf{1}) \to \bigoplus \mathbb{Z}/r_i\mathbb{Z} \to 1,$$

where E is a specific quotient group.

In general, these short exact sequences do not split but they do give us insight about the structure of $\mathscr{K}(G; \mathbf{r})$ if we know the structure of $\mathscr{K}(G_{\mathbf{r}}; \mathbf{1})$.

> **Research Project 14** What are the possible critical groups associated with a given graph G as we vary the arithmetical structures $(\mathbf{r}, \mathbf{d})$?

Answers to this question are known only in a few cases. We have already seen what happens with paths and cycles. Critical groups associated with arithmetical structures on bident graphs D_n are analyzed in [4, Section 5]. In particular, the authors show that for any $\mathbf{r}$, the matrix $L(G, \mathbf{r})$ has an $(n-2) \times (n-2)$ minor equal to 1 and use Corollary 1 to show that $\mathscr{K}(G; \mathbf{r})$ is cyclic. An analysis similar to the one leading to Corollary 6 shows that the biggest possible order will be $2n - 5$ and completely characterizes the smaller critical group orders that occur.

There are natural generalizations of many of the problems from Sect. 1 to arithmetical graphs. For example, see [16, Section 5] for results on a realization problem for arithmetical graphs.

Acknowledgements We would like to thank Luis David Garcia-Puente for initiating this project. We would further like to thank David Jensen, Pranav Kayastha, Dino Lorenzini, Sam Payne, Farbod Shokrieh, and the editors and referees for their helpful comments.

The second author is supported by NSF Grant DMS 1802281.

References[1]

1. Carlos A. Alfaro and Carlos E. Valencia, *On the Sandpile group of the cone of a graph*, Linear Algebra Appl. **436** (2012), no. 5, 1154–1176.

2. Omid Amini and Janne Kool, *A spectral lower bound for the divisorial gonality of metric graphs*, Int. Math. Res. Not. IMRN (2016), no. 8, 2423–2450.

3. Yang An, Matthew Baker, Greg Kuperberg, and Farbod Shokrieh, *Canonical representatives for divisor classes on tropical curves and the matrix-tree theorem*, Forum Math. Sigma **2** (2014), e24, 25 pp.

---

[1]We have marked papers that have at least one undergraduate coauthor in bold.

4. Kassie Archer, Abby Bishop, Alexander Diaz Lopez, Luis David García Puente, Darren Glass, and Joel Louwsma, *Arithmetical structures on bidents*, To appear in Discrete Math. https://arxiv.org/abs/1903.01393, 2019.

5. Arash Asadi and Spencer Backman, *Chip-firing and Riemann-Roch theory for directed graphs*, https://arxiv.org/abs/1012.0287v2, (2011).

6. Roland Bacher, Pierre de la Harpe, and Tatiana Nagnibeda, *The lattice of integral flows and the lattice of integral cuts on a finite graph*, Bull. Soc. Math. France **125** (1997), no. 2, 167–198.

7. Matthew Baker and Serguei Norine, *Riemann-Roch and Abel-Jacobi theory on a finite graph*, Adv. Math. **215** (2007), no. 2, 766–788.

8. Matthew Baker and Serguei Norine, *Harmonic morphisms and hyperelliptic graphs*, Int. Math. Res. Not. IMRN (2009), no. 15, 2914–2955.

9. Matthew Baker and Farbod Shokrieh, *Chip-firing games, potential theory on graphs, and spanning trees*, J. Combin. Theory Ser. A **120** (2013), no. 1, 164–182.

10. **Ryan Becker and Darren Glass, *Cyclic Critical Groups of Graphs*, Austral. Jour. of Comb. 64 (2016), 366–375.**

11. **Andrew Berget, Andrew Manion, Molly Maxwell, Aaron Potechin, and Victor Reiner, *The critical group of a line graph*, Ann. Comb. 16 (2012), no. 3, 449–488.**

12. Kenneth Berman, *Bicycles and spanning trees*, SIAM J. Algebraic Discrete Methods **7** (1986), no. 1, 1–12.

13. Norman Biggs, *Algebraic Graph Theory*. Second edition. Cambridge Mathematical Library. Cambridge University Press, Cambridge, 1993. viii+205 pp.

14. N. L. Biggs, *Chip-firing and the critical group of a graph*, J. Algebraic Combin. **9** (1999), no. 1, 25–45.

15. Anders Björner and László Lovász, *Chip-firing games on directed graphs*, J. Algebraic Combin. **1** (1992), no. 4, 305–328.

16. Siegfried Bosch and Dino Lorenzini, *Grothendieck's pairing on component groups of Jacobians*, Invent. Math. **148** (2002), no. 2, 353–396.

17. **David Brandfonbrener, Pat Devlin, Netanel Friedenberg, Yuxuan Ke, Steffen Marcus, Henry Reichard, and Ethan Sciamma, *Two-vertex generators of Jacobians of graphs*, Electr. J. Comb. 25 (2018), P1.15.**

18. Benjamin Braun, Hugo Corrales, Scott Corry, Luis David García Puente, Darren Glass, Nathan Kaplan, Jeremy L. Martin, Gregg Musiker, and Carlos E. Valencia, *Counting arithmetical structures on paths and cycles*, Discrete Math. **341** (2018), no. 10, 2949–2963.

19. Morgan V. Brown, Jackson S. Morrow, and David Zureick-Brown, *Chip-firing groups of iterated cones*, Linear Algebra Appl. **556** (2018), 46–54.

20. David B. Chandler, Peter Sin, and Qing Xiang, *The Smith and critical groups of Paley graphs*, J. Algebraic Combin. **41** (2015), no. 4, 1013–1022.

21. Sheng Chen and Sheng Kui Ye, *Critical groups for homeomorphism classes of graphs*, Discrete Mathematics **309** (2009), no. 1, 255–258.

22. Fan R. K. Chung, *Spectral graph theory*, CBMS Regional Conference Series in Mathematics, vol. 92, Published for the Conference Board of the Mathematical Sciences, Washington, DC; by the American Mathematical Society, Providence, RI, 1997.

23. Mihai Ciucu, Weigen Yan, and Fuji Zhang, *The number of spanning trees of plane graphs with reflective symmetry*, J. Combin. Theory Ser. A **112** (2005), no. 1, 105–116.

24. **Julien Clancy, Nathan Kaplan, Timothy Leake, Sam Payne, and Melanie Matchett Wood, *On a Cohen-Lenstra heuristic for Jacobians of random graphs*, J. Algebraic Combin. 42 (2015), no. 3, 701–723.**

25. **Julien Clancy, Timothy Leake, and Sam Payne, *A note on Jacobians, Tutte polynomials, and two-variable zeta functions of graphs*, Exp. Math. 24 (2015), no. 1, 1–7.**

26. Anna Comito, Jennifer Garcia, Josefina Alvarado Rivera, Natalie L. F. Hobson, and Luis David Garcia Puente, *On the Sandpile group of circulant graphs*, 2016.

27. Robert Cori and Y. Le Borgne. *The Sandpile model and Tutte polynomials*. Adv. in Appl. Math., **30** (2003), no. 1, 44–52.

28. Robert Cori and Dominique Rossin, *On the Sandpile group of dual graphs*, European J. Combin. **21** (2000), no. 4, 447–459.
29. **F. Cools, J. Draisma, S. Payne, and E. Robeva, *A tropical proof of the Brill–Noether theorem***, Adv. Math. **230** (2012), no. 2, 759–776.
30. Hugo Corrales and Carlos E. Valencia, *Arithmetical structures on graphs*, Linear Algebra Appl. **536** (2018), 120–151.
31. ———, *Arithmetical structures on graphs with connectivity one*, J. Algebra Appl. **17** (2018), no. 8, 1850147, 13.
32. Scott Corry and David Perkinson, *Divisors and Sandpiles: An introduction to chip-firing*, American Mathematical Society, Providence, RI, 2018.
33. **Josse van Dobben de Bruyn and Dion Gijswijt, *Treewidth is a lower bound on graph gonality***, https://arxiv.org/abs/1407.7055, 2014.
34. **Andrew Deveau, David Jensen, Jenna Kainic, and Dan Mitropolsky, *Gonality of random graphs***, Involve **9** (2016), no. 4, 715–720.
35. **Joshua E. Ducey, Jonathan Gerhard, and Noah Watson, *The Smith and Critical Groups of the Square Rook's Graph and its Complement***, Electr. J. Comb. **23** (2016), no. 4, P4.9.
36. **Neelav Dutta and David Jensen, *Gonality of expander graphs***, Discrete Math.. **341** (2018), no. 9, 2535–2543.
37. Alan Frieze and Michał Karoński, *Introduction to Random Graphs*, Cambridge University Press, Cambridge, 2016.
38. **Louis Gaudet, David Jensen, Dhruv Ranganathan, Nicholas Wawrykow, and Theodore Weisman, *Realization of groups with pairing as Jacobians of finite graphs***, Ann. Comb. **22** (2018), no. 4, 781–801.
39. Mark Giesbrecht, *Fast computation of the Smith normal form of an integer matrix*, Proceedings of the 1995 International Symposium on Symbolic and Algebraic Computation (New York, NY, USA), ISSAC '95, ACM, 1995, pp. 110–118.
40. Darren Glass and Criel Merino, *Critical groups of graphs with dihedral actions*, European J. Combin. **39** (2014), 95–112.
41. **Gopal Goel and David Perkinson, *Critical groups of iterated cones***, Linear Algebra Appl. **567** (2019), 138–142.
42. Fernando Q. Gouvêa, *p-adic Numbers: An introduction*, second ed., Universitext, Springer-Verlag, Berlin, 1997.
43. Phillip A Griffiths and Joseph Harris, *Principles of Algebraic Geometry*, Wiley Classics Library, Wiley, New York, NY, 1994.
44. Yaoping Hou, Chingwah Woo, and Pingge Chen, *On the Sandpile group of the square cycle $Cn2$*, Linear Algebra Appl. **418** (2006), no. 2, 457–467.
45. **Brian Jacobson, Andrew Niedermaier, and Victor Reiner, *Critical groups for complete multipartite graphs and Cartesian products of complete graphs***, Journal of Graph Theory **44** (2003), no. 3, 231–250.
46. **Sameer Kailasa, Vivian Kuperberg, and Nicholas Wawrykow, *Chip-firing on trees of loops***. Electron. J. Combin. **25** (2018), no. 1, Paper 1.19, 12 pp.
47. Edward C. Kirby, Roger B. Mallion, Paul Pollak, and Paweł J. Skrzyński, *What Kirchhoff actually did concerning spanning trees in electrical networks and its relationship to modern graph-theoretical work*, Croatica Chemica Acta **89** (2016).
48. Caroline Klivans, *The Mathematics of Chip-Firing*, Chapman and Hall/CRC, New York, 2018.
49. S. V. Konyagin, *Double exponential lower bound for the number of representations of unity by Egyptian fractions*, Math. Notes **95** (2014), no. 1-2, 277–281, Translation of Mat. Zametki **95** (2014), no. 2, 312–316.
50. Shaked Koplewitz, *Sandpile groups and the Coeulerian property for random directed graphs*, Adv. in Appl. Math. **90** (2017), 145–159.
51. ———, *Sandpile groups of random bipartite graphs*, https://arxiv.org/abs/1705.07519, 2017.
52. **Timothy Leake and Dhruv Ranganathan, *Brill–Noether theory of maximally symmetric graphs***, European J. Combin. **46** (2015), 115–125.

53. **Chang Mou Lim, Sam Payne, and Natasha Potashnik,** *A note on Brill–Noether theory and rank determining sets for metric graphs*, Int. Math. Res. Not. IMRN (2012), no. 23, 5484–5504.
54. Dino J. Lorenzini, *Arithmetical graphs*, Math. Ann. **285** (1989), no. 3, 481–501.
55. ———, *Groups of components of Néron models of Jacobians*, Compositio Math. **73** (1990), no. 2, 145–160.
56. ———, *A finite group attached to the Laplacian of a graph*, Discrete Math. **91** (1991), no. 3, 277–282.
57. ———, *Smith normal form and Laplacians*, J. Combin. Theory Ser. B **98** (2008), no. 6, 1271–1300.
58. Jessie MacWilliams, *Orthogonal matrices over finite fields*, Amer. Math. Monthly **76** (1969), 152–164.
59. A. D. Mednykh and I. A. Mednykh, *On the structure of the Jacobian group of circulant graphs*. (Russian) Dokl. Akad. Nauk 469 (2016), no. 5, 539–543; translation in Dokl. Math. 94 (2016), no. 1, 445–449
60. ———, *The number of spanning trees in circulant graphs, its arithmetic properties and asymptotic*. Discrete Math. 342 (2019), no. 6, 1772–1781.
61. András Mészáros, *The distribution of sandpile groups of random regular graphs*, https://arxiv.org/abs/1806.03736v4, 2020.
62. Rick Miranda, *Nondegenerate symmetric bilinear forms on finite abelian 2-groups*, Trans. Amer. Math. Soc. **284** (1984), no. 2, 535–542.
63. Hoi Nguyen and Melanie Matchett Wood, *Random integral matrices: universality of surjectivity and the cokernel*, https://arxiv.org/abs/1806.00596, 2018.
64. **Victor Reiner and Dennis Tseng,** *Critical groups of covering, voltage and signed graphs*, Discrete Math. **318** (2014), 10–40.
65. J. Sedláček, *On the minimal graph with a given number of spanning trees*, Canad. Math. Bull. **13** (1970), 515–517.
66. Farbod Shokrieh, *The monodromy pairing and discrete logarithm on the Jacobian of finite graphs*, J. Math. Cryptol. **4** (2010), no. 1, 43–56.
67. Daniel A. Spielman, *Graphs, vectors, and matrices*, Bull. Amer. Math. Soc. (N.S.) **54** (2017), no. 1, 45–61.
68. Richard P. Stanley, *Smith normal form in combinatorics*, J. Combin. Theory Ser. A **144** (2016), 476–495.
69. Arne Storjohann, *Near optimal algorithms for computing Smith normal forms of integer matrices*, Proceedings of the 1996 international symposium on Symbolic and algebraic computation (New York, NY, USA), ISSAC '96, ACM, 1996, pp. 267–274.
70. David G. Wagner, *The critical group of a directed graph*, https://arXiv:math/0010241, 2000.
71. C. T. C. Wall, *Quadratic forms on finite groups, and related topics*, Topology **2** (1963), 281–298.
72. Melanie Matchett Wood, *The distribution of sandpile groups of random graphs*, J. Amer. Math. Soc. **30** (2017), no. 4, 915–958.
73. ———, *Random integral matrices and the Cohen-Lenstra heuristics*, Amer. J. Math. **141** (2019), no. 2, 383–398.

Counting Tilings by Taking Walks in a Graph

Steve Butler, Jason Ekstrand, and Steven Osborne

Abstract

Given a region and a collection of basic shapes (tiles), a natural question is to look at how many ways there are to cover the region using the tiles where no pair of tiles overlaps in their interiors. We show how to transform some problems of this type into counting walks on graphs. In the latter setting, there are well-known and efficient methods to count these for small cases, and in many cases recurrences and closed-form expressions can be found. We explore variations of these problems, and get to the point where the reader can set off to explore problems of this type.

Suggested Prerequisites Rudimentary knowledge of linear algebra, graph theory, and basic combinatorics (counting and recursions) is important. To explore "larger" tiling problems knowledge of basic programming will be needed.

S. Butler (✉)
Iowa State University, Ames, IA, USA
e-mail: butler@iastate.edu

J. Ekstrand
Austin, TX, USA
e-mail: jason@jekstrand.net

S. Osborne
Ames, IA, USA

© Springer Nature Switzerland AG 2020

153

P. E. Harris et al. (eds.), *A Project-Based Guide to Undergraduate Research in Mathematics*, Foundations for Undergraduate Research in Mathematics, https://doi.org/10.1007/978-3-030-37853-0_5

1   Introduction

Tiling problems deal with covering a region (often a rectangular region formed from squares) through the use of a collection of basic shapes (tiles) with the rule that the entire region is covered and no two tiles overlap in their interiors. This naturally lends itself to many interesting types of questions including the following:

1. *Existence:* Is it possible to cover the region with the prescribed tiles? (For example: Is it possible to cover an 8×8 chessboard with opposite corners removed using only dominoes as tiles?)
2. *Enumeration:* How many ways are there to cover the region with the prescribed tiles? Can an explicit formula be given? How about a recursion?
3. *Qualitative:* What can be said about the tilings, and particularly what properties should a random tiling have? (See the Arctic Circle Theorem for tilings of the Aztec diamond using dominoes as an example (see Elkies et al. [5, 6].)
4. *Generation:* How can we generate a random tiling of the region?
5. *Interpretation:* Can the tiling be connected to another combinatorial object through bijection? What combinatorial properties of the tilings exist?

We will mainly focus on the enumeration aspect of tiling, but this list should give an indication that there is a large collection of questions to explore about tiling. Throughout this paper we will assume that our tiles can be rotated to any orientation as well as turned over (reflected), but this assumption can be readily dropped if desired.

1.1   A First Example

As a warmup, let us start with the following classical problem: Determine the number of ways to tile the $2 \times n$ board using dominoes ($\square$). For reference the first few cases are shown in Fig. 1.

Let us label the number of possible tilings of the $2 \times n$ board as F_n for $n \geq 0$ (there is one way to cover the 2×0 board, namely use no tiles). Now every tiling of a board with length at least 2 must end in one of two ways, a vertical tile or a pair of horizontal tiles. The number of such tilings is F_{n-1} and F_{n-2}, respectively, and this covers all possibilities exactly once. So we have the recurrence $F_n = F_{n-1} + F_{n-2}$. Combined with the initial conditions (e.g., in Fig. 1) we can now find the first few terms to be

$$1, 1, 2, 3, 5, 8, 13, 21, 34, 55, 89, 144, 233, 377, 610, 987, 1597, 2584, 4181, \ldots.$$

Fig. 1 Tilings of $2 \times n$ for $n = 1, 2, 3$

Fig. 2 Tilings of $4 \times n$ for $n = 1, 2$

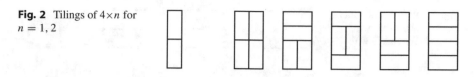

These numbers should look familiar, they are the Fibonacci numbers! The numbers so nice that they have their own journal dedicated to their study.[1]

If for some reason we did not recognize these numbers, we could have also used the On-Line Encyclopedia of Integer Sequences (OEIS) [10] to search for what they could be by entering the first few terms, and in this case find this is sequence A000045 which lists hundreds of facts and references about these numbers.[2]

1.2 A Second Example

Let us change the problem slightly: Determine the number of ways to tile the $4 \times n$ board using dominoes ($\square$). For reference the first two cases are shown in Fig. 2.

If we look back at the $2 \times n$ approach we saw that there were essentially only two ways for the structure to end. It is now easy to see that there are more possible endings (arguably infinitely many!) and so it is not immediately clear how to proceed. Indeed, this problem is nontrivial and was a problem in the Monthly in 2005 (Problem 11187). So if we are to make progress we will need a different idea than "group by how they end."

1.3 A First Example, Revisited

Let us go back and think about a different way to approach the enumeration of the tiling of $2 \times n$ board.[3] This time we will not focus on the tiling at the end, but rather shift our focus to what is happening between consecutive *columns*. *This can be thought of as a zen approach to tiling by shifting focus from the tiles to what is happening between the tiles.*[4] When the only tiles we have are dominoes, there are four possibilities of how dominoes can cross between consecutive columns: no

---

[1] *The Fibonacci Quarterly.*

[2] In general the OEIS is one of the best resources for people working on counting problems as it gives a way to see if other people have produced similar counts and opens up new avenues for exploration. We will give an example of this later.

[3] One might object that we have already done it once, and that doing it a second time will only give the same answer. But the important thing is not the answer, it is the process. A different approach might yield better insight and allow for a better generalization.

[4] Another way to think about this is akin to the character from *The Hobbit*, "Column, column,... my sweet, my love, my precious transitions-esss."

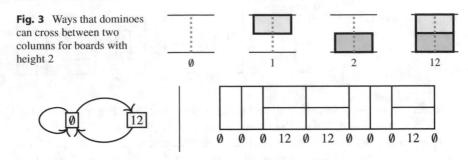

Fig. 3 Ways that dominoes can cross between two columns for boards with height 2

$\emptyset$ 1   2   12

Fig. 4 An auxiliary graph used to count tilings of $2 \times n$ board with dominoes. And an example of a $2 \times n$ tiling and corresponding sequence of column crossings

crossing, one crossing on the top, one crossing on the bottom, two crossing. These are shown in Fig. 3 along with labels (the labeling here indicates which rows cross).

When we are considering the tilings of the $2 \times n$ board with dominoes we have only two possible crossings of the type shown in Fig. 3, namely $\emptyset$ and 12. This can be seen by parity, if only one domino crossed between a particular pair of columns, then this crossing cuts the board in half and on each side of the board would leave an odd number of squares to cover using dominoes, which is impossible.

So now we come to the key idea, *look at which column crossings can occur consecutively*. Let us denote $a \to b$ to mean that column crossings corresponding to a can be immediately followed in the next column by a column crossing corresponding to b. We now have $\emptyset \to \emptyset$, $\emptyset \to 12$, and $12 \to \emptyset$. We can represent this situation by the directed graph shown in Fig. 4.[5] The figure also contains a tiling of the $2 \times n$ board and underneath each column crossing we have marked which type of crossing occurs.

Every tiling of the $2 \times n$ board begins on the left with the $\emptyset$ column crossing, then has n different column crossings ending with the $\emptyset$ column crossing. As we move from column to column in the tiling we can follow the same action in the graph (try this with Fig. 4). So we have a bijection between walks[6] with n steps that start and stop at $\emptyset$ with the tiling problem. This allows us to transform our counting problem into the following: *How many ways are there to start at $\emptyset$, take n steps (i.e., move along n edges) and end at $\emptyset$?* This can be answered with the following tool.

---

[5]For our purposes it suffices to know that a directed graph consists of a collection of objects (vertices) and connections between the objects (edges). In our case our vertices will be possible column crossings and edges will indicate which pair of columns can occur consecutively (order matters).

[6]A walk in the graph is a sequence of moves along edges, repetition is allowed and the direction of edges must be respected.

Theorem 1 (Transfer Matrix Method) *Given a directed (multi-)graph G, let T be the matrix with rows and columns indexed by the vertices of G with $T_{u,v}$ equal to the number of directed arcs from $u \rightarrow v$. Then $(T^k)_{u,v}$ is equal to the number of walks of length k which start at u and end at v.*

Exercise 1 Use induction to prove Theorem 1.

The transfer matrix method is a powerful tool for enumerating any process which can be described sequentially. For a more thorough background on the method and examples of its use see Stanley [11, Ch. 4.7].

For the tiling of the $2 \times n$ board using dominoes, the matrix would be

$$T = \begin{matrix} \emptyset \\ 12 \end{matrix} \begin{bmatrix} 1 & 1 \\ 1 & 0 \end{bmatrix}, \tag{1}$$

where on the left we have indicated the associated vertices. So by Theorem 1 we have that the number of tilings of the $2 \times n$ board is $(T^n)_{\emptyset,\emptyset}$. In a later section we will see how we can take this matrix and use this to find the recurrence relationship, generating functions, and give an explicit closed-form solution to our counting problem.

1.3.1 A Second Example, Revisited

We are now ready to take on the problem of counting the $4 \times n$ board using dominoes. So we start as in the $2 \times n$ case by looking at all sixteen possibilities that dominoes can cross between columns and seeing which columns can occur consecutively. Let us label these as some combination of $1, 2, 3, 4$ (the numbers being the rows from top to bottom), then the vertices and the possible transitions for the directed graph are as shown in Fig. 5 (verify this!).

Note that in this case we do not need the whole graph as we only need to take the vertices which we can reach in a walk starting and ending at $\emptyset$, of which there are six vertices left: $\emptyset$, 12, 23, 34, 14, 1234. This leads to the matrix

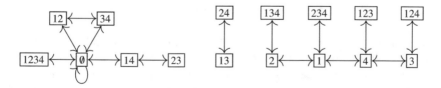

Fig. 5 An auxiliary graph used to count tilings of $4 \times n$ board with dominoes

$$
T = \begin{array}{c} \emptyset \\ 12 \\ 23 \\ 34 \\ 14 \\ 1234 \end{array}
\begin{bmatrix}
1 & 1 & 0 & 1 & 1 & 1 \\
1 & 0 & 0 & 1 & 0 & 0 \\
0 & 0 & 0 & 0 & 1 & 0 \\
1 & 1 & 0 & 0 & 0 & 0 \\
1 & 0 & 1 & 0 & 0 & 0 \\
1 & 0 & 0 & 0 & 0 & 0
\end{bmatrix} .
$$

Making a list of $(T^n)_{\emptyset,\emptyset}$ we have that the number of tilings for the $4 \times n$ board using dominoes is

$$1, 1, 5, 11, 36, 95, 281, 781, 2245, 6336, 18061, 51205, 145601, 413351, 1174500, \ldots.$$

Putting this into the OEIS [10] we see that this is sequence A005178; confirming our approach.[7]

2 General Approach

We now outline the basic underlying approach to counting tilings on $m \times n$ boards where one of the dimensions, say m, is fixed.

1. Find all possible ways that the tiles can cross between two consecutive columns of height m.
2. Construct a (directed) graph where each vertex corresponds to a crossing between two columns, and directed edges between pairs of crossings for *each* way that two columns can occur consecutively in a tiling.
3. Produce the adjacency matrix of this directed graph associated with some matrix T and look at the appropriate entry of T^n.

Exercise 2 Produce the matrix T for tiling the $2 \times n$ board where the permissible tiles consist of dominoes and bent triominoes ($\square$ and $\llcorner$; remember that we allow all possible rotations). Use this to find the number of such tilings for $n = 1, 2, \ldots, 20$. (Hint: there are eight different ways to cross columns.)

This approach reduces the problem of tiling enumeration to understanding local behavior (column crossings) and how to transition between local behavior. This also avoids the problem that can occur of having to look at the rich structure that can occur in putting tiles together and accounting for many large global possibilities. In addition, by turning the problem into counting walks, we are able to count many possibilities in *parallel* allowing us to get values for these counts that could never be achieved by going through backtracking or case-by-case analysis.

---

[7]Woohoo!

For small cases with few tiles and small fixed height, it is possible to construct the graphs and transitions by hand. As the number of tiles or height of the board grows, so to does the computational demands. So in these situations it is helpful to have a computer to automate the process. We will give an example of one such program later.

In some sense the reader can stop now, go pick a favorite set of tiles and start working on counting tilings of the rectangular board using appropriately designated tiles. We mention a few possibilities to get started.

- Tiling the $m \times n$ board with dominoes (⬛) or dominoes and monomoes (⬛ and ▫). This has been well studied as it can be rephrased in terms of matchings in a graph and there is even a closed-form expression (see Kasteleyn [9]) which involves the cosine(!) function. This can be a good warmup to develop the skills and intuitions of working on tiling problems.
- Tiling the $m \times n$ board using monomoes and bent triominoes (▫ and ⌐). (Compare with the results from Chinn et al. [4].)
- Tiling the $m \times n$ board with triominoes (⬛ and ⌐). There is less theoretical results known about this situation and for boards of a reasonable size it requires some programming. This type of tiling is the basis of Project Euler Problem 161[8] which asks for the number of tilings of the 9×12 grid using triominoes.
- There is no reason we have to assume tiles are formed by gluing squares on edges; now tackle the previous problem but now triominoes can be formed by gluing on corners giving us: ⬛, ⌐, ⌐, ⎕, and ⎕.[9]

As we start introducing triominoes we get to the situation where a single tile can span several columns. This is easily handled by keeping track of where you intersect a tile and making sure that the next column is one which intersects it one slice over.

2.1   Using Linear Algebra Tools

Before we start to explore the variations of problems that we can do with our technique, we will take a moment to see how to use linear algebra tools to give us information about our enumeration problems.

---

[8]https://projecteuler.net/problem=161. Project Euler is a collection of mathematically based problems that require computational tools to solve them. Working through the collection of these problems becomes a good way to develop mathematical programming skills.

[9]The analogous problem for dominoes where we allow gluing on edges or corners is equivalent to counting the number of perfect matchings in chess king graphs. This has been tackled by Shalosh B. Ekhad, a frequent collaborator of Doron Zeilberger.
http://sites.math.rutgers.edu/~zeilberg/tokhniot/oKamaShidukhim3.

2.1.1   Recurrence Relationships and Generating Functions

Our enumeration comes from entries in a transition matrix T, so to find a recurrence it suffices to find a recurrence for those entries. Recall that for any (square) matrix T there is a monic polynomial, $p_T(x)$, called the minimal polynomial with the property that $p_T(T) = O$ (the zero matrix). It is known that the minimal polynomial is a divisor of the characteristic polynomial.

Lemma 1 *For a matrix T let $p_T(t) = t^k - a_{k-1}t^{k-1} - \cdots - a_0$ be the minimal polynomial, and let $q_r = (T^r)_{i,j}$ (for i and j fixed). Then for $r \geq k$, we have*

$$q_r = a_{k-1}q_{r-1} + \cdots + a_0 q_{r-k}.$$

Proof For $r \geq k$ we have

$$q_r - a_{k-1}q_{r-1} - \cdots - a_0 q_{r-k} = (T^r - a_{k-1}T^{r-1} - \cdots - a_0 T^{r-k})_{i,j}$$

$$= (T^{r-k}\underbrace{(T^k - a_{k-1}T^{k-1} - \cdots - a_0 I)}_{=O})_{i,j} = O_{i,j} = 0.$$

So the minimal polynomial gives *a* recurrence, but not always the best recurrence, we will explain why in just a moment. In some sense to find the best recurrence, and so much more, we should find the generating function for our count.

A *generating function* is a way to store the values of our enumeration as coefficients of a series. Given $s_0, s_1, s_2, \ldots$ we form the series

$$S(x) = \sum_{\ell \geq 0} s_k x^k = s_0 + s_1 x + s_2 x^2 + \cdots.$$

For problems where these coefficients satisfy a linear recurrence (which from Lemma 1 includes our tiling problems) we can condense the expression for $S(x)$ into a rational function. This is done by multiplying both sides by an appropriate polynomial, in our case a modified form of the polynomial from Lemma 1. We then have

$$x^k p_T(1/x)S(x) = \sum_{\ell=0}^{k-1} t_\ell x^\ell \quad \text{so} \quad S(x) = \frac{\sum_{\ell=0}^{k-1} t_\ell x^\ell}{p_T(x)}.$$

The key to this is that the polynomial $x^k p_T(1/x)$ will combine coefficients into recurrences that zero out everything with degree k or more (we have to reverse the order of the coefficients in the polynomial to match the recurrence). So to find the t_ℓ terms it suffices to expand $(s_0 + \cdots + s_{k-1}x^{k-1})x^k p_T(1/x)$ and only keep the terms with power at most $k - 1$.

We should check to see if $S(x)$ can be simplified, e.g., common factors, and if so carry out the simplification. Note that the polynomial in the denominator for the simplified $S(x)$ gives in essence *the* recurrence.

Applying this to our examples from before we have that for tiling the $2 \times n$ strip with dominoes

$$p_T(x) = x^2 - x - 1,$$

$$x^2 p_T(1/x)(s_0 + s_1 x) = (1 - x - x^2)(1 + x) = \underbrace{1} - 2x^2,$$

$$S(x) = \frac{1}{1 - x - x^2}.$$

Doing the same for the $4 \times n$ strip with dominoes we have

$$p_T(x) = x^6 - x^5 - 6x^4 + 6x^2 + x - 1,$$

$$x^6 p_T(1/x)(s_0 + s_1 x) = (1 - x - 6x^2 + 6x^4 + x^5 - x^6)(1 + x + 5x^2 + 11x^3 + 36x^4 + 95x^5)$$

$$= \underbrace{1 - 2x^2 + x^4} - 281x^6 - 500x^7 + 222x^8 + 595x^9 + 59x^{10} - 95x^{11},$$

$$S(x) = \frac{1 - 2x^2 + x^4}{1 - x - 6x^2 + 6x^4 + x^5 - x^6} = \frac{1 - x^2}{1 - x - 5x^2 - x^3 + x^4}.$$

This last one gives a demonstration that it is possible to get some cancellation of terms when finding the generating function.

2.1.2   Closed-Form Solutions from Projection

In some cases we can use linear algebra to find closed-form solutions. Suppose that we have a matrix T which has full geometric multiplicity, in other words a full set of eigenvectors. Then the matrix can be expressed in the form

$$T = \lambda_1 P^{(1)} + \cdots + \lambda_j P^{(j)},$$

where $P^{(i)}$ is the projection matrix onto the i-th eigenspace. The key here is that the projection matrices satisfy $P^{(a)} P^{(b)} = O$ if $a \neq b$ and $P^{(a)} P^{(a)} = P^{(a)}$, in other words they are idempotent. It follows that

$$T^r = \lambda_1^r P^{(1)} + \cdots + \lambda_j^r P^{(j)}.$$

Now reading off the appropriate entry, say the $(1, 1)$ entry, in each matrix we have

$$q_r = (T_r)_{1,1} = (\lambda_1^r P_1 + \cdots + \lambda_j^r P_j)_{1,1} = \lambda_1^r P_{1,1}^{(1)} + \cdots + \lambda_j^r P_{1,1}^{(j)}.$$

In practice, this is usually not carried out because finding the explicit eigenvalues, let alone the projection matrices, is usually nontrivial. But in theory it works great!

This idea gives insight into why we might get cancellation when forming the generating function. Namely, it might be the case that $P_{1,1}^{(i)} = 0$ and in this case that means we do not need to have λ_i as part of the solution and so it reduces the degree of the polynomial.

In some cases it is possible to carry this out explicitly. For example, for the $2 \times n$ case we have

$$\begin{bmatrix} 1 & 1 \\ 1 & 0 \end{bmatrix} = \frac{1+\sqrt{5}}{2} \begin{bmatrix} \frac{1}{10}(5+\sqrt{5}) & \frac{1}{5}\sqrt{5} \\ \frac{1}{5}\sqrt{5} & \frac{1}{10}(5-\sqrt{5}) \end{bmatrix} + \frac{1-\sqrt{5}}{2} \begin{bmatrix} \frac{1}{10}(5-\sqrt{5}) & -\frac{1}{5}\sqrt{5} \\ -\frac{1}{5}\sqrt{5} & \frac{1}{10}(5+\sqrt{5}) \end{bmatrix}$$

so that a closed-form solution for the number of tilings (and hence Fibonacci numbers) is

$$\frac{1}{10}(5+\sqrt{5})\left(\frac{1+\sqrt{5}}{2}\right)^n + \frac{1}{10}(5-\sqrt{5})\left(\frac{1-\sqrt{5}}{2}\right)^n.$$

3   Using More Entries of the Matrix

It would seem that we go through the trouble of computing a large matrix to turn around and then throw most of the matrix away and grab a single entry. However, depending on the situation we can find value in more entries.

As an example, if we let T be the matrix of all possible column transitions for a fixed height m then if we look at the trace of T^n that would consist of all tilings where the left and right ends line up. In particular, we could join the left and right ends and then we have the tilings of a cylinder of height m with n units around. As an example if we go back to the tilings of the $2 \times n$ board and wanted to look at tilings when we wrap around we would have

$$T = \begin{matrix} \emptyset \\ 12 \\ 1 \\ 2 \end{matrix} \begin{bmatrix} 1 & 1 & 0 & 0 \\ 1 & 0 & 0 & 0 \\ 0 & 0 & 0 & 1 \\ 0 & 0 & 1 & 0 \end{bmatrix}$$

and then the number of tilings would be trace(T^n) which gives A068397 in the OEIS [10], a near variant of the Lucas numbers.

If instead of looking at columns of being height m (i.e., a path-like structure) we allowed the columns to be cycles of length m, then we could repeat the above and get tilings on a torus.

We can do even more weird stuff! For example, we could put some twists into our tilings. The way this is done is to look at how the ends meet up and insist that we have the reverse of what we started with (so now we really need to make sure we grab the right terms). This means we can tile Mobius strips and Klein bottles. Examples of how this was done can be found in Butler and Osborne [3].

Research Project 1 Adapt your favorite tiling problem to tiling on a Mobius strip.

4 Tiling with Statistics

One popular variation of tiling is to do statistics, and also to introduce colors. This can readily be adapted to the technique we have discussed.

As an example, suppose for each tiling of the $2 \times n$ board with monomoes and dominoes we assign it a value $x^i y^k z^\ell$ where i are the number of monomoes, k the number of vertical dominoes, and ℓ the number of horizontal dominoes. One way to interpret this is that if we assign x possible different colors to the monomoes, y possible different colors to the vertical dominoes, and z possible different colors to the horizontal dominoes, then this value is the number of different colorings that can result. Note if we set $x = y = z = 1$ then this reduces to what we have done before.

We are now going to introduce a pair of matrices, one for what happens between two column transitions (the one we have been using so far) and one that happens on a column transition (this has been implicit in our work so far as it has been the identity).

$$
\begin{array}{c}
\varnothing \\ 1 \\ 2 \\ 12
\end{array}
\underbrace{\begin{bmatrix}
x^2 + y & x & x & 1 \\
x & 0 & 1 & 0 \\
x & 1 & 0 & 0 \\
1 & 0 & 0 & 0
\end{bmatrix}}_{=A}
\quad \text{and} \quad
\begin{array}{c}
\varnothing \\ 1 \\ 2 \\ 12
\end{array}
\underbrace{\begin{bmatrix}
1 & 0 & 0 & 0 \\
0 & z & 0 & 0 \\
0 & 0 & z & 0 \\
0 & 0 & 0 & z^2
\end{bmatrix}}_{=B}.
$$

The $x^2 + y$ term in the first matrix comes from the fact that between two consecutive columns with no crossing we can either stack two monomoes (x^2) or a single vertical domino (y). The z which come from tiles which cross are in the second term which involves what happens in column crossings.

We are interested in understanding the upper left entry that comes from $B(AB)^n$. Since the first B will not change the upper left entry it suffices to understand the upper left entry of $(AB)^n$. So this gives us the following data:

n	$((AB)^n)_{1,1}$
0	1
1	$x^2 + y$
2	$x^4 + 2x^2 y + 2x^2 z + y^2 + z^2$
3	$x^6 + 3x^4 y + 4x^4 z + 3x^2 y^2 + 4x^2 yz + 4x^2 z^2 + y^3 + 2yz^2$

From this we can now get more insight into the structure of our tilings. As an example for $n = 3$ we see that the coefficient of $x^2 yz$ is 4 which means that there are four different ways to tile the 2×3 board with two monomoes, one vertical tile and one horizontal tile (verify this and the remaining coefficients).

Note that the minimal polynomial approach still can be used to get a recursion for these polynomials. For example, if we let $r_n = r_n(x, y, z)$ be the polynomial associated with the tilings for any n, then we have

$$r_n = (y + x^2)r_{n-1} + (2x^2 z + 2z^2)r_{n-2} + (x^2 z^2 - yz^2)r_{n-3} - r_{n-4}.$$

One thing that we can capture from this is how many of each tile was used. For instance, we might be interested in tilings with one square missing, and so we can have a monomoe as a tile and then keep track of the statistic of how many monomoes are used and only keep the coefficient of the term corresponding to a single monomoe being used.

Research Project 2 Given an $m \times n$ board, determine the number of vertical dominoes k and the number of horizontal dominoes ℓ that maximize the number of different tilings with exactly k vertical and ℓ horizontal dominoes. What can you say about the relationship between the values k and ℓ and the values m and n.

5   Using Other Boards for Tiling

So far we have been tiling with rectangular boards, and at first glance it seems like we are destined to continue tiling with polyominoes. The key in the process is to realize that what is important is a consistent transition. Once we understand this, there are many different boards that can be considered. One simple variation is where we have a "shifted board" and as we add on to the board we do so as a diagonal strip (see Fig. 6a); we can also consider Aztec rectangles (found by taking triangles out of a square board and rotating the result; see Fig. 6b).

Fig. 6 Variations of boards to tiles made from combining squares

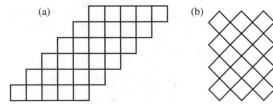

5.1 Boards Formed from Triangles

We can form a board out of equilateral triangles by stacking them together to form a parallelogram which is m triangles in height and n triangles in width. An example is shown for $m = 4$ and $n = 10$ in Fig. 7.

Bodeen et al. [2] explored the special case of the $2 \times n$ board where the tiles are a single triangle ($\triangle$) and four triangles glued together to make a larger triangle ($\triangle$). This was done by finding recursions (in the spirit of our first example), but could also have easily been done by looking at how we cross between columns. Namely, there are four possible ways to cross between columns as shown in Fig. 8 which leads to the matrix

$$T = \begin{matrix} \emptyset \\ A \\ B \\ AB \end{matrix} \begin{bmatrix} 1 & 1 & 1 & 1 \\ 1 & 0 & 1 & 0 \\ 1 & 0 & 0 & 0 \\ 1 & 0 & 0 & 0 \end{bmatrix}.$$

The matrix T has the characteristic polynomial $x^4 - x^3 - 3x^2 - x$ and so the entries of T satisfy the recursion $r_n = r_{n-1} + 3r_{n-2} + r_{n-3}$. One interesting variation that was noted in Bodeen et al. [2] is when one triangle is glued on the end as shown in Fig. 9. The number of ways to tile this board with the two types of triangles considered is the Pell numbers (sequence A000129 in the OEIS [10])!

It might not be immediately obvious that our tools can be used to count this situation in that the left and right ends do not match. But we can readily salvage the situation by looking at what happens in the first diagonal (e.g., the one past the glued triangle). We are in one of two situations, either there is no crossing or there is a crossing of type B. So we now combine both situations and conclude that the Pell numbers are $(T^n)_{\emptyset,\emptyset} + (T^n)_{B,\emptyset}$. This approach can handle any situation where we have something "sticking out the end."

Fig. 7 A board for tiling problems composed of equilateral triangles

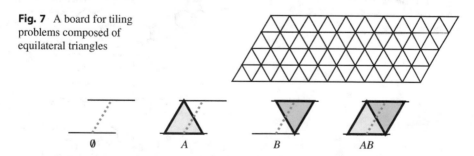

Fig. 8 The possible crossings arising from the Bodeen et al. [2] tiling

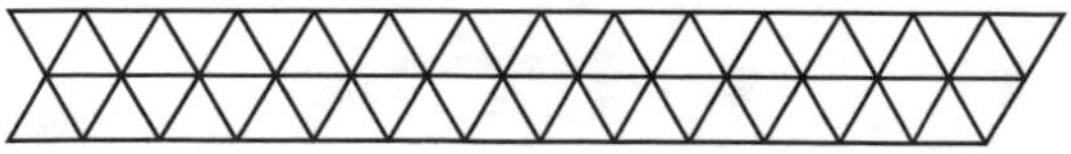

Fig. 9 A tiling board that relates to the Pell numbers

This example illustrates the advantage of playing around with small perturbations of the problem. Sometimes it might seem that the numbers being generated are random, but a small change and then it connects to previous work. In a similar fashion, look at the data and see if it has any nice properties. For example, do the numbers factor nicely, or are there nice modular conditions. If you spot a pattern, try and prove it!

Tiling based on the triangular board have been studied in the case when the tiles consist of two triangles glued together ($\square$). These are known as "lozenge tilings" and have connections to finding matchings as well as plane partitions (see the cover on Winkler's book [12]).

Research Project 3

(a) Determine the number of tilings of a board formed from triangles where the tiles are formed by gluing three triangles together ($\triangle$).
(b) Determine the number of tilings of a board formed from triangles where the tiles are formed by gluing four triangles together in all possible ways ($\square$, $\triangle$, and $\diamondsuit$).

5.2   Boards Formed from Hexagons

We can form a board out of hexagons by stacking them together with n hexagons glued in a row and then stacked m high. (Note there are two possible variations, where we consistently offset the next row in each direction and where we go back and forth.) This is shown in Fig. 10.

Fig. 10 Different boards which can be formed from hexagons

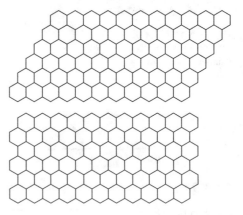

Research Project 4

(a) Determine the number of tilings of a board formed from hexagons where the tiles are single hexagons (◯) or two hexagons glued on an edge (◯◯).
(b) Determine the number of tilings of a board formed from hexagons where the tiles are formed by gluing three hexagons together in all possible ways (◯◯◯, ⬡⬡, and ◯◯).

5.3 Three Dimensional Boards

Up to this point we have looked at boards in the plane and what happens as we transition between columns. The basic idea still works in a higher dimensional setting. Namely, we look at the possible transitions between *layers*. We illustrate this by an example. Suppose we want to look at all tilings of the $2\times2\times n$ box using $1\times1\times2$ blocks (equivalent of dominoes). Then we can think of transitioning between 2×2 layers. Because of parity and consistency there end up being six possible ways to cross between two consecutive layers (shown in Fig. 11).

This leads to the following matrix

$$
T =
\begin{array}{c}
\emptyset \\
AB \\
BC \\
CD \\
AD \\
ABCD
\end{array}
\begin{bmatrix}
2 & 1 & 1 & 1 & 1 & 1 \\
1 & 0 & 0 & 1 & 0 & 0 \\
1 & 0 & 0 & 0 & 1 & 0 \\
1 & 1 & 0 & 0 & 0 & 0 \\
1 & 0 & 1 & 0 & 0 & 0 \\
1 & 0 & 0 & 0 & 0 & 0
\end{bmatrix}.
$$

Reading off $(T^n)_{\emptyset,\emptyset}$ gives

$$2, 9, 32, 121, 450, 1681, 6272, 23409, 87362, 326041, 1216800, 4541161, \ldots.$$

This is sequence A006253 in the OEIS [10] (confirming our ability to count tilings in three dimensions).

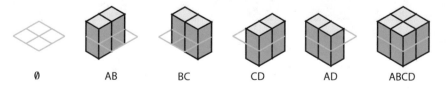

| ∅ | AB | BC | CD | AD | ABCD |

Fig. 11 Different ways to cross between two consecutive layers

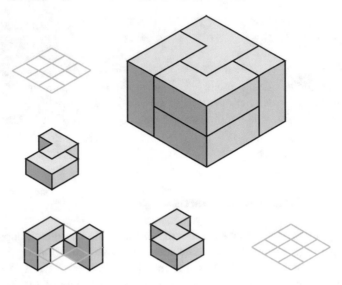

Fig. 12 A tiling of $3 \times 3 \times 2$ using "L" shapes

The assortment of tiling problems in three (and higher) dimensions has not been well mined, so almost any interesting tiling problem that involves three dimensions nontrivially has not been done. The tiling with $1 \times 1 \times 2$ tiles are good places to start.[10] Of course there are many more possibilities, for example, the number of ways to tile the $3 \times 3 \times n$ box using three cubes glued together to form an "L" shape. One of the 432 ways to do this for $n = 2$ is shown in Fig. 12, together with a decomposition going from upper left counter-clockwise where we alternate crossings between layers and a layer between crossings.

> **Research Project 5** Pick a favorite tile (or tiles) and count the number of ways to tile the $p \times q \times n$ box where p, q are fixed.

> **Research Project 6** Explore looking at tilings where the board is based on other shapes which tessellate three (or higher dimensional space. One example of where to start is the rhombic dodecahedron.

---

[10]As an example, as of this writing the number of ways to tile the $4 \times 4 \times 4$ cube with $1 \times 1 \times 2$ tiles is not in the OEIS or found via an internet search engine.

Fig. 13 (a) Examples of tilings with skinny rectangles of a step pyramid using 9 blocks. (b) Examples of shadings of a slanted pyramid with no two shaded cells sharing an edge using 10 blocks

6 Using Abstract Tiles

We begin this section with the following puzzle. In Fig. 13 we give examples of two combinatorial objects. In Fig. 13a we have examples of tilings of a step pyramid by rectangles where at least one dimension is 1 (we can call these "skinny" rectangles). In Fig. 13b we have examples of shadings of a slanted pyramid where no two shaded cells share an edge. The puzzle: are there more ways to tile the step pyramid with skinny rectangles or to shade the slanted pyramid?

We strongly encourage the reader to pause and see if they can figure this out. Bonus points for figuring out without resorting to actually counting the number of ways to do these. (A solution follows analogous to a result later in this section).

6.1 Squaring a Square

A classic problem in mathematics is "squaring the square" which refers to finding a way to take a square of integer length and to tile it with smaller squares of integer length, each square being a distinct size. The smallest number of squares needed to do this is 21 and involves decomposing a square with side length 112.[11]

When it comes to enumeration, we can drop the requirement that all squares have distinct lengths and ask instead how many ways are there to take a square of integer length and decompose it into smaller squares each of integer length. This can be counted by the method we have outlined here using as tiles all possible squares up to a given order (the key is to remember that when we look at how a column intersects a square we also need to understand how far into the square we are intersecting). The counts for small cases are given as A219924 and A045846 in the OEIS [10].

> **Research Project 7** Determine the number of ways to triangle a triangle. That is, take an equilateral triangle of integer length and count how many ways there are to decompose it into smaller equilateral triangles, each of integer length.

---

[11] For more information on squaring the square visit the website http://www.squaring.net/.

6.2   Rectangling a Rectangle

Let us consider a simple variation where squares become rectangles, so now we are dividing a rectangle with integer lengths into small rectangles of integer length. In principle we can approach the same as squaring the square, namely we load in all possible rectangles as our tiles and then we keep track of how far into the rectangle we are slicing. This, however, will lead us into a fundamental problem, namely that the number of possible tiles will continue to grow even as we keep one dimension fixed; meaning *every* board we could consider would become a new problem to perform (while this might be fun, it becomes prohibitive in enumerating).

Alternatively let us change our view on how to approach the problem. We no longer think of our tiles as rectangles of fixed height *and* width, but rather of a rectangle with only fixed height. By this we mean that we look at how rectangles of various heights can cross between columns and make sure that we get consistency of heights between columns (if we have a rectangle of a given height at a column crossing, then in the next column crossing it must be either a rectangle of the same height at the same location, or no crossing occurring at all). So this changes the problem to working with a collection of abstract tile shapes. Using this approach leads to the sequences A116694 and A182275 in the OEIS [10].

> **Research Project 8** Find the number of ways to tile $m \times m \times n$ box with rectangular boxes with integer lengths for fixed m.

6.3   Tiling with Skinny Rectangles

Let us take a moment to look at the special case of tiling with skinny rectangles (rectangles where at least one of the dimensions is one). We begin with the following code written in sage.

```
def M(L1,L2,n):
    tally = 0
    for i in range(n-1):
        if (not i in L1) and (not i in L2):
            if (not i+1 in L1) and (not i+1 in L2):
                tally += 1
    return 2^tally

def count_tilings(n,m):
    ind = {}
    counter = 0
    for C in Combinations(n):
        ind[counter] = C
```

```
        counter += 1
  v = [0]*counter
  v[0] = 1
  for t in range(m):
      new_v = [0]*counter
      for i in range(counter):
          for j in range(counter):
              new_v[i] += v[j]*M(ind[i],ind[j],n)
      v = new_v
  return v[0]
```

Running the code count_tilings(n,m) will produce the number of tilings of the $m \times n$ board with skinny rectangles. So it will be useful to understand how this works to see how similar problems can be explored.

The first thing to note is that in many of these tiling problems, the size of the matrices grows very fast and gets to the point where it becomes unmanageable to store the matrix.[12] Now recall that our approach not only needs the matrix, but often we care about the matrix to a high power which will have *large* entries. So this would appear to be an unavoidable obstacle in our enumeration problem. But we have one small saving grace: we do not need the entire matrix, we only need one entry (or a few entries). If we let $e_1 = (1\ 0\ 0\ \cdots\ 0)^T$, then we have

$$(A^n)_{1,1} = e_1^* A^n e_1 = e_1^* \big(A\big(A\big(\cdots A\big(Ae_1\big)\cdots\big)\big)\big).$$

Notice that this allows us to transform the task of computing a matrix to the n-th power to performing a series of n matrix-vector multiplications which are simpler. Moreover, we do not even need to ever store A in memory! Instead we can find a way to produce entries on the fly as they are needed. So we can reduce the amount of storage tremendously, down to a single vector.

With the preceding in mind we can see how the code works. Namely, we look at all possible ways that we can cross columns and look at these as sets (i.e., which rows are crossed) and use this as an indexing set for reference. We now initiate a vector v and carry out matrix multiplication m times, every time we need an entry from the matrix we generate it in a separate routine (so the matrix is never stored). Once we have finished our multiplication we then grab the first entry of the vector (this corresponds to the no crossing between columns) and return the result.

One other thing to notice is that the entries in the matrix are powers of 2, and particularly some entries are large. This can happen when we have two consecutive columns with coinciding runs of no crossing occurring, in particular this means

---

[12] For example, when the authors were using this approach to count the number of ways to fill up a 10×20 board using Tetris pieces (the board sized used in the game) they had 447426747 different crossings to keep track of which would create an exceedingly huge matrix that even the largest computers would have a hard time working with. On a side note, the number of such tilings is: 29105323812018491321183537645 6587574.

that there are multiple ways to connect what is happening between two column crossings.

We now have a way to generate the data, which leads to the important question: Has this been done before? Thanks to the OEIS [10] and powerful search engines it is relatively easy to answer this question. This leads to three possibilities:

1. *This has never been counted before.* Time to double check our work, and then we can stake our claim to immortality by recording it in the OEIS and/or writing up the result.[13]
2. *This particular tiling problem has been counted before.* This means we are doing a great job of counting stuff, and now we start thinking about ways to modify/enhance our counts to get something new. (This is often easier than it sounds, e.g., add colors to the tiles (statistics), modify the board, add restrictions to pieces, and so on).
3. *These numbers have been involved in a counting problem before, but not by tiling.* This is often the most interesting because it leads us to figuring out why these objects are related, and in particular looking for a bijective proof between these objects. (We should still make a note of this in the OEIS, once a proof has been found of correctness, to help those who follow after. This is also the most publishable of outcomes).

For this particular problem it turned out that it was in the third case; these numbers had been counted before in a paper by Zhang [13], but not in terms of a tiling (see A254414 and A254127 in the OEIS [10]). This particular counting problem dealt with counting *independent sets in Aztec rectangle graphs.* Let $G_{m,n}$ be the graph with vertex set $V(G) = \{(i, j) : 1 \le i \le 2m - 1, 1 \le j \le 2n - 1\}$ and edge set $E(G) = \{\{(i, j), (i', j')\} : i + j \text{ odd}, |i - i'| = |j - j'| = 1\}$. The graph $G_{6,3}$ is shown in Fig. 14a. Moreover, in the graph we have filled some vertices which are an independent set (an independent set of vertices is one in which no two are connected by an edge).

Theorem 2 *The number of independent sets in $G_{m,n}$ is the number of ways to tile an $m \times n$ board with skinny rectangles.*

Proof We show how to go back and forth between these two objects (a bijective proof). The key is to think of tiling not as putting pieces down on the board, but rather to think of it as removing edges out of our picture. For each internal vertex of the $m \times n$ board we associate it with a vertex of a graph, and we connect two edges if removing both of them would result in something which is not a skinny rectangle. The resulting graph is $G_{m,n}$. Now we see that every independent set in $G_{m,n}$ is

---

[13]For publication you will be expected to go beyond the enumeration, look for interesting characteristics and properties that you can prove about these numbers.

Fig. 14 An example of $G_{6,3}$ with a marked independent set and a tiling of the 6×3 board with skinny rectangles. To see the relationship between them, try crossing your eyes

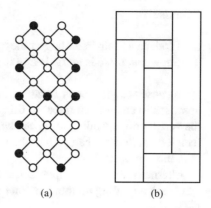

(a)   (b)

associated with a unique tiling of the $m\times n$ board and conversely every tiling of the $m\times n$ board produces a unique independent set.

This is illustrated in Fig. 14.

Research Project 9 Find the number of ways to tile the $m\times m\times n$ box with "skinny" boxes (where at least two of the three dimensions is 1). Find and describe the graph where the number of independent sets in the graph is equal to the number of tilings.

What happens if we only require one of the dimensions to be 1?

Research Project 10 For any tiling board that was discussed before, repeat the process for counting with skinny strips. For the triangular board, is there an interpretation in terms of independent sets? What about other possible interpretations?

7   A Global Constraint on Tiling

Our approach on focusing on counting by looking at what happens when we cross columns gives a simple local way to carry out the computation. The trade-off to this local approach is that sometimes there is a global constraint which our local approach can have a difficult time capturing.

We will call a tiling of an $m\times n$ board *fault-free* if between any two consecutive columns *and* any two consecutive rows some tile crosses. These were introduced by Graham [7].

Exercise 3

(a) Produce a fault-free tiling of the 5×6 board using dominoes ($\square$).
(b) Show that there is no fault-free tiling of the 6×6 board using dominoes ($\square$).

Because we have been focusing on column crossings it is simple to adapt our approach to count the number of tilings with no vertical faults. Namely, we remove the vertex corresponding to no crossings ($\emptyset$) and take $n - 1$ steps on the graph. We then look at how we can go into the graph from $\emptyset$ and from the graph go to the $\emptyset$ to grab the appropriate entries.

Alternatively, one can approach this via generating functions by recognizing we can break any tiling up into a union of consecutive tilings with no vertical faults. If we let

$$F(x) = \sum_{n \geq 0} f_n x^n, \qquad \text{and} \qquad G(x) = \sum_{n \geq 1} g_n x^n$$

where f_n is all tilings for the $m \times n$ board and where g_n is all vertical fault-free tilings of the $m \times n$ board, then we have

$$F(x) = 1 + G(x)F(x).$$

In other words if we want to count all tilings (the $F(x)$) we either have the empty tiling (the "1") or we split it into the first vertical fault-free part (the "$G(x)$") and then the remainder which is some tiling (the $F(x)$). From this we conclude that

$$G(x) = \frac{F(x) - 1}{F(x)}.$$

Note that for tiling problems where we can find the characteristic polynomial of the matrix used to generate the data that we have ways to get the generating function, and so this method gives a quick approach (and as added bonus we already have the generating function). A good introduction to the manipulation of generating functions is in Graham et al. [8, Ch. 7], which also contains an interesting discussion about generating functions and tilings.

There are many questions about fault-free tilings that can be asked, and we lift the following from the paper by Graham [7].

> It is typical of this business that one answer leads to n more questions. For example, *how many fault-free tilings* does a rectangle have? What if we can use two sizes of tiles instead of just one? What about these same questions in 3 (or more) dimensions? ... We encourage the interested reader to explore this fascinating byway of geometry and discover the gems which must surely lie waiting to be discovered.

Research Project 11 Pick a board and a tile (or set of tiles) and determine the number of fault-free tilings of the board.

Some enumeration (without details) of the fault-free tilings of the square with dominoes is given as A124997 in the OEIS [10]. There has also been work done in the enumeration of fault-free tilings using triominoes (⌐), see A084477, A084479 and A084481 in the OEIS [10] and references therein.

8   Concluding Remarks

A surprising amount of mathematics can be connected to tilings. The wonderful book by Benjamin and Quinn [1] give examples of how to use tilings to establish properties of Fibonacci and Lucas numbers through interpreting tilings.[14]

So tilings are not just fun to play with, they can have mathematical power. Even in the answering of some of these "toy" questions we often develop tools that can in turn be used to solve other problems. The main method used here (the transfer matrix method) has applications all throughout combinatorics and can be adapted to many settings.

We are *counting* on you to help further push our knowledge about tilings.

References

1. Benjamin, A., Quinn, J.: Proofs that Really Count: The Art of Combinatorial proof, MAA, 193 p. (2003)
2. Bodeen, J., Butler, S., Kim, T., Sun, X., Wang, S.: Tiling a strip with triangles. Electronic Journal of Combinatorics **21**, P1.7, 15 p. (2014)
3. Butler, S., Osborne, S.: Counting tilings by taking walks. The Journal of Combinatorial Mathematics and Combinatorial Computing **88**, 83–94 (2014)
4. Chinn, P., Grimaldi, R., Heubach, S.: Tiling with L's and squares. Journal of Integer Sequences **10**, 07.2.8, 17 p. (2007)
5. Elkies, N., Kuperberg, G., Larsen, M., Propp, J.: Alternating sign matrices and domino tilings I, Journal of Algebraic Combinatorics **1**, 111–132 (1992)
6. Elkies, N., Kuperberg, G., Larsen, M., Propp, J.: Alternating sign matrices and domino tilings II, Journal of Algebraic Combinatorics **2**, 219–234 (1992)
7. Graham, R.: Fault-free tilings of rectangles. The Mathematical Gardner, D. Klarner ed., Wadsworth, 120–126 (1981)
8. Graham, R., Knuth, D., Patashnik, O.: Concrete Mathematics; second edition, Wiley, 657 p. (1994)
9. Kasteleyn, P.M.: The statistics of dimers on a lattice. Physica **27**, 1209–1225 (1961)
10. N. J. A. Sloane, The On-Line Encyclopedia of Integer Sequences, http://oeis.org, 2019.

---

[14]With many colorful examples.

11. Stanley, R.: Enumerative Combinatorics, Volume I, second edition. Cambridge, New York (2012)
12. Winkler, P.: Mathematical Puzzles: A Connoisseur's Collection, CRC Press, Boca Raton (2004)
13. Zhang, Z.: Merrifield-Simmons index of generalized Aztec diamond and related graphs. MATCH Communications in Mathematical and in Computer Chemistry **56**, 625–636 (2006)

Beyond Coins, Stamps, and Chicken McNuggets: An Invitation to Numerical Semigroups

Scott Chapman, Rebecca Garcia, and Christopher O'Neill

Abstract

We give a self-contained introduction to numerical semigroups and present several open problems centered on their factorization properties.

Suggested Prerequisites Linear algebra, Number theory

1   Introduction

Many difficult mathematics problems have extremely simple roots. For instance, suppose you walk into your local convenience store to buy that candy bar you more than likely should not eat. Suppose the candy bar costs X cents and you have c_1 pennies, c_2 nickels, c_3 dimes, and c_4 quarters in your pocket (half dollars are of course too big to carry in your pocket). Can you buy the candy bar? You can if there are non-negative integers $x_1, \ldots, x_4$ such that

$$x_1 + 5x_2 + 10x_3 + 25x_4 \geq X$$

with $0 \leq x_i \leq c_i$ for each $1 \leq i \leq 4$. Obviously, this is not difficult mathematics; it is a calculation that almost everyone goes through in their heads multiple times

S. Chapman (✉) · R. Garcia
Mathematics and Statistics Department, Sam Houston State University, Huntsville, TX, USA
e-mail: scott.chapman@shsu.edu; mth_reg@shsu.edu

C. O'Neill
Mathematics and Statistics Department, San Diego State University, San Diego, CA, USA
e-mail: cdoneill@sdsu.edu

© Springer Nature Switzerland AG 2020   177
P. E. Harris et al. (eds.), *A Project-Based Guide to Undergraduate Research in Mathematics*, Foundations for Undergraduate Research in Mathematics,
https://doi.org/10.1007/978-3-030-37853-0_6

a week. Now, suppose the cashier indicates that the register is broken, and that the store can only accept the exact amount of money necessary in payment for the candy bar. This changes the problem to

$$x_1 + 5x_2 + 10x_3 + 25x_4 = X$$

with the same restrictions on the x_i's.

Given our change system, the two equations above are relatively easy with which to deal, but changing the values of the coins involved can make the problem much more difficult. For instance, instead of our usual change, suppose you have a large supply of 3-cent pieces and 7-cent pieces. Can you buy the 11-cent candy bar? With a relatively gentle calculation, even your English major roommate concludes that you cannot. There is no solution of $3x_1 + 7x_2 = 11$ in the non-negative integers. But with a little more tinkering, you can unearth a deeper truth.

Big Fact: *In a 3–7 coin system, you can buy any candy bar costing above 11 cents.*

The Big Fact follows since $12 = 3 \cdot 4$, $13 = 1 \cdot 7 + 2 \cdot 3$, $14 = 2 \cdot 7$, and any integer value greater than 14 can be obtained by adding the needed number of 3 cent pieces to one of these sums. But why limit the fun to coins? Analogous problems can be constructed using postage stamps and even Chicken McNuggets. The key to what we are doing involves an interesting mix of linear algebra, number theory, and abstract algebra, and quickly leads to some simply stated mathematics problems that are very deep and remain (over a long period of time) unsolved. Moreover, these problems have been the basis of a wealth of undergraduate research projects, many of which led to publication in major mathematics research journals. In order to discuss these research level problems, we will now embark on a more technical description of the work at hand. As our pages unfold, the reader should keep in mind the humble beginnings of what will become highly challenging work.

A *numerical semigroup* is a subset $S \subset \mathbb{Z}_{\geq 0}$ of the non-negative integers that

1. is closed under addition, i.e., whenever $a, b \in S$, we also have $a + b \in S$ and
2. has finite complement in $\mathbb{Z}_{\geq 0}$.

The two smallest examples of numerical semigroups are $S = \mathbb{Z}_{\geq 0}$ and $S = \mathbb{Z}_{\geq 0} \backslash \{1\}$. As is usually tacit in the literature, we assume that $0 \in S$ which in fact makes S an additive monoid. We will drift between the use of the term semigroup and monoid throughout the remainder of this work.

Often, the easiest way to specify a numerical semigroup is by providing a list of *generators*. For instance,

$$\langle n_1, \ldots, n_k \rangle = \{a_1 n_1 + \cdots + a_k n_k : a_1, \ldots, a_k \in \mathbb{Z}_{\geq 0}\}$$

equals the set of all non-negative integers obtained by adding copies of $n_1, \ldots, n_k$ together. The smallest nontrivial numerical semigroup $S = \mathbb{Z}_{\geq 0} \setminus \{1\}$ can then also be written as $S = \langle 2, 3 \rangle$, since every non-negative even integer can be written as $2k$ for some $k \geq 0$, and every odd integer greater than 1 can be written as $2k + 3$ for some $k \geq 0$. It is clear that generating systems are not unique (for instance, $\langle 2, 3 \rangle = \langle 2, 3, 4 \rangle$), but we will argue later that each numerical semigroup has a unique generating set of minimal cardinality. Note that problems involving the 3–7 coin system take place in the numerical semigroup $\langle 3, 7 \rangle$.

Example 1 Although numerical semigroups may seem opaque, there are some very practical ways to think about them. For many years, McDonald's sold Chicken McNuggets in packs of 6, 9, and 20, and as such, it is possible to buy exactly n Chicken McNuggets using only those three pack sizes precisely when $n \in \langle 6, 9, 20 \rangle$. For this reason, the numerical semigroup $S = \langle 6, 9, 20 \rangle$ is known as the *McNugget semigroup* (see [9]). It turns out that it is impossible to buy exactly 43 Chicken McNuggets using only packs of 6, 9, and 20, but for any integer $n > 43$, there is some combination of packs that together contain exactly n Chicken McNuggets.

By changing the quantities involved (be it with coins or Chicken McNuggets) yields what is known in the literature as the *Frobenius coin-exchange problem*. To Frobenius, each generator of a numerical semigroup corresponds to a coin denomination, and the largest monetary value for which one cannot make even change is the *Frobenius number*. In terms of numerical semigroups, the *Frobenius number* of S is given by

$$F(S) = \max(\mathbb{Z} \setminus S).$$

Sylvester proved in 1882 (see [22]) that in the 2-coin problem (i.e., if $S = \langle a, b \rangle$ with gcd $(a, b) = 1$), the Frobenius number is given by $F(S) = ab - (a+b)$. To date, a general formula for the Frobenius number of an arbitrary numerical semigroup (or even a "fast" algorithm to compute it from a list of generators) remains out of reach.

While deep new results concerning the Frobenius number are likely beyond the scope of a reasonable undergraduate research project, a wealth of problems related to numerical semigroups have been a popular topic in REU programs for almost 20 years. To better describe this work, we will need some definitions. Assume that $n_1, \ldots, n_k$ is a set of generators for a numerical semigroup S. For $n \in S$, we refer to

$$\mathsf{Z}(n) = \left\{ (x_1, \ldots, x_k) : n = \sum_{i=1}^{k} x_i n_i \right\}$$

as the *set of factorizations* of $n \in S$, and to

$$\mathsf{L}(n) = \left\{ \sum_{i=1}^{k} x_k : (x_1, \ldots, x_k) \in \mathsf{Z}(n) \right\}$$

as the *set of factorization lengths* of $n \in S$. Each element of $Z(n)$ represents a distinct *factorization* of n (that is, an expression

$$n = x_1 n_1 + \cdots + x_k n_k$$

of n as a sum of $n_1, \ldots, n_k$, wherein each x_i denotes the number of copies of n_i used in the expression). The local descriptors $Z(n)$ and $L(n)$ can be converted into global descriptors of S by setting

$$\mathscr{Z}(S) = \{Z(n) : n \in S\}$$

to be the *complete set of factorizations* of S and

$$\mathscr{L}(S) = \{L(n) : n \in S\}$$

to be the *complete set of lengths* of S (note that these are both sets of sets).

Hence, while we started by exploring the membership problem for a numerical semigroup (i.e., given $m \in \mathbb{Z}_{\geq 0}$, is $m \in S$?), we now focus on two different questions.

1. Given $n \in S$ what can we say about the set $Z(n)$?
2. Given $n \in S$, what can we say about the set $L(n)$?

We note that the set of factorizations of an element $n \in S$ is known as the *denumerant* of n. This notion is related to the idea of a Hilbert series and the interested reader can find more information in [19]. We start with a straightforward but important observation.

Exercise 1 If S is a numerical semigroup and $n \in S$, then $Z(n)$ and $L(n)$ are both finite sets.

Example 2 Calculations of the above sets tend to be nontrivial and normally require some form of a computer algebra system. To demonstrate this, we return to the elementary example $S = \langle 2, 3 \rangle$ mentioned earlier. As previously noted, any integer $n \geq 2$ is in S. In Table 1, we give $Z(n)$ and $L(n)$ for some basic values of $n \in S$.

Example 3 Patterns in the last example are easy to identify (and we will return to Example 2 in our next section), but the reader should not be too complacent, as the two-generator case is the simplest possible. We demonstrate this by producing in Table 2 the same sets, now for the semigroup $S = \langle 7, 10, 12 \rangle$. We make special note that while the length sets in Table 1 are sets of consecutive integers, $L(42) = \{4, 6\}$ in Table 2 breaks this pattern. We will revisit the concept of "skips" in length sets at the end of the next section.

Table 1 Some basic values of $Z(n)$ and $L(n)$ where $S = \langle 2, 3\rangle$

n	$Z(n)$	$L(n)$	n	$Z(n)$	$L(n)$
2	$\{(1,0)\}$	$\{1\}$	11	$\{(1,3),(4,1)\}$	$\{4,5\}$
3	$\{(0,1)\}$	$\{1\}$	12	$\{(6,0),(3,2),(0,4)\}$	$\{4,5,6\}$
4	$\{(2,0)\}$	$\{2\}$	13	$\{(5,1),(2,3)\}$	$\{5,6\}$
5	$\{(1,1)\}$	$\{2\}$	14	$\{(7,0), (4,2), (1,4)\}$	$\{5,6,7\}$
6	$\{(3,0), (0,2)\}$	$\{2,3\}$	15	$\{(6,1),(3,3),(0,5)\}$	$\{5,6,7\}$
7	$\{(2,1)\}$	$\{3\}$	16	$\{(8,0),(5,2),(2,4)\}$	$\{6,7,8\}$
8	$\{(4,0),(1,2)\}$	$\{3,4\}$	17	$\{(7,1),(4,3),(1,5)\}$	$\{6,7,8\}$
9	$\{(3,1),(0,3)\}$	$\{3,4\}$	18	$\{(9,0), (6,3), (3,4), (0,6)\}$	$\{6,7,8,9\}$
10	$\{(5,0),(2,2)\}$	$\{4,5\}$	19	$\{(8,1), (5,3), (2,5)\}$	$\{7,8,9\}$

Table 2 Some basic values of $Z(n)$ and $L(n)$ where $S = \langle 7, 10, 12\rangle$

n	$Z(n)$	$L(n)$	n	$Z(n)$	$L(n)$
7	$\{(1,0,0)\}$	$\{1\}$	30	$\{(0,3,0)\}$	$\{3\}$
10	$\{(0,1,0)\}$	$\{1\}$	31	$\{(3,1,0),(1,0,2)\}$	$\{3,4\}$
12	$\{(0,0,1)\}$	$\{1\}$	32	$\{(0,2,1)\}$	$\{3\}$
14	$\{(2,0,0)\}$	$\{2\}$	33	$\{(3,0,1)\}$	$\{4\}$
17	$\{(1,1,0)\}$	$\{2\}$	34	$\{(2,2,0),(0,1,2)\}$	$\{3,4\}$
19	$\{(1,0,1)\}$	$\{2\}$	35	$\{(5,0,0)\}$	$\{5\}$
20	$\{(0,2,0)\}$	$\{2\}$	36	$\{(2,1,1),(0,0,3)\}$	$\{3,4\}$
21	$\{(3,0,0)\}$	$\{3\}$	37	$\{(1,3,0)\}$	$\{4\}$
22	$\{(0,1,1)\}$	$\{2\}$	38	$\{(4,1,0), (2,0,2)\}$	$\{4,5\}$
24	$\{(2,1,0),(0,0,2)\}$	$\{2,3\}$	39	$\{(1,2,1)\}$	$\{4\}$
26	$\{(2,0,1)\}$	$\{3\}$	40	$\{(0,4,0),(4,0,1)\}$	$\{4,5\}$
27	$\{(1,2,0)\}$	$\{3\}$	41	$\{(3,2,0),(1,1,2)\}$	$\{4,5\}$
28	$\{(4,0,0)\}$	$\{4\}$	42	$\{(6,0,0), (0,3,1)\}$	$\{4,6\}$
29	$\{(1,1,1)\}$	$\{3\}$	43	$\{(3,1,1), (1,0,3)\}$	$\{4,5\}$

Much of the remainder of this paper will focus on the study of $\mathscr{Z}(S)$ and $\mathscr{L}(S)$, the complete systems of factorizations and factorization lengths of S. The next section presents a crash course on definitions and basic results. We will review some of the significant results in this area, with an emphasis on those obtained in summer and yearlong REU projects. Section 3 explores the computation tools available to embark on similar studies, and Sects. 4 and 5 contain actual student level projects which we hope will pique students' minds and interests.

2   A Crash Course on Numerical Semigroups

We start with a momentary return to the notion of minimal generating sets alluded to in Sect. 1. Let S be a numerical semigroup, and $m > 0$ its smallest positive element. We call a generating set W for S *minimal* if $W \subseteq T$ for any other generating set

T of S. We claim any minimal generating set for S has at most m elements (and in particular is finite). Indeed, for each i with $0 \leq i < m$, set

$$M_i = \{n \in S : n > 0 \text{ and } n \equiv i \bmod m\}.$$

Note that by the definition of S, each $M_i \neq \emptyset$ (in fact, each is infinite). Moreover, for each $n \in M_i$, we must have $n + m \in M_i$ as well since S is closed under addition. Hence, setting

$$n_i = \min M_i$$

for each i, we can write each $M_i = \{n_i + qm : q \geq 0\}$. This implies $N = \{n_0, \ldots, n_{m-1}\}$ is a generating set for S, i.e.,

$$S = \langle N \rangle = \langle n_0, \ldots, n_{m-1} \rangle,$$

since the remaining elements of S can each be obtained from an element of N by adding $m = n_0$ sufficiently many times (note that this is precisely the argument used to justify the Big Fact at the beginning of Sect. 1). As a consequence, any minimal generating set for S must be a subset of N. The generating set is often referred to in the literature as the *Apéry basis*.

Example 4 While the elements of N are chosen with respect to minimality modulo m, the generating set N may not be minimal. For instance, if

$$S = \{0, 5, 8, 10, 13, 15, 16, 18, 20, 21, 23, 24, 25, 26, 28, 29, 30, 31, 32, \ldots\},$$

then $N = \{5, 16, 32, 8, 24\}$, and while $S = \langle 5, 8, 16, 24, 32 \rangle$, the fact that $16 = 2 \cdot 8$, $24 = 3 \cdot 8$, and $32 = 4 \cdot 8$ yields $S = \langle 5, 8 \rangle$.

Using Example 4, we can reduce N to a minimal generating set by setting

$$\widehat{N} = \{n \in N : n \notin \langle N \setminus \{n\} \rangle\}.$$

We note that the minimal generating set can also be expressed as $(S^* + S^*) \setminus S^*$ where $S^* = S \setminus \{0\}$. The following is a good exercise, and implies that every numerical semigroup has a unique minimal generating set.

Exercise 2 If S is a numerical semigroup and T is any generating set of S, then $\widehat{N} \subseteq T$. In particular, $\widehat{N}$ is the unique minimal generating set of S.

Exercise 2 and the argument preceding it establish some characteristics of a numerical semigroup S that are widely used in the mathematics literature. The smallest positive integer in S is called the *multiplicity* of S and denoted by $m(S)$. The cardinality of $\widehat{N}$ above is called the *embedding dimension* of S and is denoted by $e(S)$. By the argument preceding Exercise 2, $e(S) \leq m(S)$, and additionally the elements of $\widehat{N}$ are pairwise incongruent modulo $m(S)$. Moreover, note that $\gcd(\widehat{N}) = 1$, as otherwise $\mathbb{Z}_{\geq 0} \setminus S$ would be infinite.

Example 5 Using Exercise 2, we can set up several obvious classes of numerical semigroups which have garnered research attention. Suppose that $m \geq 2$ and $d \geq 1$ are integers with $\gcd(m, d) = 1$. Given k with $1 \leq k \leq m - 1$, set

$$A_k = \{m, m + d, \ldots, m + kd\}.$$

Using elementary number theory, it is easy to see that A_k is the minimal generating set of the numerical semigroup $\langle A_k \rangle$, which is called an *arithmetical numerical semigroup* (since its minimal generating set is an arithmetical sequence). This is a very large class of numerical semigroups, which contains many important subclasses:

- all 2-generated numerical semigroups (i.e., $k = 1$);
- all numerical semigroups generated by consecutive integers (i.e., $d = 1$); and
- numerical semigroups consisting of all positive integers greater than or equal to a fixed positive integer m (i.e., $d = 1$, $k = m - 1$, and $S = \langle m, m+1, \ldots, 2m-1 \rangle$). In the literature, these numerical semigroups are referred to as *ordinary*.

The latter subclass consists of all numerical semigroups for which $F(S) < m(S)$.

Just as we factor integers as products of primes, or polynomials as products of irreducible factors, we now factor elements in a numerical semigroup S in terms of its minimal generators (in this context, "factorization" means an expression of an element of S as a sum of generators, and as we will see, many elements have multiple such expressions). In terms of S, we have already defined the notation $Z(n)$, $L(n)$, $\mathscr{Z}(S)$, and $\mathscr{L}(S)$. Let us consider some further functions that concretely address structural attributes of these sets. We denote the maximum and minimum factorization lengths of an element $n \in S$ by

$$\ell(n) = \min L(n) \quad \text{and} \quad L(n) = \max L(n).$$

These functions satisfy the following recurrence for sufficiently large semigroup elements; we state this result now and revisit it in much more detail in Sect. 3.

Theorem 1 ([2, Theorems 4.2 and 4.3]) *If $S = \langle n_1, \ldots, n_k \rangle$ with $n_1 < \cdots < n_k$, then*

$$\ell(n + n_k) = \ell(n) + 1 \quad \text{and} \quad L(n + n_1) = L(n) + 1$$

for all $n > n_{k-1} n_k$.

The *elasticity* of a nonzero element $n \in S$, denoted $\rho(n)$, measures the deviation between $\ell(n)$ and $L(n)$ and is defined as

$$\rho(n) = L(n)/\ell(n).$$

The *elasticity* of S is then defined as

$$\rho(S) = \sup\{\rho(n) : n \in S\}.$$

The elasticity of a semigroup element measures the "spread" of its factorization lengths. One of the advantages of defining the elasticity of an element n as the quotient of the maximum and minimum lengths (as opposed to, say, their difference) is that one cannot obtain larger elasticity values "for free" by simply taking multiples of n. Indeed, if $\ell(n) = 3$ and $L(n) = 5$, then $2n$ has factorizations of length 6 and 10 obtained by concatenating factorizations of n of minimum and maximum length, respectively. The only way for $\rho(2n)$ to exceed $\rho(n)$ is for $2n$ to have "new" factorizations not obtained from those of n.

We introduce two more terms before exploring an in-depth example. When the supremum in this expression is attained (i.e., there exists $n \in S$ with $\rho(n) = \rho(S)$) we call the elasticity of S *accepted*. We say that S is *fully elastic* if for every rational $q \in \mathbb{Q} \cap [1, \rho(S))$ (or $[1, \infty)$ if the elasticity is infinite), there exists a nonzero $n \in S$ such that $\rho(n) = q$.

Example 6 We return to the basic semigroup $S = \langle 2, 3 \rangle$ in Example 2 to offer some examples of the calculations thus far suggested. Hence, each factorization of $n \in S$ has the form

$$n = 2x_1 + 3x_2.$$

Table 1 suggests that factorizations of a given element of S are far from unique in general. Notice that in S, the longest factorization of an element $n \in S$ contains the most possible copies of 2 and the shortest the most possible copies of 3. This is the intuition behind Theorem 1: for large semigroup elements, a maximum length factorization for $n + 2$ can be obtained a maximum length factorization for n by adding a single copy of 2.

Using this fact and some elementary number theory, explicit formulas for all the invariants discussed to this point can be worked out for arbitrary elements of S. For instance, for all $n \in S$ we have that

$$\ell(n) = \left\lceil \frac{n}{3} \right\rceil \quad \text{and} \quad L(n) = \left\lfloor \frac{n}{2} \right\rfloor,$$

and thus

$$\rho(n) = \left\lfloor \frac{n}{2} \right\rfloor / \left\lceil \frac{n}{3} \right\rceil.$$

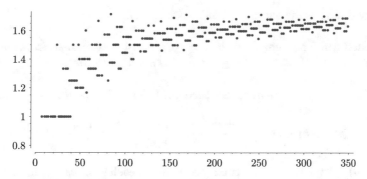

Fig. 1 Elasticity values for elements of $S = \langle 7, 10, 12 \rangle$

Using the fact that 2 copies of the generator 3 can be exchanged for 3 copies of the generator 2 in any factorization in S, we obtain for $n \geq 4$ that

$$L(n) = \left\{ \left\lceil \frac{n}{3} \right\rceil, \left\lceil \frac{n}{3} \right\rceil + 1, \ldots, \left\lfloor \frac{n}{2} \right\rfloor - 1, \left\lfloor \frac{n}{2} \right\rfloor \right\}.$$

Using the notation $[x, y] = \{z \in \mathbb{Z} \mid x \leq z \leq y\}$ for $x \leq y$ integers, we conclude

$$\mathscr{L}(S) = \left\{ \{1\}, \left[\lceil \tfrac{4}{3} \rceil, \lfloor \tfrac{4}{2} \rfloor \right], \left[\lceil \tfrac{5}{3} \rceil, \lfloor \tfrac{5}{2} \rfloor \right], \ldots \right\}.$$

Moreover, it is easy to verify that $\rho(n) \leq 3/2$ for all $n \in S$ and that $\rho(n) = 3/2$ if and only if $n \equiv 0 \mod 6$. Thus $\rho(S) = 3/2$ and the elasticity is accepted.

Though a comparable analysis of elasticities for $S = \langle 7, 10, 12 \rangle$ is out of reach, we offer in Fig. 1 a graph of the elasticity values for $S = \langle 7, 10, 12 \rangle$ to give the reader a feel for how the elasticity behaves for large elements of S.

Many basic results concerning elasticity in numerical semigroups are worked out in the paper [6], which was a product of a summer REU program. We review, with proof, three of that paper's principal results, the first of which yields an exact calculation for the elasticity of S.

Theorem 2 ([6, Theorem 2.1]) *Let $S = \langle n_1, \ldots, n_k \rangle$ be a numerical semigroup, where $n_1 < n_2 < \cdots < n_k$ is a minimal set of generators for S. Then $\rho(S) = n_k/n_1$ and the elasticity of S is accepted.*

Proof If $n \in S$ and $n = x_1 n_1 + \ldots + x_k n_k$, then

$$\frac{n}{n_k} = \frac{n_1}{n_k} x_1 + \ldots + \frac{n_k}{n_k} x_k \leq x_1 + \ldots + x_k \leq \frac{n_1}{n_1} x_1 + \ldots + \frac{n_k}{n_1} x_k = \frac{n}{n_1}.$$

Thus $L(n) \leq n/n_1$ and $l(n) \geq n/n_k$ for all $n \in S$, so we can conclude $\rho(S) \leq n_k/n_1$. Also, $\rho(S) \geq \rho(n_1 n_k) = n_k/n_1$, yielding equality and acceptance. $\qquad \square$

We now answer the question of full elasticity in the negative.

Theorem 3 ([6, Theorem 2.2]) *If $S = \langle n_1, \ldots, n_k \rangle$ is a numerical semigroup, where $2 \le n_1 < \cdots < n_k$ and $k \ge 2$, then S is not fully elastic.*

Proof Let $N = n_{k-1}n_k + n_1 n_k$. By Theorem 1, for each $n > n_{k-1}n_k$, we have

$$\rho(n + n_1 n_k) = \frac{L(n + n_1 n_k)}{\ell(n + n_1 n_k)} = \frac{L(n) + n_k}{\ell(n) + n_1} \ge \frac{L(n)}{\ell(n)} = \rho(n)$$

since $\rho(n) \le n_k/n_1$ by Theorem 2. As such, for each $r < n_k/n_1$, there are only finitely many elements with elasticity less than r, so S cannot be fully elastic. $\square$

The proof of Theorem 3 can be used to prove a result which is of its own interest. For a numerical semigroup S, let

$$R(S) = \{\rho(n) : n \in S\}.$$

Corollary 1 ([6, Corollary 2.3]) *For any numerical semigroup S, the only limit point of $R(S)$ is $\rho(S)$.*

Proof Let $n_1 < n_2 < \cdots < n_k$ be a minimal set of generators for the numerical semigroup $S = \langle n_1, \ldots, n_k \rangle$, where $k \ge 2$. If $n = a(n_1 n_k) + n_1$ for $a \in \mathbb{Z}_{\ge 0}$, then

$$\rho(n) = \frac{L(n)}{l(n)} = \frac{an_k + 1}{an_1 + 1}.$$

It follows that $\rho(n) < n_k/n_1$ for all $a \in \mathbb{Z}_{\ge 0}$ and $\lim_{a \to \infty} \rho(n) = n_k/n_1$, making n_k/n_1 a limit point of the set $R(S)$. Additionally, by Theorem 1, for $n > n_{k-1}n_k$ and $a \ge 1$ we have

$$\rho(n + an_1 n_k) = \frac{L(n + an_1 n_k)}{l(n + an_1 n_k)} = \frac{L(n) + an_k}{L(n) + an_1},$$

meaning $R(S)$ is the union of a finite set (elasticities of the elements less than $n_{k-1}n_k + n_1 n_k$) and a union of $n_1 n_k$ monotone increasing sequences approaching n_k/n_1. As such, we conclude n_k/n_1 is the only limit point. $\square$

The original proofs in [6] did not use Theorem 1 and were much more technical. The proofs given above are a consequence of a complete description of $R(S)$ in [2], a recent paper with an undergraduate co-author in which Theorem 1 first appeared.

The elasticity does lend us information concerning the structure of the length set, but only limited information. While it deals with the maximum and minimum length values, it does not explore the finer structure of $L(n)$ (or more generally of $Z(n)$). There are several invariants studied in the theory of non-unique factorizations that yield more refined information—we introduce one such measure here, which is known as the delta set.

Let $S = \langle n_1, \ldots, n_t \rangle$ be a numerical semigroup, where $n_1, \ldots, n_k \in \mathbb{N}$ minimally generate S and $k \geq 2$. If $\mathsf{L}(n) = \{\ell_1, \ldots, \ell_t\}$ with the ℓ_i's listed in increasing order, then set

$$\Delta(n) = \{\ell_i - \ell_{i-1} : 2 \leq i \leq t\}$$

and

$$\Delta(S) = \bigcup_{0 < n \in S} \Delta(n).$$

Our hypothesis that $t \geq 2$ ensure $\Delta(S) \neq \emptyset$, since (for instance) if $n = n_1 n_2$, then both n_1 and $n_2 \in \mathsf{L}(n)$. Also, for each $n \in S$, since $|\mathsf{L}(n)| < \infty$ by Exercise 2, we clearly have $|\Delta(n)| < \infty$ as well.

Example 7 We use the calculations already presented in Example 6. For $S = \langle 2, 3 \rangle$, our formula for $\mathsf{L}(n)$ yields for all $n \in S$ that

$$\Delta(n) = \{1\} \quad \text{and thus} \quad \Delta(S) = \{1\}.$$

Calculations for $S = \langle 7, 10, 12 \rangle$ in Example 3 require advanced techniques. From Table 2 we have that $\Delta(24) = \{1\}$ while $\Delta(42) = \{2\}$. Thus

$$\Delta(S) \supseteq \{1, 2\}.$$

Using [3, Corollary 2.3], we obtain $\max \Delta(S) \leq 2$, which yields equality.

Many basic results concerning the structure of the delta set of a numerical semigroup can be found in [3] (another paper that is the product of an REU project). The publication of [3] led to a long series of papers devoted to the study of delta sets and related properties in numerical semigroups, which approach delta sets from both theoretical and computational standpoints. In our bibliography, we offer a subset of this list of papers that include undergraduate co-authors [1, 4, 5, 7–9, 20].

Before proceeding, we will need two fundamental results. While we state these results in terms of numerical semigroups, they are actually valid for any affine semigroup (i.e., a subset $S \subset \mathbb{Z}_{\geq 0}^d$ closed under vector addition and finitely generated). We omit the proofs, but invite interested readers to construct proofs specifically for the numerical semigroup setting. The first merely establishes the finiteness of $\Delta(S)$; proofs can be found in both [3, Proposition 2.3] or [5, Theorem 2.5]. Another proof can be constructed using our still to come Theorem 5.

Proposition 1 *If S is a numerical semigroup, then $|\Delta(S)| < \infty$.*

The second result is a deeper structure theorem concerning delta sets, due to Geroldinger, and a general proof can be found in [18, Lemma 3].

Proposition 2 *If S is a numerical semigroup, then*

$$\min \Delta(S) = \gcd \Delta(S).$$

Hence, if $d = \gcd \Delta(S)$, then

$$\Delta(S) \subseteq \{d, 2d, \ldots, ad\}$$

for some $a \in \mathbb{Z}_{\geq 0}$.

Proposition 2 raises two interesting questions, both of which were addressed by the authors of [3].

- Given positive integers d and k, can one construct a numerical semigroup S with $\Delta(S) = \{d, 2d, \ldots, kd\}$?
- Must the set containment in Proposition 2 be an equality?

Prior to [3], all examples in the literature of delta sets (albeit in different settings— primarily in Krull domains and monoids) consisted of a set of consecutive multiples of a fixed positive integer d. For numerical semigroups, on the other hand, the answer to the first question is yes, but the answer to the second is no.

Proposition 3 ([3, Corollary 4.8]) *For each $n \geq 3$ and $k \geq 1$ with $\gcd(n, k) = 1$, the numerical semigroup*

$$S = \langle n, n + k, (k + 1)n - k \rangle,$$

is minimally 3-generated and

$$\Delta(S) = \left\{ k, 2k, \ldots, \left\lfloor \frac{n + k - 1}{k + 2} \right\rfloor k \right\}.$$

Hence, for any positive integers k and t, there exists a 3-generated numerical semigroup S such that $\Delta(S) = \{k, 2k, \ldots, tk\}$.

Proposition 4 ([3, Proposition 4.9]) *For each $n \geq 3$, the numerical semigroup*

$$S = \langle n, n + 1, n^2 - n - 1 \rangle,$$

is minimally 3-generated and

$$\Delta(S) = \{1, \ldots, n - 2\} \cup \{2n - 5\}.$$

The semigroups in Propositions 3 and 4 have fairly intuitive minimal generators. For instance, in Proposition 4, $n^2 - n - 1$ is the Frobenius number of $\langle n, n + 1 \rangle$, and in Proposition 3, S reduces to $\langle n, n + 1, 2n - 1 \rangle$ when $k = 1$. Note also that Proposition 4 gives a "loud" no to the second question, as it shows that one can construct as large a "gap" as desired in the set $\{d, 2d, \ldots, kd\}$. After the publication of [3], the delta sets for all numerical semigroups of embedding dimension three were determined. The interested reader can find the details in [15] and [16].

While some fairly deep results have been obtained, a good grasp on the general form for the delta set remains out of reach. We will do such a computation for arithmetical numerical semigroups (i.e., when $S = \langle a, a + d, \ldots, a + kd \rangle$ for $0 \leq k < a$ and $\gcd(a, d) = 1$). This result was originally proved in [3, Theorem 3.9], but we present a much shorter self-contained proof which later appeared in [1], another product of an REU. We begin with a lemma.

Lemma 1 ([1, Lemma 2.1]) *Let S be an arithmetical numerical semigroup with a, d, and k defined as above. If $n \in S$, then $n = c_1 a + c_2 d$ with $c_1, c_2 \in \mathbb{Z}_{\geq 0}$ and $0 \leq c_2 < a$.*

Proof Any $n \in S$ can be written in the form $c_1 a + c_2 d$ for some $c_1, c_2 \in \mathbb{Z}_{>0}$. Write $c_2 = qa + r$ with $0 \leq r < a$. Now $n = c_1 a + c_2 d = a(c_1 + qr) + rd$. □

Theorem 4 ([1, Theorem 2.2]) *If S is an arithmetical numerical semigroup with a, d, and k defined as above, $n = c_1 a + c_2 d \in S$ with $0 \leq c_2 < a$, and*

$$K = \frac{c_2 - c_1 k}{a + kd},$$

then we have

$$\mathsf{L}(n) = \{c_1 + jd : K \leq j \leq 0\}.$$

Proof Suppose $l \in \mathsf{L}(n)$. Now $la \equiv n \equiv c_1 a \pmod{d}$, and thus $\mathsf{L}(n) \subset c_1 + d\mathbb{Z}$. Writing $l = c_1 + jd$ for $j \in \mathbb{Z}$, we see

$$a(c_1 + jd) = al \leq n \leq (a + kd)l = (a + kd)(c_1 + jd),$$

so

$$K = \frac{c_2 - c_1 k}{(a + kd)} = \frac{n - c_1(a + kd)}{(a + kd)d} \leq j \leq \frac{n - c_1 a}{ad} = \frac{c_2}{a} < 1.$$

This means $\mathsf{L}(n) \subset \{c_1 + jd : K \leq j \leq 0\}$.

It remains to locate a factorization of length $c_1 + jd$ for each $j \in \mathbb{Z}$ with $K \leq j \leq 0$. Write $c_2 - j = qk + r$ for $q, r \in \mathbb{Z}$ with $0 \leq r < k$. We have

$$n = a(c_1 + jd) + d(c_2 - j) = a(q + 1 + c_1 + jd - 1 - q) + d(qk + r)$$
$$= q(a + kd) + (a + rd) + (c_1 + jd - 1 - q)a,$$

which is a factorization of n of length $c_1 + jd$. Thus $c_1 + jd \in \mathsf{L}(n)$, as desired.   □

An obvious corollary to this theorem follows.

Corollary 2 *If S is an arithmetical numerical semigroup and a, d, and k are as defined above, then $\Delta(S) = \{d\}$. Moreover,*

- *if $n_2 > n_1 > 1$ are relatively prime integers, then $\Delta(\langle n_1, n_2 \rangle) = \{n_2 - n_1\}$*
- *if $n > 1$ and k are integers with $1 \leq k \leq n - 1$, then for $S = \langle n, n+1, \ldots, n+k \rangle$ we have that $\Delta(S) = \{1\}$.*

Additionally, Theorem 4 can be used to show that $\mathscr{L}(S)$ is not a perfect invariant, that is, one cannot in general recover a given numerical semigroup S from $\mathscr{L}(S)$.

Challenge Problem 1 Use Theorem 4 to find two numerical semigroups S_1 and S_2 so that $\mathscr{L}(S_1) = \mathscr{L}(S_2)$ but $S_1 \neq S_2$.

We close this section with another REU related result that appears in [7]. Writing the elements of a numerical semigroup S in order as $s_1, s_2, \ldots$, where $s_i < s_{i+1}$ for all $i \geq 1$, we now consider the sequence of sets

$$\Delta(s_1), \Delta(s_2), \Delta(s_3), \ldots$$

In the case where S is arithmetical, then for large i this sequence is comprised solely of $\{k\}$, which is not too interesting. Using Table 2, one can construct the beginning of this sequence for $S = \langle 7, 10, 12 \rangle$:

$$\emptyset, \emptyset, \ldots, \emptyset, \{1\}, \emptyset, \emptyset, \emptyset, \emptyset, \emptyset, \{1\}, \emptyset, \emptyset, \{1\}, \emptyset, \{1\}, \emptyset, \{1\}, \emptyset, \{1\}, \{1\}, \{2\}, \{1\}, \ldots$$

While the beginning behavior of these sequences is in some sense "chaotic," in the long run, they are much more well behaved. This can be better demonstrated with some graphs. Figure 2 represents the sequence of delta sets for $S = \langle 7, 10, 12 \rangle$, while Fig. 3 does so for the Chicken McNugget semigroup. On these graphs, a point is plotted at (n, d) if $d \in \Delta(n)$.

Using data such as the above, it was conjectured shortly after the publication of [3] that this sequence of sets is eventually periodic. Three years later, this problem was solved, again as part of an REU project.

Theorem 5 ([7, Theorem 1]) *Given a numerical semigroup $S = \langle n_1, \ldots, n_k \rangle$ with $n_1 < n_2 < \cdots < n_k$ and $N = 2kn_2 n_k^2 + n_1 n_k$, we have*

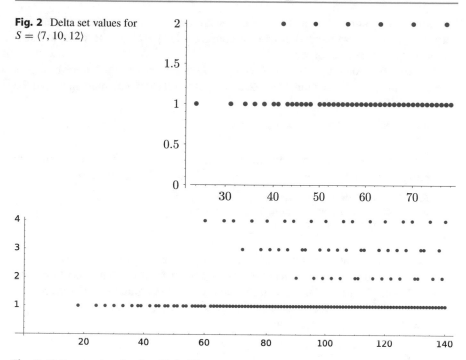

Fig. 2 Delta set values for $S = \langle 7, 10, 12 \rangle$

Fig. 3 Delta set values for $S = \langle 6, 9, 20 \rangle$

$$\Delta(n) = \Delta(n - n_1 n_k)$$

for every $n \geq N$. Hence,

$$\Delta(S) = \bigcup_{n \in S, \, n < N} \Delta(n).$$

The importance of Theorem 5 cannot be overstated, as it turns the problem of computing $\Delta(S)$ into a finite time exercise. The bound N given in Theorem 5 has been drastically improved in [14] (Table 1 in that paper shows exactly how drastic this improvement is). An alternate view of the computation of $\Delta(S)$ using the *Betti numbers* of S can be found in [5], which is also an REU product.

3 Using Software to Guide Mathematical Inquisition

One of the most reliable tools when working with numerical semigroups is computer software. We will give an overview of using the GAP package numericalsgps. GAP (Groups, Algorithms, Programming) is a computer algebra system used in a variety of discrete mathematical areas, and numericalsgps is a package for working specifically with numerical semigroups, including over 400 preprogrammed functions to compute numerous invariants and properties of numerical semigroups. Full documentation can be found in [11] and on the GAP website. https://www.gap-system.org/.

We begin by providing a brief overview of the functionality related to the topics covered in the previous section. Once GAP is up and running, you must first load the numericalsgps package.

```
gap> LoadPackage("numericalsgps");
true
```

Once this is done, you can begin to compute information about the numerical semigroups you are interested in examining, such as the Frobenius number.

```
gap> McN:= NumericalSemigroup(6,9,20);
<Numerical semigroup with 3 generators>
gap> FrobeniusNumber(McN);
43
```

Many of the quantities the numericalsgps package can compute center around factorizations and their lengths. Given how central the functions that compute $Z(n)$ and $L(n)$ are, these functions have undergone numerous improvements since the early days of the numericalsgps package, and now run surprisingly fast even for reasonably large input.

```
gap> Factorizations(50, McN);
[ [ 5, 0, 1 ], [ 2, 2, 1 ] ]
gap> Factorizations(60, McN);
[ [ 10, 0, 0 ], [ 7, 2, 0 ], [ 4, 4, 0 ],
  [ 1, 6, 0 ], [ 0, 0, 3 ] ]
gap> LengthsOfFactorizationsElementWRTNumericalSemigroup(60,
                                                          McN);
[ 3, 7, 8, 9, 10 ]
gap> LengthsOfFactorizationsElementWRTNumericalSemigroup(150,
                                                          McN);
[ 10, 11, 13, 14, 15, 16, 17, 18, 19, 20, 21, 22, 23, 24, 25 ]
```

The numericalsgps package can also compute delta sets, both of numerical semigroups and of their elements. The original implementation of the latter function used Theorem 5 to compute the delta set of every element up to N, and only more recently was a more direct algorithm developed [17].

```
gap> DeltaSet(60, McN);
[ 1, 4 ]
gap> DeltaSetOfNumericalSemigroup(McN);
[ 1, 2, 3, 4 ]
```

One of the primary goals is to use these observations to formulate meaningful conjectures among these concepts, and even to aide in the development of a proof. In what follows, we hope to give a better sense of this process by walking through a specific example.

Suppose we decide to study maximum factorization length. We begin by using the numericalsgps package to compute maximum factorization lengths for elements of $S = \langle 7, 10, 12 \rangle$ from Example 3.

```
gap> S := NumericalSemigroup(7,10,12);
<Numerical semigroup with 3 generators>
gap> Factorizations(60,S);
[ [ 0, 6, 0 ], [ 4, 2, 1 ], [ 2, 1, 3 ], [ 0, 0, 5 ] ]
gap> Maximum(
> LengthsOfFactorizationsElementWRTNumericalSemigroup(60,S));
7
```

Using built-in GAP functions, maximum factorization length can be computed for several semigroup elements in one go. We first compute a list of the initial elements of S (this avoids an error message when attempting to compute $Z(n)$ when $n \notin S$).

```
gap> elements := Intersection([1..60], S);
[ 7, 10, 12, 14, 17, 19, 20, 21, 22, 24, 26, 27, 28, 29, 30,31,
  32, 33, 34, 35, 36, 37, 38, 39, 40, 41, 42, 43, 44, 45, 46,47,
  48, 49, 50, 51, 52, 53, 54, 55, 56, 57, 58, 59, 60 ]
```

Next, we compute the values of $L(n)$ for semigroup elements $n \le 100$.

```
gap> List(elements, n -> Maximum(
> LengthsOfFactorizationsElementWRTNumericalSemigroup(n,S)));
[ 1, 1, 1, 2, 2, 2, 2, 3, 2, 3, 3, 3, 4, 3, 3, 4, 3, 4, 4, 5,
  4, 4, 5, 4, 5, 5, 6, 5, 5, 6, 5, 6, 6, 7, 6, 6, 7, 6, 7, 7,
  8, 7, 7, 8, 7 ]
```

Well-chosen plots can be an incredibly effective tool for visualizing such data. Figure 4 depicts the values output above; the repeating pattern in the right half of the plot is undeniable. This is what we call a *quasilinear* function, that is, a linear function with periodic coefficients. For our particular S and $n \ge 26$, we have

$$L(n) = \tfrac{1}{7}n + a(n),$$

where $a(n)$ is a periodic function with period 7 (for instance, $a(n) = -\frac{11}{7}$ whenever $n \equiv 4 \bmod 7$, so that $L(60) = \frac{1}{7}(60) - \frac{11}{7} = 7$). The resulting plot resembles 7 parallel lines, each with slope $\frac{1}{7}$. Another way to express a quasilinear function with constant linear coefficient is via constant successive differences, i.e.,

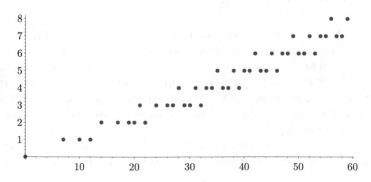

Fig. 4 Maximum factorization length values for elements of $S = \langle 7, 10, 12 \rangle$

$$L(n + 7) - L(n) = 1$$

for all $n \geq 26$. Notice that this is precisely the relation claimed in Theorem 1.

In addition to elucidating the quasilinear pattern, computations can also be used to help us work towards a proof. We begin by asking numericalsgps to compute the factorizations of maximum length for $n = 44$, 51, and 58 (each 7 apart).

```
gap> LengthsOfFactorizationsElementWRTNumericalSemigroup(44,S);
[ 4, 5 ]
gap> LengthsOfFactorizationsElementWRTNumericalSemigroup(51,S);
[ 5, 6 ]
gap> LengthsOfFactorizationsElementWRTNumericalSemigroup(58,S);
[ 5, 6, 7 ]
gap> Filtered(Factorizations(44,S),
>                    f -> (Sum(f)=5));
[ [ 2, 3, 0 ] ]
gap> Filtered(Factorizations(51,S),
>                    f -> (Sum(f)=6));
[ [ 3, 3, 0 ] ]
gap> Filtered(Factorizations(58,S),
>                    f -> (Sum(f)=7));
[ [ 4, 3, 0 ] ]
```

Notice the only change in the factorizations is the first coordinate, which increases by exactly 1 each time the element n increases by exactly 7. Intuitively, this is because longer factorizations should use more small generators. This identifies where the period of 7 and the leading coefficient of $\frac{1}{7}$ originate. However, this does not yet explain why the quasilinear pattern does not begin until $n = 26$. After testing our conjecture on several more numerical semigroups, we come across an example that provides some insight behind this final piece of the puzzle.

```
gap> S2 := NumericalSemigroup(9,10,21);
<Numerical semigroup with 3 generators>
gap> Factorizations(41,S2);
[ [ 0, 2, 1 ] ]
gap> Factorizations(50,S2);
[ [ 0, 5, 0 ], [ 1, 2, 1 ] ]
gap> Factorizations(59,S2);
[ [ 1, 5, 0 ], [ 2, 2, 1 ] ]
```

Here, we see the longest factorization of $n = 50$ in $S_2 = \langle 9, 10, 21 \rangle$ does not use any copies of the smallest generator. As it turns out, this phenomenon can only happen for small semigroup elements, as once n is large enough, any factorization with no copies of the smallest generator can be "traded" for a longer factorization that does. This highlights the key to proving Theorem 1: determining how large n must be to ensure all of its factorizations of maximal length have at least one copy of the smallest generator.

We invite the reader to use the ideas discussed above to obtain a rigorous proof of Theorem 1 (indeed, the proof appearing in [2] utilizes these ideas).

Challenge Problem 2 Prove Theorem 1.

4   Research Projects: Asymptotics of Factorizations

The length of a factorization coincides with the ℓ_1-norm of the corresponding point. Much like Theorem 1, several other norms that arise in discrete optimization appear to have EQP behavior. The following was observed during the 2017 San Diego State University Mathematics REU and motivates the research project that follows.

In what follows, for $\mathbf{a} \in \mathbb{Z}^k$ and $r \in \mathbb{Z}_{\geq 1}$, let

$$\|\mathbf{a}\|_r = (a_1^r + \cdots + a_k^r)^{1/r}$$

and

$$\|\mathbf{a}\|_\infty = \max(a_1, \ldots, a_k),$$

which are known as the ℓ_r- and ℓ_∞-norm, respectively.

Challenge Problem 3 Let $S = \langle n_1, \ldots, n_k \rangle$. Prove the function

$$\ell_\infty(n) = \min\{\|\mathbf{a}\|_\infty : \mathbf{a} \in \mathsf{Z}_S(n)\}$$

is eventually quasilinear with period $n_1 + \cdots + n_k$.

Research Project 1 Determine for which fixed $r \in [2, \infty)$ the functions

$$\mathsf{M}_r(n) = \max\{(\|\mathbf{a}\|_r)^r : \mathbf{a} \in \mathsf{Z}_S(n)\}$$

and

$$\mathsf{m}_r(n) = \min\{(\|\mathbf{a}\|_r)^r : \mathbf{a} \in \mathsf{Z}_S(n)\}$$

are eventually quasipolynomial.

Given a numerical semigroup $S = \langle n_1, \ldots, n_k \rangle$ with $n_1 \leq \cdots \leq n_k$, one can define

$$N(\ell) = |\{n \in S : \ell \in \mathsf{L}(n)\}|,$$

which counts the number of elements of S with a given length ℓ in their length set. Unlike many functions discussed above, which take semigroup elements as input, N takes factorization lengths as input. Since each semigroup element counted by $N(\ell)$ must lie between $n_1\ell$ and $n_k\ell$, we see that

$$N(\ell) \le (n_k - n_1)\ell,$$

so $N(\ell)$ grows at most linearly in ℓ. This yields the following natural question.

Research Project 2 Fix a numerical semigroup S. Determine whether

$$N(\ell) = |\{n \in S : \ell \in L(n)\}|$$

is eventually quasilinear in $\ell \ge 0$.

One of the running themes of results in the numerical semigroups literature is that the factorization structure is "chaotic" for small elements, but "stabilizes" for large elements. Typically, the latter is easier to describe, as evidenced by the word "eventually" in several of the results presented above. Broadly speaking, it would be interesting to determine how much of a numerical semigroup's structure can be recovered from that of its "large" elements. The following project is an initial step in this direction, and at its heart is the question "does the eventual behavior of a given factorization invariant uniquely determine its behavior for the whole semigroup?"

Research Project 3 Given a numerical semigroup $S = \langle n_1, \ldots, n_k \rangle$ satisfying $n_1 < \cdots < n_k$, Theorem 1 implies

$$\mathsf{M}_S(n) = \tfrac{1}{n_1} n + a_S(n) \qquad \text{and} \qquad \mathsf{m}_S(n) = \tfrac{1}{n_k} n + b_S(n)$$

for some periodic functions $a_S(n)$ and $b_S(n)$. Characterize the functions $a_S(n)$ and $b_S(n)$ in terms of the generators of S. For distinct numerical semigroups S and T, is it possible that $a_S(n) = a_T(n)$ or $b_S(n) = b_T(n)$ for all n?

Most of the invariants introduced thus far (and indeed, most in the literature) are derived from "extremal" factorizations. In a recent REU project, "medium" factorization lengths were studied. More precisely, the *length multiset*

$$\mathsf{L}_S[\![n]\!] = \{\{|\mathbf{a}| : \mathbf{a} \in Z_S(n)\}\}$$

was defined, wherein factorization lengths are considered with repetition, and the following quantities were considered:

- $\mu_S(n)$, the *mean* of the elements of $\mathsf{L}_S[\![n]\!]$ and
- $\eta_S(n)$, the *median* of the elements of $\mathsf{L}_S[\![n]\!]$.

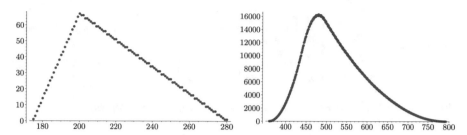

Fig. 5 Histograms of the length multiset $L_S[\![1400]\!]$ for $S = \langle 5, 7, 8\rangle$ (left) and the length multiset $L_S[\![3960]\!]$ for $S = \langle 5, 8, 9, 11\rangle$ (right)

Notice that $L_S[\![n]\!]$ has the same cardinality as $Z_S(n)$, since factorization lengths are counted with repetition in $L_S[\![n]\!]$.

Example 8 The length multiset of $n = 1400$ in $S = \langle 5, 7, 8\rangle$ is depicted in Fig. 5, wherein a point at (ℓ, m) indicates the length ℓ appears exactly m times in $L[\![n]\!]$. The lengths in $L[\![n]\!]$ range from 175 to 280 (as predicted by Theorem 2), and the mode length(s) occur around $n/7$ (note that 7 is the middle generator of S). In this case, as $n \to \infty$, the histogram approaches a *triangular distribution*. The second histogram in Fig. 5 is for the length multiset of an element of $S = \langle 5, 8, 9, 11\rangle$ and has a visually different shape. Indeed, the limiting distribution of the length multiset is only triangular for 3-generated numerical semigroups.

The following result will appear in a forthcoming paper, and implies that although $\mu(n)$ is not itself (eventually) quasipolynomial, it can be expressed in terms of quasipolynomial functions.

Theorem 6 *Fix a numerical semigroup $S = \langle n_1, \ldots, n_k\rangle$. The function $\mu(n)$ equals the quotient of two quasipolynomial functions, and*

$$\lim_{n\to\infty} \frac{\mu_S(n)}{n} = \frac{1}{k}\left(\frac{1}{n_1} + \cdots + \frac{1}{n_k}\right).$$

Median factorization length has proven more difficult to describe in general. The limiting distribution of $L[\![n]\!]$ is characterized for 3-generated numerical semigroups in [13], yielding the following theorem regarding the asymptotic growth rate of median factorization length in this case.

Theorem 7 *Fix a numerical semigroup $S = \langle n_1, n_2, n_3\rangle$, and let*

$$F = \frac{n_1(n_3 - n_2)}{n_2(n_3 - n_1)}$$

(called the fulcrum constant*). We have*

$$\lim_{n\to\infty} \frac{\eta_S(n)}{n} = \begin{cases} \dfrac{1}{n_1}\left(1-\sqrt{\dfrac{1-F}{2}}\right) + \dfrac{1}{n_3}\sqrt{\dfrac{1-F}{2}} & \text{if } F \le \tfrac{1}{2}, \\[3ex] \dfrac{1}{n_1}\sqrt{\dfrac{F}{2}} + \dfrac{1}{n_3}\left(1-\sqrt{\dfrac{F}{2}}\right) & \text{if } F \ge \tfrac{1}{2}, \end{cases}$$

the value of which is irrational for some, but not all, numerical semigroups.

Theorem 7 is a stark contrast to many of the invariants discussed above, since if the limit therein is irrational, then $\eta(n)$ cannot possibly coincide with a quasipolynomial for large n (indeed, this follows from the fact that any linear function sending at least 2 rational inputs to rational outputs must have rational coefficients). As such, studying the asymptotic behavior of median factorization length requires different techniques than previously studied invariants.

As the histograms in Fig. 5 demonstrate, the limiting distribution of the length multiset for 3-generated numerical semigroup elements differs drastically from semigroups with more generators. Students interested in the following project are encouraged to begin by reading [13], wherein the limiting distribution is carefully worked out in the 3-generated case.

Research Project 4 Fix a numerical semigroup S. Find a formula for

$$\lim_{n\to\infty} \frac{\eta(n)}{n},$$

the asymptotic growth rate of the median factorization length of n.

5   Research Projects: Random Numerical Semigroups

Suppose someone walks up to you on the street and hands you a "random" numerical semigroup. What do we expect it to look like? Is it more likely to have a lot of minimal generators, or only a few? How large do we expect its Frobenius number to be? How many gaps do we expect it to have? Such questions of "average" or "expected" behavior arise frequently in discrete mathematics, and often utilize tools from probability and real analysis that are otherwise uncommon in discrete settings.

The general strategy is to define a *random model* that selects a mathematical object "at random," and then determine the probability that the chosen object has a particular property. One prototypical example comes from graph theory: given a fixed integer n and probability p, select a random graph G on n vertices by deciding, with independent probability p, whether to draw an edge between each pair of vertices v_1 and v_2. A natural question to ask is "what is the probability G is connected?" (note that the larger p is, the more edges one expects to draw, and thus

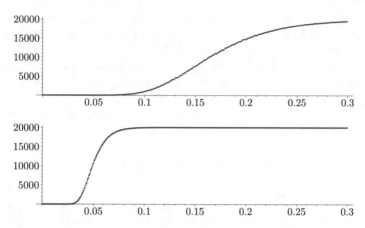

Fig. 6 Scatterplots recording the number of connected graphs sampled as a function of p for $n = 20$ and $n = 100$, respectively. Sample size is 20,000 for each plotted value of p

the higher chance the resulting graph is connected). This is a difficult question to obtain an exact answer for, although estimates can be obtained for small n (with the help of computer software) since there are only finitely many graphs with n vertices. That said, it turns out that for very large n, there is an $\epsilon > 0$ so that

- if $p < \log(n)/n - \epsilon$, then G has low probability of being connected and
- if $p > \log(n)/n + \epsilon$, then G has high probability of being connected,

where $\epsilon \to 0$ as $n \to \infty$. Here, the phrases "low probability" and "high probability" mean probability tending to 0 and 1, respectively, as $n \to \infty$. This kind of bifurcation (illustrated in Fig. 6 for varying values of n) is a phenomenon known as a *threshold function*, and occurs frequently when answering probabilistic questions in discrete mathematics.

The authors of [10] introduce a model of selecting a numerical semigroup at random that is similar to the above model for random graphs. Their model takes two inputs $M \in \mathbb{Z}_{\geq 1}$ and $p \in [0, 1]$, and randomly selects a numerical semigroup by selecting a generating set A that includes each integer $n = 1, 2, \ldots, M$ with independent probability p. For example, if $M = 40$ and $p = 0.1$, then one possible set is $A = \{6, 9, 18, 20, 32\}$ (this is not unreasonable, as one would expect 4 to be selected on average). However, only 3 elements of A are minimal generators, since $18 = 9 + 9$ and $32 = 20 + 6 + 6$. As such, the selected semigroup $S = \langle A \rangle = \langle 6, 9, 20 \rangle$ has embedding dimension 3.

Regarding the expected properties of numerical semigroups selected in this way, two main results are proven in [10]. First, the threshold function for whether or not the resulting numerical semigroup S has finite complement is proven to be $1/M$. More precisely, for each large M, there exists $\epsilon > 0$ (with $\epsilon \to 0$ as $M \to \infty$) so

- if $p < 1/M - \epsilon$, then $\mathbb{Z}_{\geq 0} \setminus S$ is finite with low probability and
- if $p > 1/M + \epsilon$, then $\mathbb{Z}_{\geq 0} \setminus S$ is finite with high probability.

One expects $|A| = 1$ on average when $p = 1/M$, meaning for large M, the selected numerical semigroup is most likely to either have finite complement (if $p > 1/M$) or equal the trivial semigroup $\{0\}$ (if $p < 1/M$).

The remaining results in [10] provide lower and upper bounds on the expected number of minimal generators of the selected semigroup S. More precisely, a formula is obtained for $\mathbb{E}[e(S)]$ in terms of p and M (though it is computationally infeasible for large M), and derives from it lower and upper bounds on $\lim_{M \to \infty} \mathbb{E}[e(S)]$ for fixed p. It is also shown that

$$\lim_{M \to \infty} \mathbb{E}[e(S)] = \frac{p}{1 - p} \lim_{M \to \infty} \mathbb{E}[g(S)],$$

thereby providing lower and upper bounds on the expected number of gaps as well.

Asymptotic estimates of this nature can be useful, for instance, in testing conjectures in semigroup theory. Suppose a researcher has a conjecture regarding numerical semigroups with exactly 150 gaps. They could test their conjecture on a small number of "larger" numerical semigroups selected using a random model, choosing the parameters so as to maximize the chances of selecting a numerical semigroup with 150 gaps.

This random model is just one of many possible models for randomly selecting numerical semigroups, and using different models is likely to yield different expected behavior for the resulting semigroups. Given below are some alternative models that have yet to be explored. The first adds a new parameter to the existing model, namely the multiplicity of the semigroup, yielding more control over which semigroups are selected. The second selects *oversemigroups* (that is, semigroups containing a given semigroup) instead of generators, and takes their intersection.

Research Project 5 Study random numerical semigroups selected using the following model: given $M, m \in \mathbb{Z}_{\geq 1}$ and $p \in [0, 1]$, select the semigroup

$$S = \langle \{m\} \cup A \rangle \cup ([M + 1, \infty) \cap \mathbb{Z})$$

by selecting a random subset $A \subset [m + 1, M] \cap \mathbb{Z}$ that includes each integer the original model discussed above.

Research Project 6 Study random numerical semigroups selected using the following model: given $N \in \mathbb{Z}_{\geq 1}$ and $p \in [0, 1]$, select the semigroup

$$S = \bigcap_{2 \leq a < b \leq N} \langle a, b \rangle,$$

(continued)

> where each numerical semigroup $\langle a, b \rangle$ with $\gcd(a, b) = 1$ is included in the intersection with independent probability p.

For each of these projects, a natural starting place would be to use computer software to produce a large sample of numerical semigroups and compute the average number of minimal generators, Frobenius number, etc. as estimates of their expected value for varying choices of the parameters.

Acknowledgements The authors would like to thank an unknown referee for detailed comments that greatly improved this manuscript. They would also thank Nathan Kaplan for his discussions and input. All plots created using Sage [21], and all computations involving numerical semigroups utilize the GAP package numericalsgps [12].

References

1. Amos, J., Chapman, S., Hine, N., Paixao, J.: Sets of lengths do not characterize numerical monoids. Integers, 7(1): Paper-A50 (2007).
2. Barron, T., O'Neill, C., Pelayo, R.: On the set of elasticities in numerical monoids. Semigroup Forum **94**: 37–50 (2017). Available at arXiv:math.CO/1409.3425.
3. Bowles, C., Chapman, S., Kaplan, N., Reiser, D.: On delta sets of numerical monoids. J. Algebra Appl. **5**: 695–718 (2006).
4. Chapman, S., Daigle, J., Hoyer, R., Kaplan, N.: Delta sets of numerical monoids using nonminimal sets of generators. Comm. Algebra. **38**: 2622–2634 (2010).
5. Chapman, S., García-Sánchez, P., Llena, D., Malyshev, A., Steinberg, D.: On the delta set and the Betti elements of a BF-monoid. Arabian Journal of Mathematics. **1**: 53–61 (2012).
6. Chapman, S., Holden, M., Moore, T.: Full elasticity in atomic monoids and integral domains. The Rocky Mountain Journal of Mathematics. **36**:1437–1455 (2006).
7. Chapman, S., Hoyer, R., Kaplan, N.: Delta sets of numerical monoids are eventually periodic. Aequationes mathematicae. **77**: 273–279 (2009).
8. Chapman, S., Kaplan, N., Lemburg, T., Niles, A., Zlogar, C.: Shifts of generators and delta sets of numerical monoids. International Journal of Algebra and Computation. **24**: 655–669 (2014).
9. Chapman, S., O'Neill, C.: Factorization in the Chicken McNugget Monoid. Math. Mag. **91**(5): 323–336 (2018).
10. De Loera, J., O'Neill, C., Wilbourne, D.: Random numerical semigroups and a simplicial complex of irreducible semigroups. Electronic Journal of Combinatorics. **25**: #P4.37 (2018).
11. Delgado, M., García-Sánchez, P.: numericalsgps, a GAP package for numerical semigroups. ACM Commun. Comput. Algebra 50 (2016), no. 1, 12–24.
12. Delgado, M., García-Sánchez, P., Morais, J.: NumericalSgps, A package for numerical semigroups, Version 1.2.0 (2019), (Refereed GAP package), https://gap-packages.github.io/numericalsgps.

13. García, S., O'Neill, C., Yih, S.: Factorization length distribution for affine semigroups I: numerical semigroups with three generators. to appear, European Journal of Combinatorics. Available at arXiv:1804.05135.
14. García-García, J., Moreno-Frías, M., Vigneron-Tenorio, A.: Computation of delta sets of numerical monoids. Monatshefte für Mathematik. **178**: 457–472 (2015).
15. García-Sánchez, P., Llena, D., Moscariello, A.: Delta sets for symmetric numerical semigroups with embedding dimension three. Aequationes Math. **91**(2017), 579–600.
16. García-Sánchez, P., Llena, D., Moscariello, A.: Delta sets for nonsymmetric numerical semigroups with embedding dimension three. Forum Math. **30**(2018), 15–30.
17. García-Sánchez, P., O'Neill, C., Webb, G.: On the computation of factorization invariants for affine semigroups. Journal of Algebra and its Applications. **18**: 1950019, 21 pp (2019).
18. Geroldinger, A.: On the arithmetic of certain no integrally closed noetherian integral domains. Comm. Algebra. **19**: 685–698 (1991).
19. Glen, J., O'Neill, C., Ponomarenko, V., Sepanski, B.: Augmented Hilbert series of numerical semigroups. arXiv:1806.11148.
20. Colton, S., Kaplan, N.: The realization problem for delta sets of numerical semigroups. Journal of Commutative Algebra. **9**: 313–339 (2017).
21. The Sage Developers: SageMath, the Sage Mathematics Software System (Version 7.2). (2016) http://www.sagemath.org.
22. Sylvester, J.: On subinvariants, i.e. Semi-Invariants to Binary Quantics of an Unlimited Order. American Journal of Mathematics. **5**: 134 (1882).

Lateral Movement in Undergraduate Research: Case Studies in Number Theory

Stephan Ramon Garcia

Abstract

We explore the thought processes, strategies, and pitfalls involved in entering new territory, developing novel projects, and seeing them through to publication. We propose twenty-one general principles for developing a sustainable undergraduate research pipeline and we illustrate those ideas in three case studies.

Suggested Prerequisites Number theory

1   Introduction

This paper is an account of how an operator theorist began supervising undergraduate research projects in number theory, a field far removed from his original research area. We explore the thought processes, strategies, and pitfalls involved in entering new territory (with no formal training), developing novel projects with students, and seeing them through to publication. We propose twenty-one general principles (Sect. 2) for developing a sustainable undergraduate research pipeline and we illustrate those ideas in action through three case studies (Sects. 3–5). We hope that the following account will be accessible enough for readers to translate the author's experiences and ideas into their own personal domains.

S. R. Garcia (✉)
Pomona College, Claremont, CA, USA
e-mail: stephan.garcia@pomona.edu; http://pages.pomona.edu/~sg064747

© Springer Nature Switzerland AG 2020   203
P. E. Harris et al. (eds.), *A Project-Based Guide to Undergraduate Research in Mathematics*, Foundations for Undergraduate Research in Mathematics,
https://doi.org/10.1007/978-3-030-37853-0_7

The author was trained in function-related operator theory, the study of certain operators on Banach spaces of holomorphic functions. The questions studied in that field require graduate-level analysis, complex analysis, and functional analysis to state. Although the author had supervised a few undergraduate-research projects in operator theory and matrix analysis [5, 18, 19, 38–43, 45], he felt the need to branch out in order to meet the demands of his department's mandatory senior thesis requirement (Pomona College is an elite liberal arts college at which around 10% of the student body majors in mathematics). Given the background of the typical senior thesis student (no exposure to complex variables or advanced linear algebra and only a first course in analysis), dabbling in number theory made sense. Questions in number theory are often less abstruse and more easily stated than in other areas and they are sometimes amenable to computation. Both of these properties were crucial for developing new research threads capable of supporting a pipeline of undergraduate research students.

We track the development of three independent research threads in number theory. Each program consists of several interrelated papers written by the author and his collaborators (most of whom were students at the time) over a period of several years. A simple flowchart accompanies each program (Fig. 1).

A few important disclaimers are required before we begin.

- We do not provide the complete details of every individual project, although we do this to the extent necessary to illustrate our guiding principles.
- The author is a pure mathematician and his experiences might not translate perfectly outside of this sphere. Nevertheless, we hope that the guiding principles set forth below have some universal value for readers in the mathematical sciences.

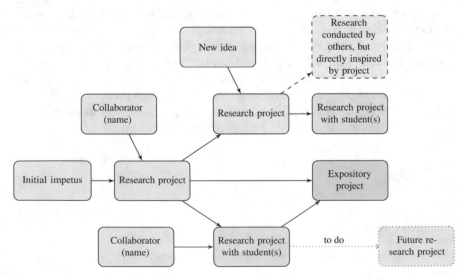

Fig. 1 Sample project flowchart. Time flows approximately from left to right. Initial problems or external inflows of knowledge are in yellow. Collaborators coming on board are in purple

- Most of the projects discussed stem from academic-year research, often as part of a senior exercise. Readers focused on summer-research programs with short timeframes (but perhaps more time to work on mathematics) may encounter different issues than those considered by the author.
- The author's philosophy and goals might not correlate exactly with those of the reader. The author has focused on producing papers for publication in respectable journals. However, there are other markers of success in undergraduate research. These range from poster presentations and technical reports to retention and recruitment of students from historically underrepresented groups. Research experience can also help prepare students for a variety of careers.
- There is a human dimension to undergraduate research. The author is not qualified to provide advice upon how to handle psychological or emotional adversity. This article focuses exclusively on the mathematical side of the equation: how, as a supervisor of undergraduate research, you can direct the student to fruitful research topics and shepherd subsequent results through to publication.
- The following case studies concern the personal recollections and experiences of the author. Consequently, we often lapse into the first person ("I"). We trust that the reader will not find the overall tone too conversational or informal.
- Although it is neither egalitarian nor consistent, I will often refer to students by their first names and to professional mathematicians by their last names.

With these caveats and disclaimers out of the way, we may proceed to our guiding principles (Sect. 2). These are general precepts suggested by the author's personal experience and reflections. The three case studies (Sects. 3–5) that follow it are written in a loose narrative format that we hope is amenable to the illustration of our principles while also remaining true to the evolution and chronology of events.

2   Guiding Principles

The author has not significantly reflected, until the present moment, on his precise strategy for supervising undergraduate research. His philosophy grew organically from long experience mingled with trial and error. We would like to take the opportunity to share what has been distilled from our reflections on this endeavor.

1. **Time is a luxury you don't have.** A Ph.D. student trains for years in a specific area assigned by the advisor, a leading expert on the subject. A supervisor of undergraduate research is not so fortunate. Their preferred subject matter might be unrealistic for a student to grasp in the time allotted. You must be flexible.
2. **You are not an old dog.** You can learn new tricks. Graduate school is not the only time to learn new math. Be open to new ideas and areas. Attend workshops, seminars, and colloquia. You might get new ideas or meet new collaborators.
3. **You know more than they do.** You have an advanced degree in the mathematical sciences. With your training and experience, you can stay a few steps ahead

of your students, even in an area that you are not an expert in. Your ideas and perspectives may be useful in a new area.

4. **You can be human.** You do not have to be an expert in the subject that you and your students are investigating. This shows students how a mathematician learns new mathematics. We are human and it is not bad for students to see that.

5. **Follow their passions.** Students have more energy and enthusiasm for projects they love than for projects they view as tedium. Let them work on projects they want to work on, rather than doggedly assigning them a single, fixed problem.

6. **Search for fertile ground.** Find a topic that has not been combed over by experts. Students need fertile ground that spawns low-hanging fruit. Competing against generations of dedicated experts is difficult. Find your own niche and master it.

7. **Your students are not Andrew Wiles.** Students lack our experience and instinct. They cannot tell which problems are too difficult and which are realistic for them. Be prepared to pivot and shift to more tractable problems if necessary.

8. **Focus more broadly.**[1] Have your students skim through a few papers in the area. Ask them to return with a list of ten or twenty questions, perhaps only vaguely related to the subject at hand. Even if only 10% of them pan out, it is a success.

9. **Everything is negotiable.** No question is set in stone. Feel free to change the context or the hypotheses. Turn your problems around and invent new variations. You only need traction on one problem before the results start pouring in.

10. **Complement your research.** Undergraduate research need not be distinct from "real" research. Your problems might have versions suitable for students. Conversely, student projects may suggest new problems for your own research.

11. **The computer is your friend.** A few moments on the computer can save a lot of pencil-and-paper time. If counterexamples exist, the computer might find them. Symbolic manipulation takes care of tedious algebra. Moreover, computational tasks can get students started on a project almost immediately.

12. **Build upon previous success.** An apparently straightforward generalization may lead to unexpected results. Usually some new complications turn up that make things less transparent and more difficult. Look at this as good news: complications make for more interesting follow-up projects.

13. **Turn lemons into lemonade.** A disproved conjecture or a failed proof does not mean that the situation is unsalvageable. Perhaps the counterexamples are more interesting than the conjecture itself. Perhaps the proof broke down in an interesting way. Always seek ways to turn a negative into a positive, failure into success.

14. **Feel free to hand wave.** Heuristic arguments and informal reasoning can lead to better student intuition. A skeleton of an argument can be fleshed out later.

---

[1] A grant reviewer once admonished a colleague to "focus more broadly." Although oxymoronic and unintentionally humorous, this feels like the appropriate phrase.

Perhaps you are close to the finish line but don't know it. A colleague or collaborator may see how to turn an informal argument into a rigorous one.

15. **You don't have to go it alone!** Mathematicians love to talk about their work. I "met" many of my frequent collaborators via e-mail. Most mathematicians appreciate it when someone asks about one of their papers. Feel free to write to someone out of the blue if their work is relevant. In the worst case, they won't answer. In the best case, you might learn something or gain a future collaborator.

16. **Know your audience.** Who works on problems like yours? What journals do they publish in? If your paper is more of a curiosity or an oddball result, the chances of acceptance are lower. Relate your work to the existing literature.

17. **Do not drag your feet.** Make sure that the paper gets written, posted on the arXiv, and submitted to a journal. Do not sit on it for months. Graduate-school applications may hinge on whether or not a publication comes through in time.

18. **Is there an opportunity for exposition?** Have you run into a topic for which key results are strewn throughout dozens of obscure articles? There may be an opportunity to write a much-appreciated survey article (or monograph) on the subject. These can be excellent opportunities for student research since a proper exposition may require detailed examples, fully flesh-out proofs, computer investigations, and so forth. Each piece might provide a small project for a student.

19. **Modularity principle.** An instructive example might not be publishable, nor might be a minor improvement on a known result. But both might form a "module" in a larger work. A paper need not be the product of a single undergraduate research collaboration. It may take several years and a few "generations" of students before enough results can be assembled into a compelling research paper.

20. **Reach out to new communities.** If projects pull you in new directions, make an effort to meet people in the area. Attend local seminars and meet some experts. Sign up for talks in appropriate special sessions and get your name out there.

21. **Getting recognized.** Perhaps your students published in a good journal. Maybe a student won a prize? If there is a compelling story to tell, your institution's communications staff wants to know. They are always on the lookout for good stories that highlight student research. Get to know your communications staff.

The author's personal experiences as an undergraduate-research supervisor have been distilled and abstracted, to the extent possible, into the preceding list. We hope that the following three case studies illustrate these principles in action.

3   Case Study I: Quotient Sets

Our first case study begins with a senior thesis that set the author on the path to number theory. This research line eventually involved a dozen students and several mathematicians from across the globe (Fig. 2).

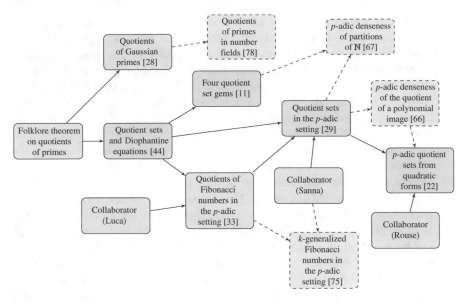

Fig. 2 Flowchart for the quotient set project

A Folklore Theorem An old folk theorem asserts that the set of quotients of prime numbers dense in $\mathbb{R}^+$, the set of positive real numbers (see [20, Ex. 218], [25, Ex. 4.19], [44, Cor. 5], [58, Thm. 4], [69, Ex. 7, p. 107], [70, Thm. 4], [79, Cor. 2]). In other words, if $0 < a < b$, there exist prime numbers p, q such that $a < p/q < b$. The proof depends on the prime number theorem:

$$\lim_{x \to \infty} \frac{\pi(x)}{x / \log x} = 1,$$

in which $\pi(x)$ denotes the number of primes at most x. If $0 < a < b$, then

$$\lim_{q \to \infty} \left(\pi(bq) - \pi(aq) \right) = \infty,$$

in which q is any sequence of primes that tends to infinity. In particular, for q large enough there is a prime $p \in (aq, bq)$; that is, $a < p/q < b$.

I had asked "is the set of quotients of prime numbers dense in the positive real numbers?" as a bonus problem in a real analysis class. This was a question that I had seen somewhere, perhaps in the MONTHLY. My thesis student, Noah Simon, learned of the "folk theorem" from students in my class. "I want to do my thesis on this!" he declared, abandoning all interest in the linear algebra project he was working on.

It is often best to go where a student's passion leads, even if it requires one to learn new material. I agreed to Noah's request, even though there was not enough substance in the folk theorem itself to constitute a proper senior thesis. It was more of a one-off result, so we'd have to improvise and find new angles to explore.

Quotient Sets and Diophantine Equations A search through a stack of number theory books eventually turned up a few references. Ratio sets

$$R(A) = \{a/a' : a, a' \in A\},$$

in which $A \subseteq \mathbb{N} = \{1, 2, 3, \ldots\}$, had been studied over the years and the quotients-of-primes result rediscovered several times. Most importantly, we had identified some key publications on the subject and journals that were open to the topic.

I encouraged Noah, and a few other students who later became involved, to come up with a host of questions of their own. Some were easy, some were hard, and some (we would learn) had already been solved. Nevertheless, they were a place to start. Working through them gave us experience and instructive examples.

It took a couple years and extensive use of the "modularity principle," but the quotient-set investigations spurred by Noah came to fruition. With the combined efforts of several students, we soon had amassed a compelling array of examples and general theorems. Moreover, we had an "application" (if it could be so called) of quotient sets to certain Diophantine equations. The key observation arose from an incidental discussion about Möbius transformations and their mapping properties [44, Lem. 2]. If $U, V \subseteq \mathbb{N}$, $a, b, c, d \in \mathbb{N}$, and $ad - bc = 1$, then the system[2]

$$ax + by = u,$$
$$cx + dy = v,$$

has a solution $(x, y, u, v) \in \mathbb{N} \times \mathbb{N} \times U \times V$ if and only if there exists $(u, v) \in U \times V$ such that $\frac{b}{d} < \frac{u}{v} < \frac{a}{c}$. With this elementary observation, we could cook up some diabolical examples. For example, we could show that the system

$$21x + 17y = 16p^4 + 5p^2q + 13p^2 + 24pq^2 + 9pq + 7q^4 + 7,$$

$$58x + 47y = 44p^4 + 15p^2q + 36p^2 + 66pq^2 + 25pq + 21q^4 + 20,$$

in which $p \equiv 83 \pmod{97}$ and $q \equiv 59 \pmod{103}$ are primes and $x, y \in \mathbb{N}$, has infinitely many solutions [44, Example 15].

Since the MONTHLY already had a history of publishing material on quotients sets [56, 58, 68, 79], it seemed the natural place to try our luck. The paper had a new angle on an old topic and plenty of cute examples, so it was accepted [44]. All of this was spawned from the original "folk theorem" about quotients of primes.

---

[2]The requirement that $x, y \in \mathbb{N}$ (instead of $x, y \in \mathbb{Z}$) makes this problem more difficult than it first appears. This system can be thought of as a simultaneous Frobenius coin problem.

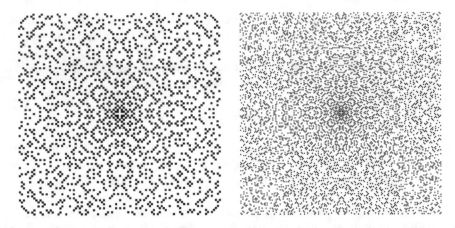

Fig. 3 Gaussian primes $a + bi$ satisfying $|a|, |b| \leq 50$ and $|a|, |b| \leq 100$, respectively

Gaussian Primes and Number Fields A *Gaussian prime* is a prime in the ring $\mathbb{Z}[i] = \{a + bi : a, b \in \mathbb{Z}\}$ of *Gaussian integers*; here $i^2 = -1$ (Fig. 3). The short note [28] proves that the set of quotients of Gaussian primes is dense in the complex plane. Although this was a solo piece, it is a project that I would not have undertaken had not my students pushed me to consider quotient sets. Moreover, the background reading required for the original project [44] had exposed me to the variants of the prime number theorem necessary to tackle the Gaussian integer case. This led to further work by Brian Sittinger [78], a lecturer at Cal State Channel Islands, who extended the folklore theorem to general number fields. Student research can suggest new research avenues for you and your colleagues.

Four Quotient Set Gems My senior thesis student, Michael Dairyko, wanted to do a senior thesis in analysis. Since the paper [44] left open a lot of loose ends and since many additional questions emerged over the years, I sent Michael off with a couple papers and asked him to return with a list of questions inspired by the reading.

The set of all natural numbers whose base-10 representation begins with the digit 1 is "fractionally dense," in the sense that its ratio set is dense in $\mathbb{R}^+$ [44, Example 19]. Michael asked "what happens for other bases?" Some initial work on Mathematica suggested a peculiar phenomenon: the set of all natural numbers whose base-b representation begins with the digit 1 is fractionally dense for $b = 2, 3, 4$, but not for $b \geq 5$ (Fig. 4). Once we "knew" the answer, the rest was a matter of working through the details. However, we would not have known about this interesting phenomenon without first checking it out on the computer. A little preparatory work on the computer goes a long way.

This result, of course, was a one-off curiosity. The modularity principle was necessary to package it with other results into something that could be published in

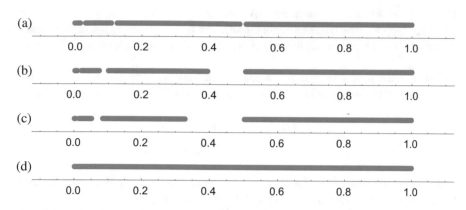

Fig. 4 (**a**)–(**c**) Approximate depiction of the quotient sets for $b = 4, 5, 6$, respectively. (**d**) Quotient set for $b = 4$ with more terms added; compare with (**a**)

a journal such as the MONTHLY. Another student worked through the details of delicate proofs from a rich central European literature (including a paper coauthored by Paul Erdős) that had been overlooked in the MONTHLY [14, 15, 71, 72, 83, 84]. This was an opportunity for exposition. Two other students worked on problems related to arithmetic progressions and asymptotic density.[3] The final paper, published in the MONTHLY, eventually contained the following four quotient set "gems" [11].

1. The set of all natural numbers whose base-b representation begins with the digit 1 is fractionally dense for $b = 2, 3, 4$, but not for $b \geq 5$.
2. For each $\delta \in [0, \frac{1}{2})$, there exists a set $A \subset \mathbb{N}$ with $\underline{d}(A) = \delta$ that is not fractionally dense. On the other hand, if $\underline{d}(A) \geq \frac{1}{2}$, then A is fractionally dense [83].
3. One can partition $\mathbb{N}$ into three sets, each of which is not fractionally dense. However, such a partition is impossible using only two sets [15].
4. There are subsets of $\mathbb{N}$ which contain arbitrarily long arithmetic progressions, yet that are not fractionally dense. On the other hand, there exist fractionally dense sets that have no arithmetic progressions of length ≥ 3.

Items (2) and (3) are due to Strauch and Tóth [83, Thm. 1] and Bukor et al. [15], respectively. Two new results and the exposition of two beautiful, but underappreciated theorems, made our paper a perfect fit for the MONTHLY [11].

---

[3]The *lower asymptotic density* of $A \subseteq \mathbb{N}$ is $\underline{d}(A) = \liminf_{n \to \infty} |A(n)|/n$, in which $A(x) = A \cap [1, x]$ and $|A(x)|$ denotes the number of elements in A that are at most x.

Quotients of Fibonacci Numbers   A summer-research student, Evan Schechter, was searching for new questions about quotient sets. "What about p-adic numbers?" he asked, "I just heard about them in analysis and they seem interesting".[4]

Fix a prime p. If $x = p^n a/b$ is a nonzero rational number, in which $n, a, b \in \mathbb{Z}$ and a, b, p are pairwise relatively prime, then the *p-adic absolute value* of x is

$$
\|x\|_p = \begin{cases} 0 & \text{if } x = 0, \\ p^{-n} & \text{if } x = p^n a/b \text{ as above.} \end{cases}
$$

The *p-adic metric* on $\mathbb{Q}$ is $d_p(x, y) = \|x - y\|_p$. The field $\mathbb{Q}_p$ of *p-adic numbers* is the completion of $\mathbb{Q}$ with respect to the p-adic metric [50, 63].

One of the instructive examples from [44, Example 17] concerns the Fibonacci numbers, defined by the recurrence $F_{n+2} = F_{n+1} + F_n$ with initial conditions $F_0 = 0$ and $F_1 = 1$. Binet's formula (see [53, X.10.14])

$$
F_n = \frac{1}{\sqrt{5}} \left(\left(\frac{1 + \sqrt{5}}{2} \right)^n - \left(\frac{1 - \sqrt{5}}{2} \right)^n \right) \tag{1}
$$

and the fact that $|\frac{1}{2}(1 - \sqrt{5})| < 1$ ensure that the only accumulation points of $\{F_n/F_m \ : \ m, n \ \geq \ 1\}$ are the integral powers of the golden ratio $\frac{1}{2}(1 + \sqrt{5})$. Consequently, the set of quotients of nonzero Fibonacci numbers is not dense in the positive real numbers. On the other hand, Florian Luca[5] and I proved that the set of quotients of nonzero Fibonacci numbers is dense in $\mathbb{Q}_p$ for each prime p [28].

This first foray into the p-adic setting inspired several other authors to explore the topic [66, 67, 75]. Although [28] did not have undergraduate coauthors, it demonstrated that the p-adic quotient set avenue was viable for further research.

p-Adic Quotient Sets   In 2016, I assembled a team of three students to work on p-adic quotient sets for a semester. I instructed them to skim through the existing literature on quotient sets and then formulate twenty questions that we might attack in the p-adic setting. These are the sorts of results and observations that emerged:

---

[4]Unforeseen circumstances forced Evan to withdraw from research that summer, but his idea was a great one. He did prove a couple useful lemmas that appeared in [29], which he is a coauthor of.

[5]Don't be afraid to ask questions! Several years ago, I had a project in which some rather specific results about Lucas numbers (a Fibonacci-like sequence generated by the recurrence $L_0 = 2$, $L_1 = 1$, and $L_{n+2} = L_{n+1} + L_n$ [86]) were required [16, Thm. 19]. My students and I needed to prove that if $p \geq 5$ is an odd prime, then $p - 1$ divides L_p. I asked a few colleagues who work with Fibonacci-flavored number theory, and they all referred me to Florian Luca. This led to our first coauthored article [16]. Since then Florian and I have collaborated on five or six papers on various topics [16, 29, 33–35]. Our collaboration has been entirely online: we have never met in person.

1. For each set P of prime numbers, there is an $A \subseteq \mathbb{N}$ such that $R(A)$ is dense in $\mathbb{Q}_p$ if and only if $p \in P$.
2. A concrete example exists for each of the four statements of the form "$R(A)$ is (dense/not dense) in every $\mathbb{Q}_p$ and (dense/not dense) dense in $\mathbb{R}^+$."
3. There exists an $A \subseteq \mathbb{N}$ that contains no arithmetic progression of length three and such that $R(A)$ is dense in each $\mathbb{Q}_p$.
4. For each $\alpha \in [0, 1)$, there is an $A \subseteq \mathbb{N}$ such that $\underline{d}(A) \geq \alpha$ and $R(A)$ is dense in no $\mathbb{Q}_p$.
5. We determined precisely when the ratio set of $A_{m,n} = \{a \in \mathbb{N} : a = x_1^n + x_2^n + \cdots + x_m^n,\ x_i \geq 0\}$ is dense in $\mathbb{Q}_p$ for $n = 2$ and 3.
6. We extended the Fibonacci result [33] to a wide variety of second-order linear recurrences.[6] For example, the set of quotients of nonzero Lucas numbers is dense in $\mathbb{Q}_p$ if and only if $p \neq 2$ and $p \mid L_n$ for some $n \geq 1$.
7. Let p be an odd prime, let b be a nonzero integer, and let

$$A = \{p^j : j \geq 0\} \cup \{b^j : j \geq 0\}.$$

Then $R(A)$ is dense in $\mathbb{Q}_p$ if and only if b is a primitive root modulo p^2. For example, if $A = \{5^j : j \geq 0\} \cup \{7^j : j \geq 0\}$, then $R(A)$ is dense in $\mathbb{Q}_7$ but not in $\mathbb{Q}_5$ since 5 is a primitive root modulo 7^2 but 7 is not a primitive root modulo 5^2.

This last item was elaborated on by Florian Luca, who proved that there are infinitely many pairs of primes (p, q) such that p is not a primitive root modulo q and q is a primitive root modulo p^2. This yields infinitely many prime pairs (p, q) such that the ratio set of $\{p^j : j \geq 0\} \cup \{q^k : k \geq 0\}$ is dense in $\mathbb{Q}_p$ but not in $\mathbb{Q}_q$. The final paper was coauthored with four students (Evan included), Sanna, and Luca [29].

The "modularity principle" was on full display here. The students had tackled a variety of wide-ranging problems. Their collected work, combined with the observations of my colleagues and I, was greater than the sum of the parts. We had observed a variety of interesting phenomena and laid the groundwork for future explorations of p-adic quotient sets [66, 67].

Quadratic Forms Chris Donnay had just spent the summer working on quadratic forms at an REU and he was eager to continue exploring the topic for his senior thesis. A result from the previous project begged for a generalization [29, Problem 4.4]. If $A = \{x^2 + y^2 : x, y \in \mathbb{Z}\}\backslash\{0\}$, then $R(A)$ is dense in $\mathbb{Q}_p$ if and only if $p \equiv 1 \pmod 4$. When is the quotient set

$$R(Q) = \{Q(\mathbf{x})/Q(\mathbf{y}) : \mathbf{x}, \mathbf{y} \in \mathbb{Z}^r,\ Q(\mathbf{y}) \neq 0\}$$

generated by a quadratic form Q dense in $\mathbb{Q}_p$?

---

[6]Carlo Sanna, who wrote a paper [75] generalizing the Fibonacci result [33], read the `arXiv` preprint and extended our arguments. Thus, he came aboard as a coauthor on the final paper.

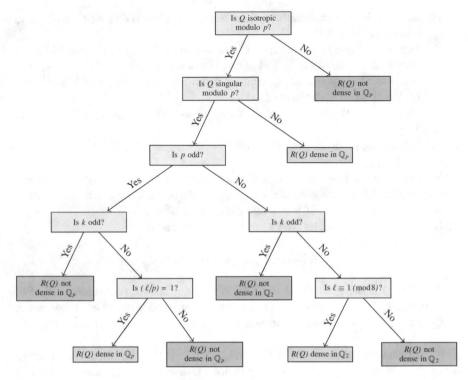

Fig. 5 How to decide if $R(Q)$ is dense in $\mathbb{Q}_p$. Here Q is an integral, binary, and primitive quadratic form of discriminant $p^k \ell$, in which $\gcd(p, \ell) = 1$. Here (ℓ/p) denotes a Legendre symbol

Chris had not worked over the p-adic numbers before and I had never gotten my hands on quadratic forms, so this would be a learning experience for both of us. Over the course of the year, Chris and I managed to prove a complicated web of results that mostly solved the problem (Fig. 5). Our proofs were long and tedious and contained some repetition. They worked, but they were inelegant. Moreover, the prime $p = 2$ stuck out as an annoying special case. Chris' former REU advisor, Jeremy Rouse, later helped us unify our approach and taught us a few facts from Serre [76] that simplified things immensely. The three of us submitted a paper that provided a complete answer to the quotient set problem for quadratic forms [22].

Future Work The study of quotient sets provided my students and I with a host of accessible projects [11, 22, 28, 29, 33, 44] Although there is much to be said for moving on before a subject becomes stale, we cannot completely close the door on returning to quotient sets if new inspiration arrives.

4 Case Study II: Primitive Roots, Prime Pairs, and the Bateman–Horn Conjecture

Our second case study stems from a question posed by a student in the author's Spring 2017 introductory number-theory course (teaching a subject is often the best way to learn it). This launched a research program that resulted in several papers, some publicity, and an award for the student (Fig. 6).

An Innocent Question The Diffie–Hellman key exchange protocol is a standard topic in elementary number-theory courses, the author's being no exception. Although the technical details do not concern us, we should say that the method permits two entities, traditionally named Alice and Bob, to agree on a secret key without meeting "in person" and while communicating over an insecure channel. To perform this feat, Alice and Bob require the use of primitive roots [53, 59, 63].

A *primitive root* modulo a prime p is a generator g of the multiplicative group $(\mathbb{Z}/p\mathbb{Z})^{\times}$; that is, $g, g^2, \ldots, g^{p-1}$ are congruent modulo p to $1, 2, \ldots, p-1$, in some order. An old theorem of Gauss ensures that each prime p possesses exactly $\phi(p-1)$ primitive roots (modulo p), in which

$$\phi(n) = \left|\left\{a \in \{1, 2, \ldots, n\} : \gcd(a, n) = 1\right\}\right|$$

is the Euler totient function.

In highlighting the Diffie–Hellman key exchange protocol and the practical value of primitive roots, I displayed a table of data to illustrate the unpredictability of primitive roots (Fig. 7). What happened next launched an entire research thread.

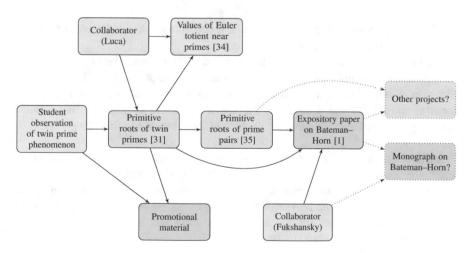

Fig. 6 Flow chart for the prime-pairs project

p	primitive roots modulo p
2	1
3	2
5	2,3
7	3,5
11	2,6,7,8
13	2,6,7,11
17	3,5,6,7,10,11,12,14
19	2,3,10,13,14,15
23	5,7,10,11,14,15,17,19,20,21
29	2,3,8,10,11,14,15,18,19,21,26,27
31	3,11,12,13,17,21,22,24
37	2,5,13,15,17,18,19,20,22,24,32,35
41	6,7,11,12,13,15,17,19,22,24,26,28,29,30,34,35
43	3,5,12,18,19,20,26,28,29,30,33,34
47	5,10,11,13,15,19,20,22,23,26,29,30,31,33,35,38,39,40,41,43,44,45
53	2,3,5,8,12,14,18,19,20,21,22,26,27,31,32,33,34,35,39,41,45,48,50,51
59	2,6,8,10,11,13,14,18,23,24,30,31,32,33,34,37,38,39,40,42,43,44,47,50,52,54,55,56
61	2,6,7,10,17,18,26,30,31,35,43,44,51,54,55,59 ·
67	2,7,11,12,13,18,20,28,31,32,34,41,44,46,48,50,51,57,61,63
71	7,11,13,21,22,28,31,33,35,42,44,47,52,53,55,56,59,61,62,63,65,67,68,69

Fig. 7 A student noticed that for most twin-prime pairs $(p, p + 2)$, the first prime has at least as many primitive roots as the second (the original table shown in class was black and white)

After looking at the data for a few seconds, Elvis Kahoro asked "is it true that for twin primes, apart from 3 and 5, the first one has at least many primitive roots as the second?" In other words, if $(p, p + 2)$ is a pair of twin primes with $p \geq 5$, is

$$\phi(p - 1) \geq \phi(p + 1)? \tag{2}$$

The limited numerical data on hand made the conjecture seem plausible. I said that I would have to give the matter more thought.

I wrote a `Mathematica` program to test the conjecture. The computer found several counterexamples, the first of which is the twin-prime pair $(2381, 2383)$.[7] The results were convincing. Asymptotically, 98% of twin-prime pairs $(p, p + 2)$ seem to satisfy (2) and a stubborn 2% seem to violate it (Fig. 8). This was an opportunity to make lemonade [new theorems] from lemons [a false conjecture].

Primitive Roots for Twin Primes Numerical evidence suggested that the overwhelming majority of twin-prime pairs satisfy (2). However, we do not know if infinitely many twin primes exist. The assertion that there are the fame twin prime conjecture. Sometimes the best remedy is to be näive, wave your hands, and invite some friends along. Over the course of a few days and a flurry of e-mails with Florian Luca, a path forward emerged. The product formula

---

[7]By chance, my office number is 2383.

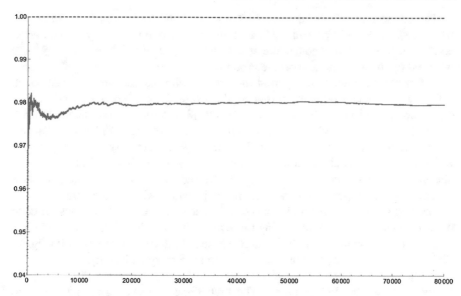

Fig. 8 The horizontal axis denotes the number of twin-prime pairs surveyed. The vertical axis is the running ratio of twin-prime pairs $(p, p+2)$ for which $\phi(p-1) \geq \phi(p+1)$. Apart from some initial fluctuations, the ratio settles down to somewhere around 98%

$$\frac{\phi(n)}{n} = \prod_{q \mid n} \left(1 - \frac{1}{q}\right), \tag{3}$$

in which q runs over all primes that divide n, holds the key. If p and $p+2$ are prime, then $2|(p-1)$, $3 \nmid (p-1)$, and $6|(p+1)$. Thus,

$$\frac{\phi(p-1)}{p-1} = \frac{1}{2} \prod_{\substack{q \mid (p-1) \\ q \geq 5}} \left(1 - \frac{1}{q}\right) \quad \text{and} \quad \frac{\phi(p+1)}{p+1} = \frac{1}{3} \prod_{\substack{q \mid (p+1) \\ q \geq 5}} \left(1 - \frac{1}{q}\right)$$

$$\tag{4}$$

and hence

$$\frac{\phi(p-1)}{p-1} \geq \frac{\phi(p+1)}{p+1} \tag{5}$$

should hold most of the time. Why? Experience suggests that the primes behave "randomly" and hence the two products in (4) should be comparable in size. Thus, the first expression in (4) should tend to be larger than the second because, quite simply, $1/2$ is larger than $1/3$. If p is sufficiently large, then replacing the denominators in (5) with p should cause no harm and hence we predict that the inequality (2) holds most of the time. This heuristic argument, although plausible, is no proof. Moreover, it does not explain why a stubborn percentage (approximately 2%) of twin-prime pairs satisfy the reverse inequality $\phi(p-1) < \phi(p+1)$.

The key to putting this heuristic reasoning on solid footing was the *Bateman–Horn conjecture*, an important conjecture in analytic number theory (discussed in detail later on). Indeed, a conjecture at least as strong as the twin prime conjecture is necessary to get the project off the ground.

A few months after Elvis spurred the project with his question, we had found a family of counterexamples, reframed the original question, partnered with a distant colleague, and produced our main results [31]. Assuming the Bateman–Horn conjecture, the set of twin-prime pairs $(p, p + 2)$ for which (2) holds has lower density (as a subset of twin primes) at least 65.13%; the set of twin-prime pairs for which the reverse inequality holds has lower density at least 0.47%. Thus, a definite bias exists and the dominant inequality is reversed for a small percentage of twin primes. The paper spurred several other projects, which we describe below.

Elvis' story was a compelling one, so I contacted our communications office to see if they were interested in hearing more about it. They loved the story and ran a web-banner and an article in the Pomona College Magazine based upon it (Fig. 9). Elvis later earned third place in a poster session at Emory University.

Primitive Roots for Prime Pairs   The first prime in a twin-prime pair tends to have more primitive roots than does the second (if we assume the Bateman–Horn conjecture). It is natural to consider whether other prime pairs $(p, p + k)$ exhibit a similar bias. What about $k = 4$ (cousin primes), $k = 6$ (sexy primes), and so forth?

Tim Schaaff, a community-college transfer student, had recently graduated from Pomona and was looking for a summer-research project before heading off into the "real world." Although he had no number theory background, he had taken several analysis courses and abstract algebra. He was at the right level to wade through and adapt the arguments from the twin-prime project [31] to the prime-pair case.

Although there are some similarities, many new complications arise when passing from $(p, p + 2)$ to $(p, p + k)$ with $k \geq 4$. Several complicated parameters needed to be introduced and a difficult asymptotic lemma was required. Moreover, the tolerances are tight for certain k: among the first twenty million primes, each prime pair $(p, p + 70)$ satisfies $\phi(p - 1) < \phi(p + 69)$. Nevertheless, a positive proportion (at least 1.81×10^{-20}) of such pairs satisfy the reverse inequality!

Fig. 9 (Left) Promotional material from the Pomona College website. (Right) Pomona College Magazine article, Spring 2018

In reality, these complications were good news. A simple-minded, straight-forward generalization of an existing result is usually not worth publishing in a reputable journal. However, a broad generalization that requires new approaches and greater technical skill is often publishable. So obstacles are not necessarily bad.

Tim and I worked through most of the details over the summer, occasionally checking in with Florian Luca. This resulted in the paper [35]. This tale is noteworthy for at least two reasons. First of all, it illustrates how an apparently straightforward generalization need not be simple in practice. Second, the added complications inherent in some generalizations can be turned into a positive. Simply put, the project would not have resulted in a decent paper if it were simply a matter of changing every occurrence of $+2$ in the first paper to $+k$.

Behavior of the Euler Totient Near Primes Were these sorts of results truly about prime pairs? Or was there a bias inherent in the Euler totient function near prime values? Would the bias disappear if we assume only that p is prime?

Through a complicated argument, Florian Luca and I were able to prove unconditionally that for each $\ell \geq 1$, the difference $\phi(p - \ell) - \phi(p + \ell)$ is positive for 50% of odd primes p and negative for 50% [34]. Although there was no undergraduate collaborator on this paper, it would not have been possible had not the original line of research been initiated by Elvis. Undergraduate research, which need not be distinct from "research" itself, can provide new problems for one's own research. Questions posed by students can lead to new avenues for your own research.

One Conjecture to Rule Them All The Bateman–Horn conjecture is a far-reaching conjecture, widely supported by numerical evidence, that provides asymptotic predictions for the number of prime values simultaneously assumed by families of polynomials. The precise statement is quite a mouthful [1, 6, 7].

Bateman–Horn Conjecture *Let $f_1, f_2, \ldots, f_k \in \mathbb{Z}[x]$ be distinct irreducible polynomials with positive leading coefficients and let*

$$Q(f_1, f_2, \ldots, f_k; x) = \#\{n \leq x : f_1(n), f_2(n), \ldots, f_k(n) \text{ are prime}\}.$$

Suppose that $f = f_1 f_2 \cdots f_k$ does not vanish identically modulo any prime. Then

$$Q(f_1, f_2, \ldots, f_k; x) \sim \frac{C(f_1, f_2, \ldots, f_k)}{\prod_{i=1}^k \deg f_i} \int_2^x \frac{dt}{(\log t)^k},$$

in which

$$C(f_1, f_2, \ldots, f_k) = \prod_p \left(1 - \frac{1}{p}\right)^{-k} \left(1 - \frac{\omega_f(p)}{p}\right) \tag{6}$$

and $\omega_f(p)$ is the number of solutions to $f(x) \equiv 0 \pmod{p}$.

We content ourselves here with a worked example. Let $f_1(t) = t$ and $f_2(t) = t + 2$. Then $f_1(t)$ and $f_2(t)$ are simultaneously prime if and only if t is the lesser element of a twin-prime pair. Let $f = f_1 f_2$. Then $f(t) \equiv 0 \pmod{p}$ if and only if $t(t - 2) \equiv 0 \pmod{p}$, and hence

$$\omega_f(p) = \begin{cases} 1 & \text{if } p = 2, \\ 2 & \text{if } p \geq 3. \end{cases}$$

The corresponding Bateman–Horn constant (6) is $C(f_1, f_2) = 2C_2$, in which

$$C_2 = \prod_{p \geq 3} \frac{p(p - 2)}{(p - 1)^2} \approx 0.660161815$$

is the *twin primes constant*. The Bateman–Horn conjecture predicts that

$$Q(f_1, f_2; x) \sim 2C_2 \int_2^x \frac{dt}{(\log t)^2},$$

an asymptotic prediction that was first proposed by Hardy and Littlewood [52].

When Florian introduced me to the Bateman–Horn conjecture, I could guess at how many of the seminal theorems and famous conjectures in number theory would follow from it, but I never chased down the precise details. For example, the Bateman–Horn conjecture should tell us something about the Ulam spiral (Fig. 10)

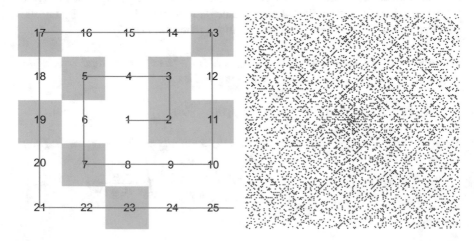

Fig. 10 (Left) The Ulam spiral is produced by enumerating the natural numbers in a spiral pattern. Squares corresponding to prime numbers are colored. (Right) Curious diagonal features are suggested by the Ulam spiral when one looks at larger scales

and Euler's "prime-producing" polynomial $n^2 + n + 41$. It should imply the Green–Tao theorem (the primes contain arbitrarily long arithmetic progressions [51]) and the prime number theorem for arithmetic progressions. This seemed like the perfect topic for a senior thesis. The Bateman–Horn conjecture was just too interesting, broad-reaching, and exciting not to study some more.

I assigned one of my senior thesis students, Soren Aletheia-Zomlefer, to the task. We found few accounts of the conjecture in the literature, the most notable being Lang's brief survey [65]. This lack of sources was a blessing in disguise. There was a hole in the literature that needed to be filled. This was another opportunity to turn lemons [lack of information] into lemonade [a much-needed survey paper].

As an expository paper on the conjecture began forming in early 2018, another problem emerged. Why does the infinite product (6) converge? Soren and I never tackled this question since we had been focused on deriving consequences of the conjecture. The convergence of the product was assumed in all sources, most of which pointed to each other for explanations. The sketch provided in [7] was referred to in most subsequent papers, although Bateman and Horn left many details out. Although I have no doubt that leading experts could reconstruct the argument in detail, a thorough exposition was in order.

Although I was able to work out most of the argument, there were some algebraic number theory issues that still puzzled me. Fortunately, my neighbor and frequent collaborator, Lenny Fukshanky, is a number theorist who teaches a topics course in algebraic number theory on occasion. The three of us eventually completed the expository paper [1], building upon Soren's host of examples.

This illustrates the "modularity principle": the student does not need to be involved in all aspects of the project. It was sufficient for Soren to work through dozens of applications, without dealing with the technical convergence issue. We received valuable feedback on the arXiv preprint from several top number theorists. This sharpened the manuscript and indicated that the survey was widely read. It also put me in touch with top names in the field, something valuable for a "newbie."

Future Work The primitive roots project, initiated by a student's question, spawned a minor industry. Several additional projects inspired by the project are in the works.

1. We know $\phi(p - 1) < \phi(p + 1)$ is possible, although rare, for twin-prime pairs $(p, p + 2)$. How extreme can this inequality be for twin primes? Assuming a standard conjecture, a recent preprint coauthored with two undergraduates shows that $\{\phi(p + 1)/\phi(p - 1) : p, p + 2 \text{ prime}\}$ is dense in the positive reals [36].
2. What happens for prime triples, such as $(p, p+2, p+6)$? Does one of the primes tend to have more primitive roots than the others? How much more?
3. Can similar results be proved about primitive roots modulo odd prime powers?

4. Other arithmetic functions, such as the sum-of-divisors function $\sigma(n) = \sum_{d|n} d$, enjoy product representations reminiscent of (3). How do these functions behave near prime arguments?
5. Fukshansky and I plan are writing a monograph on the Bateman–Horn conjecture, using the expository paper [1] as a skeleton.

The primitive-roots program is alive and well. A good question, numerical experimentation, and a frequent collaborator got the initial project off the ground. Once the ball was rolling, new questions emerged and provided work for students while also inspiring a couple projects for my collaborators and I.

5   Case Study III: Supercharacters and Exponential Sums

Our final case study involves the theory of supercharacters, a novel generalization of classical character theory that was developed by P. Diaconis and I.M. Isaacs (building upon seminal work of C. André [2–4]) to explore the character theory of certain intractable groups [21]. When applied to finite abelian groups, connections to analytic number theory, discrete Fourier analysis, and additive combinatorics arise. Many of these links furnished undergraduate research projects (Fig. 11).

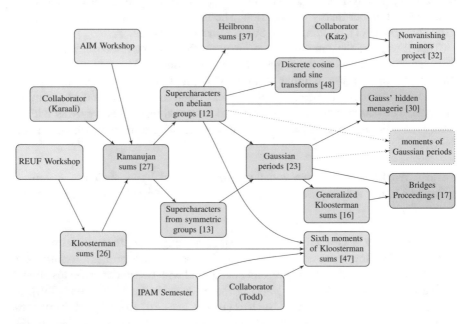

Fig. 11 Flowchart for the supercharacter theory project

A Research Experience In 2009, the author had the good fortune to attend the Research Experiences for Undergraduate Faculty (REUF) program at the American Institute of Mathematics (AIM). This program brings faculty members, many of whom are from groups underrepresented in the mathematical sciences, together "to provide faculty participants with a research experience investigating open questions in the mathematical sciences" and "to equip participants to engage in research with undergraduate students at their home institutions."

Several topics were proposed by the four team leaders. I gravitated towards the one described by Phil Kutzko, from whom I would learn a great deal. The sum

$$K_u = \sum_{x=1}^{p-1} e\left(\frac{x + ux^{-1}}{p}\right),$$

in which p is a prime, x^{-1} denotes the inverse of x modulo p, and $e(x) = \exp(2\pi i x)$, is a *Kloosterman sum*. Exponential sums of this form play an important role in analytic number theory [60–62]. Phil proposed an elementary method, inspired by one of his early papers [64], to prove the *Weil bound*: $|K_u| \leq 2\sqrt{p}$ for $p \nmid u$ [80, 81, 87]. Although we did not accomplish this,[8] we found a simple proof of a nontrivial bound, obtained new identities, and found a connection with Ramanujan multigraphs [26].

How does this relate to undergraduate research? First, it highlights the importance of the REUF program for invigorating research programs. Although not quite an "old dog" at that point, I was able to learn "new tricks." Most importantly, I acquired new tools that would later spawn several undergraduate research projects.

Ramanujan Sums In 2010, my wife (Gizem Karaali) attended an AIM workshop on supercharacters and combinatorial Hopf algebras. She became convinced that supercharacters would simplify the messy computations from our REUF project [26]. Instead of considering characters and conjugacy classes of a finite group, as in classical character theory, one studies certain sums of characters (supercharacters) and compatible unions of conjugacy classes (superclasses) [8, 12, 21, 27, 57]. Just as in classical character theory, where one has a character table that enjoys certain orthogonality relations, in supercharacter theory one has a supercharacter table (a "compression" of the original character table) with similar orthogonality properties.

Chris Fowler, an undergraduate student co-advised by Karaali and I, computed the supercharacters on $\mathbb{Z}/n\mathbb{Z}$ that arose from the multiplication action of the unit group $(\mathbb{Z}/n\mathbb{Z})^\times$. The intriguing expressions produced were *Ramanujan sums*

---

[8] A colleague and I recently made a key step in this direction using supercharacter theory [47].

$$c_n(x) = \sum_{\substack{j=1 \\ \gcd(j,n)=1}}^{n} e\left(\frac{jx}{n}\right).$$

The three of us developed the basics for supercharacter theory on abelian groups in the process. In order to set the stage for what follows, we need to jump forward a bit to a later incarnation [12, Thm. 2] of one of the early results [27, Thm. 4.2].

Theorem 1 *Let $\Gamma = \Gamma^T$ be a subgroup of $GL_d(\mathbb{Z}/n\mathbb{Z})$, the invertible $d \times d$ matrices over $\mathbb{Z}/n\mathbb{Z}$, let $\{X_1, X_2, \ldots, X_N\}$ denote the set of Γ-orbits in $G = (\mathbb{Z}/n\mathbb{Z})^d$ induced by the action of Γ, and let $\sigma_1, \sigma_2, \ldots, \sigma_N$ denote the corresponding supercharacters*

$$\sigma_i(\mathbf{y}) = \sum_{\mathbf{x} \in X_i} e\left(\frac{\mathbf{x} \cdot \mathbf{y}}{n}\right), \tag{7}$$

in which $\mathbf{x} \cdot \mathbf{y}$ denotes the formal dot product of two elements of $(\mathbb{Z}/n\mathbb{Z})^d$ and $e(x) = \exp(2\pi i x)$. For each fixed $\mathbf{z}$ in X_k, let $c_{i,j,k}$ denote the number of solutions $(\mathbf{x}_i, \mathbf{y}_j) \in X_i \times X_j$ to the equation $\mathbf{x} + \mathbf{y} = \mathbf{z}$.

1. *$c_{i,j,k}$ is independent of the representative $\mathbf{z}$ in X_k that is chosen.*
2. *Each σ_i is a superclass function: $\sigma_i(\mathbf{x})$ depends only upon the X_j that contains $\mathbf{x}$.*
3. *The $N \times N$ matrix*

$$U = \frac{1}{\sqrt{n^d}} \left[\frac{\sigma_i(X_j)\sqrt{|X_j|}}{\sqrt{|X_i|}} \right]_{i,j=1}^{N} \tag{8}$$

   is complex symmetric ($U = U^T$) and unitary. It satisfies $U^4 = I$.
4. *The identity*

$$\sigma_i(X_\ell)\sigma_j(X_\ell) = \sum_{k=1}^{N} c_{i,j,k}\sigma_k(X_\ell)$$

   holds for $1 \leq i, j, k, \ell \leq N$.
5. *The matrices $T_1, T_2, \ldots, T_N$, whose entries are given by*

$$[T_i]_{j,k} = \frac{c_{i,j,k}\sqrt{|X_k|}}{\sqrt{|X_j|}},$$

   each satisfy

$$T_i U = U D_i,$$

in which

$$D_i = \text{diag}\left(\sigma_i(X_1), \sigma_i(X_2), \ldots, \sigma_i(X_N)\right).$$

In particular, the T_i are simultaneously unitarily diagonalizable.

6. *Each T_i is a normal matrix ($T_i^* T_i = T_i T_i^*$) and the set $\{T_1, T_2, \ldots, T_N\}$ forms a basis for the commutative algebra of all $N \times N$ complex matrices T such that $U^* T U$ is diagonal.*

Although we did not have this exact result available, we were able to use something similar [27, Thm. 4.2] to convert statements about certain exponential sums, such as Ramanujan sums, into statements in linear algebra. We provided a unified treatment of Ramanujan-sum identities and proved many new identities too [27].

The Graphic Nature of the Symmetric Group   In the summer of 2012, I ran an 8-week, Claremont-based REU that involved four undergraduate students and one graduate student. I gave an introductory lecture on the representation theory of finite groups and a brief survey of supercharacter theory. There was no fixed agenda and I encouraged the students to come up with questions of their own. This was fertile territory and almost any question they asked would be novel and unexplored.

We investigated multiple threads to keep the group operating at peak efficiency and to increase our chances of making a breakthrough. A couple students looked at $G = (\mathbb{Z}/n\mathbb{Z})^d$ with S_d, the symmetric group on d letters, acting upon G by permuting entries. When we plotted the values of the corresponding supercharacters $\sigma_X : G \to \mathbb{C}$, defined by (7), a host of beautiful images appeared (Fig. 12). Many of the images suggested higher-dimensional phenomena. Others seemed to enjoy a high degree of symmetry but were, in fact, not so symmetric after all.

We dropped everything for a couple weeks in order to study the complex relationship between the parameters, the combinatorics of the symmetric group, and the qualitative appearance of the resulting images. This led to the paper [13].

Supercharacters on Abelian Groups   The Ramanujan-sum project required the development of a substantial portion of supercharacter theory for abelian groups [27]. Building upon this, the REU group abstracted and formalized the approach, which crystalized into Theorem 1. The students worked through many illustrative examples and important computations too. The resulting paper [12] had six student coauthors from several "generations" along with my collaborators from [27].

Gaussian Periods   The Ramanujan sum and symmetric group projects indicated that even the most elementary groups were fair game. I assigned Bob Lutz, my senior thesis student, to investigate another family of examples. If $G = \mathbb{Z}/n\mathbb{Z}$ and

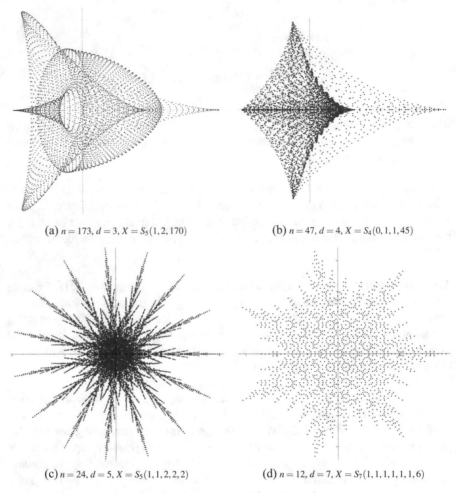

(a) $n = 173, d = 3, X = S_5(1, 2, 170)$ (b) $n = 47, d = 4, X = S_4(0, 1, 1, 45)$

(c) $n = 24, d = 5, X = S_5(1, 1, 2, 2, 2)$ (d) $n = 12, d = 7, X = S_7(1, 1, 1, 1, 1, 1, 6)$

Fig. 12 Images in $\mathbb{C}$ of supercharacters $\sigma_X : (\mathbb{Z}/n\mathbb{Z})^d \to \mathbb{C}$ arising from the permutation action of S_d on $(\mathbb{Z}/n\mathbb{Z})^d$ for various moduli n, dimensions d, and orbits X. Observe that (**c**) and (**d**) do not enjoy the full rotational symmetry suggested by their large-scale structure. The symmetry groups of (**c**) and (**d**) are the dihedral group D_3 and $\mathbb{Z}/2\mathbb{Z}$, respectively

$\Gamma = \langle \omega \rangle$ is a cyclic subgroup of $(\mathbb{Z}/n\mathbb{Z})^\times$, then the corresponding supercharacters (7) assume values that are *Gaussian periods* [9, 30]. For example, if $n = p$ is an odd prime and ω has order d in $(\mathbb{Z}/p\mathbb{Z})^\times$, then (7) produces expressions like

$$\sum_{j=0}^{d-1} e\left(\frac{\omega^j y}{p}\right).$$

If $d = (p - 1)/2$, then these are, more or less, quadratic Gauss sums.

The supercharacter plots were even more amazing than those produced in the symmetric group project (Fig. 13). Bob had the great idea to paint each point $\sigma_X(\xi)$ according to the residue class of ξ (mod c), in which c is a small divisor of the modulus n. This revealed subtle structures and suggested a host of new theorems, which we eventually proved. Still, there were some deeper properties that eluded us.

In 2013, I gave a talk on supercharacter theory and exponential sums at the UCLA combinatorics seminar. One of the attendees was number theorist William Duke. He realized that the right way to approach the emergent properties Bob and I were witnessing (Fig. 14) was with equidistribution theory, a technique that my students

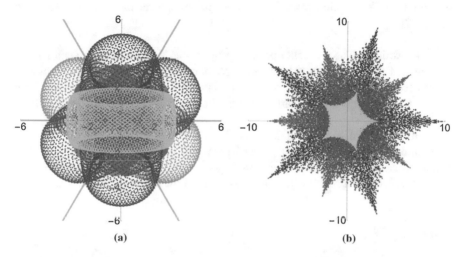

(a)   (b)

Fig. 13 Images in $\mathbb{C}$ of supercharacters $\sigma_X : \mathbb{Z}/n\mathbb{Z} \to \mathbb{C}$ arising from the action of a cyclic subgroup $\Gamma \subseteq (\mathbb{Z}/n\mathbb{Z})^\times$. Here $X = \Gamma 1$ is the orbit of 1 and $\sigma_X(\xi)$ is colored according to ξ (mod c). (**a**) $n = 3 \cdot 5 \cdot 17 \cdot 29 \cdot 37$, $\Gamma = \langle 184747 \rangle$, $c = 3 \cdot 17$. (**b**) $n = 5 \cdot 251 \cdot 281$, $\Gamma = \langle 54184 \rangle$, $c = 5$

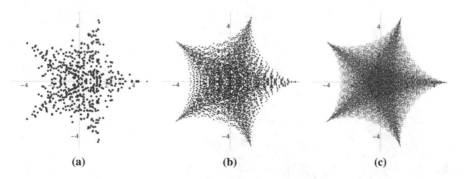

(a)   (b)   (c)

Fig. 14 Images in $\mathbb{C}$ of supercharacters $\sigma_X : \mathbb{Z}/p\mathbb{Z} \to \mathbb{C}$ arising from the action of a cyclic subgroup $\Gamma \subseteq (\mathbb{Z}/p\mathbb{Z})^\times$. A hypocycloid gets "filled out" if the parameters are chosen appropriately. Observe that the images are not rotationally symmetric. (**a**) $p = 2791$, $\Gamma = \langle 800 \rangle$. (**b**) $p = 27011$, $\Gamma = \langle 9360 \rangle$. (**c**) $p = 202231$, $\Gamma = \langle 61576 \rangle$

and I later used to prove similar results about "generalized Kloosterman sums" (see below). This collaboration resulted in [23], largely based on Bob's senior thesis. In particular, this fruitful opportunity highlights the importance of presenting one's work. New opportunities for collaboration can arise from chance meetings.

Heilbronn Sums Bob worked on another supercharacter project before leaving for graduate school. If $G = (\mathbb{Z}/p^2\mathbb{Z})$, then $\Gamma = \{1^p, 2^p, \ldots, (p-1)^p\}$, the set of perfect pth powers modulo p^2, is a subgroup of $(\mathbb{Z}/p^2\mathbb{Z})^\times$. The corresponding supercharacter theory produces Heilbronn sums [54, 55, 77], a curious exponential sum connected to Fermat congruences $ax^p + by^p \equiv cz^p \pmod{p^2}$. A few months of work provided explicit formulas for the first few power moments of Heilbronn sums [37].

Getting Noticed The Gaussian-periods project was a good story: a senior thesis student had discovered remarkable new phenomena in centuries-old expressions introduced by Gauss. I alerted our communications office about the story. They interviewed Bob and wrote an article for the Pomona College Magazine (Fig. 15).

Editors of popular mathematics outlets are always on the lookout for well-written, eye-catching articles of general interest. I had already published a short historical piece in the *Notices of the AMS*, coauthored with an undergraduate, on a problem solved by Erdős [46]. Thus, I had my "foot in the door" with the editor

Fig. 15 Fall 2013 Pomona College Magazine article about Bob Lutz' senior thesis

of the *Notices*, Steven Krantz. I wrote to him and proposed an article on Gaussian periods. He liked the images and said that the idea had promise. Bob and I got to work right away (Duke became department chair around this time and did not participate).

The computational aspects of the project were daunting. We wanted bigger pictures, with more color than those in [23]. Bob (now at Michigan) said that his fellow graduate student, Trevor Hyde, was good with computers and number theory in general. I was happy to have another eager student join the project and the three of us put together the expository article [30], which appeared in the *Notices* in 2015.

Building Bridges I soon had a new generation of undergraduate students to work with. We met once a week to bounce around ideas until something stuck. Typically students would take turns explaining what they worked on, with appropriate feedback and questions from myself and the other students. We would often crowd over a laptop to see the latest code in action. After a brainstorming session, we found another family of exponential sums that exhibited new phenomena. Suffice it to say that [16], coauthored with four students, was my first encounter with Florian Luca, who we met in the previous case studies (p. 212).

Around this time, another opportunity for exposition arose. A good deal of energy had been spent cooking up visually appealing examples [13, 16, 23, 30], so I gave a talk at the 2015 Bridges conference, which celebrates "mathematical connections to art, music, architecture, education and culture." Several students and I submitted a proceedings paper highlighting some of the images we had produced [17].

Discrete Cosine Transform If we apply Theorem 1 to $G = \mathbb{Z}/n\mathbb{Z}$ and $\Gamma = \{1\}$, then the matrix U, defined by (8), is the conjugate of the discrete Fourier matrix

$$\frac{1}{\sqrt{n}} \begin{bmatrix} 1 & 1 & 1 & \cdots & 1 \\ 1 & \zeta & \zeta^2 & \cdots & \zeta^{n-1} \\ 1 & \zeta^2 & \zeta^4 & \cdots & \zeta^{2(n-1)} \\ \vdots & \vdots & \vdots & \ddots & \vdots \\ 1 & \zeta^{n-1} & \zeta^{2(n-1)} & \cdots & \zeta^{(n-1)^2} \end{bmatrix}, \qquad \zeta = \exp(2\pi i/n),$$

of order n. Theorem 1 recovers the fact that a matrix is diagonalized by the discrete Fourier transform (DFT) if and only if it is a *circulant* matrix:

$$\begin{bmatrix} c_0 & c_{n-1} & \cdots & c_2 & c_1 \\ c_1 & c_0 & c_{n-1} & & c_2 \\ \vdots & c_1 & c_0 & \ddots & \vdots \\ c_{n-2} & & \ddots & \ddots & c_{n-1} \\ c_{n-1} & c_{n-2} & \cdots & c_1 & c_0 \end{bmatrix}.$$

If we let $\Gamma = \{-1, 1\}$, then U is a discrete cosine transform (DCT) matrix

$$
\frac{2}{\sqrt{n}}
\begin{bmatrix}
\frac{1}{2} & \frac{1}{\sqrt{2}} & \frac{1}{\sqrt{2}} & \cdots & \frac{1}{\sqrt{2}} \\
\frac{1}{\sqrt{2}} & \cos\frac{2\pi}{n} & \cos\frac{4\pi}{n} & \cdots & \cos\frac{(n-1)\pi}{n} \\
\frac{1}{\sqrt{2}} & \cos\frac{4\pi}{n} & \cos\frac{8\pi}{n} & \cdots & \cos\frac{2(n-1)\pi}{n} \\
\vdots & \vdots & \vdots & \ddots & \vdots \\
\frac{1}{\sqrt{2}} & \cos\frac{(n-1)\pi}{n} & \cos\frac{2(n-1)\pi}{n} & \cdots & \cos\frac{(n-1)^2\pi}{2n}
\end{bmatrix} ; \tag{9}
$$

here depicted for n odd [12, Sect. 4.3]. The DCT is a staple in modern software. For example, the MP3 and JPEG file formats both make use of it [49].

What is the analogue of the circulant result for the DCT? Undergraduate Sam Yih and I used Theorem 1 to produce a novel description of the algebra diagonalized by the DCT that has a simple combinatorial interpretation. In addition to recapturing known results (see [10, 24, 73, 74]) in a systematic manner, we were also able to treat the discrete sine transform (DST) as well [48].

Nonvanishing Minors All of this work on signal processing and number theory had exposed me to a lot of finite Fourier analysis. I also tried to keep up, to the extent possible, with some of the exciting developments in additive number theory.

In 2006, Terence Tao provided a new proof of a beautiful result of Chebotärev: every minor of the $n \times n$ Fourier matrix is nonzero if and only if n is prime [82, 85]. He used this to provide a new proof of the Cauchy–Davenport inequality

$$
|A + B| \geq \min\{|A| + |B| - 1, p\} \quad \text{for nonempty } A, B \subseteq \mathbb{Z}/p\mathbb{Z}
$$

from additive combinatorics. After spending so much time on the discrete cosine transform with Sam, it was natural to investigate whether a similar Chebotärev-type theorem holds for the DCT or DST. A few numerical experiments suggested that this was the case. A year of on-and-off work with two collaborators eventually provided a broad generalization of Chebotärtev's result, which encompassed the DCT, DST, and many other examples, along with applications to additive combinatorics [32].

Although this work involved no undergraduate students, it would not have been possible without the experience I had built up guiding many undergraduate research projects. I would never have been in a position to think of, or work on, such problems, without having supervised many student research projects on the area.

Future Work The supercharacter program is still up and running. For example, my current senior thesis student, Brian Lorenz, is applying Theorem 1 to investigate the moments of the Gaussian periods modulo a prime p. Another group of students, jointly supervised by Karaali and I, are in the beginning stages of their research. There is no doubt that supercharacter theory is fertile ground for student research and we have no plans to let up while so many theorem wait to be proved.

6 Conclusion

We have outlined twenty-one general principles for fostering long-term, sustainable undergraduate research programs (2). The author hopes that the three case studies in number theory presented above (Sects. 3–5) have illustrated these principles. In particular, we hope that the reader will find the prospect of guiding students into new territory a little less frightening.

References

1. Soren Aletheia Zomlefer, Stephan Ramon Garcia, and Lenny Fukshansky. One conjecture to rule them all: Bateman–Horn. *Expo. Math.* (in press). https://arxiv.org/abs/1807.08899.
2. Carlos A. M. André. Basic characters of the unitriangular group. *J. Algebra*, 175(1):287–319, 1995.
3. Carlos A. M. André. The basic character table of the unitriangular group. *J. Algebra*, 241(1):437–471, 2001.
4. Carlos A. M. André. Basic characters of the unitriangular group (for arbitrary primes). *Proc. Amer. Math. Soc.*, 130(7):1943–1954 (electronic), 2002.
5. Levon Balayan and Stephan Ramon Garcia. Unitary equivalence to a complex symmetric matrix: geometric criteria. *Oper. Matrices*, 4(1):53–76, 2010.
6. Paul T. Bateman and Roger A. Horn. A heuristic asymptotic formula concerning the distribution of prime numbers. *Math. Comp.*, 16:363–367, 1962.
7. Paul T. Bateman and Roger A. Horn. Primes represented by irreducible polynomials in one variable. In *Proc. Sympos. Pure Math., Vol. VIII*, pages 119–132. Amer. Math. Soc., Providence, R.I., 1965.
8. Samuel G. Benidt, William R. S. Hall, and Anders O. F. Hendrickson. Upper and lower semimodularity of the supercharacter theory lattices of cyclic groups. *Comm. Algebra*, 42(3):1123–1135, 2014.
9. Bruce C. Berndt, R.J. Evans, and K.S. Williams. *Gauss and Jacobi sums*. Canadian Mathematical Society series of monographs and advanced texts. Wiley, 1998.
10. Dario Bini and Milvio Capovani. Spectral and computational properties of band symmetric Toeplitz matrices. *Linear Algebra Appl.*, 52/53:99–126, 1983.
11. Bryan Brown, Michael Dairyko, Stephan Ramon Garcia, Bob Lutz, and Michael Someck. Four quotient set gems. *Amer. Math. Monthly*, 121(7):590–599, 2014.
12. J. L. Brumbaugh, Madeleine Bulkow, Patrick S. Fleming, Luis Alberto Garcia German, Stephan Ramon Garcia, Gizem Karaali, Matt Michal, Andrew P. Turner, and Hong Suh. Supercharacters, exponential sums, and the uncertainty principle. *J. Number Theory*, 144:151–175, 2014.
13. J. L. Brumbaugh, Madeleine Bulkow, Luis Alberto Garcia German, Stephan Ramon Garcia, Matt Michal, and Andrew P. Turner. The graphic nature of the symmetric group. *Exp. Math.*, 22(4):421–442, 2013.
14. J. Bukor, P. Erdős, T. Šalát, and J. T. Tóth. Remarks on the (R)-density of sets of numbers. II. *Math. Slovaca*, 47(5):517–526, 1997.
15. Jozef Bukor, Tibor Šalát, and János T. Tóth. Remarks on R-density of sets of numbers. *Tatra Mt. Math. Publ.*, 11:159–165, 1997. Number theory (Liptovský Ján, 1995).
16. Paula Burkhardt, Alice Zhuo-Yu Chan, Gabriel Currier, Stephan Ramon Garcia, Florian Luca, and Hong Suh. Visual properties of generalized Kloosterman sums. *J. Number Theory*, 160:237–253, 2016.
17. Paula Burkhardt, Gabriel Currier, Mathieu de Langis, Stephan Ramon Garcia, Bob Lutz, and Hong Suh. An exhibition of exponential sums: visualizing supercharacters. *Proceedings of Bridges 2015: Mathematics, Music, Art, Architecture, Culture*, pages 475–478, 2015.

18. Alice Zhuo-Yu Chan, Luis Alberto Garcia German, Stephan Ramon Garcia, and Amy L. Shoemaker. On the matrix equation $XA + AX^T = 0$, II: Type 0-I interactions. *Linear Algebra Appl.*, 439(12):3934–3944, 2013.

19. Jeffrey Danciger, Stephan Ramon Garcia, and Mihai Putinar. Variational principles for symmetric bilinear forms. *Math. Nachr.*, 281(6):786–802, 2008.

20. Jean-Marie De Koninck and Armel Mercier. *1001 Problems in Classical Number Theory*. American Mathematical Society, Providence, RI, 2007.

21. Persi Diaconis and I. M. Isaacs. Supercharacters and superclasses for algebra groups. *Trans. Amer. Math. Soc.*, 360(5):2359–2392, 2008.

22. Christopher Donnay, Stephan Ramon Garcia, and Jeremy Rouse. p-adic quotient sets II: Quadratic forms. *J. Number Theory*, 201:23–39, 2019.

23. William Duke, Stephan Ramon Garcia, and Bob Lutz. The graphic nature of Gaussian periods. *Proc. Amer. Math. Soc.*, 143(5):1849–1863, 2015.

24. Ephraim Feig and Michael Ben-Or. On algebras related to the discrete cosine transform. *Linear Algebra Appl.*, 266:81–106, 1997.

25. Benjamin Fine and Gerhard Rosenberger. *Number Theory: An Introduction via the Distribution of Primes*. Birkhäuser, Boston, 2007.

26. Patrick S. Fleming, Stephan Ramon Garcia, and Gizem Karaali. Classical Kloosterman sums: representation theory, magic squares, and Ramanujan multigraphs. *J. Number Theory*, 131(4):661–680, 2011.

27. Christopher F. Fowler, Stephan Ramon Garcia, and Gizem Karaali. Ramanujan sums as supercharacters. *Ramanujan J.*, 35(2):205–241, 2014.

28. Stephan Ramon Garcia. Quotients of Gaussian Primes. *Amer. Math. Monthly*, 120(9):851–853, 2013.

29. Stephan Ramon Garcia, Yu Xuan Hong, Florian Luca, Elena Pinsker, Carlo Sanna, Evan Schechter, and Adam Starr. p-adic quotient sets. *Acta Arith.*, 179(2):163–184, 2017.

30. Stephan Ramon Garcia, Trevor Hyde, and Bob Lutz. Gauss's hidden menagerie: from cyclotomy to supercharacters. *Notices Amer. Math. Soc.*, 62(8):878–888, 2015.

31. Stephan Ramon Garcia, Elvis Kahoro, and Florian Luca. Primitive root bias for twin primes. *Exp. Math.*, 28(2):151–160, 2019.

32. Stephan Ramon Garcia, Gizem Karaali, and Daniel J. Katz. On Chebotarëv's nonvanishing minors theorem and the Biró–Meshulam–Tao discrete uncertainty principle. (submitted). https://arxiv.org/abs/1807.07648.

33. Stephan Ramon Garcia and Florian Luca. Quotients of Fibonacci numbers. *Amer. Math. Monthly*, 123(10):1039–1044, 2016.

34. Stephan Ramon Garcia and Florian Luca. On the difference in values of the Euler totient function near prime arguments. In *Irregularities in the distribution of prime numbers*, pages 69–96. Springer, Cham, 2018.

35. Stephan Ramon Garcia, Florian Luca, and Timothy Schaaff. Primitive root biases for prime pairs I: Existence and non-totality of biases. *J. Number Theory*, 185:93–120, 2018.

36. Stephan Ramon Garcia, Florian Luca, Kye Shi, and Gabe Udell. Primitive root bias for twin primes II: Schinzel-type theorems for totient quotients and the sum-of-divisors function. *J. Number Theory*, 208:400–417, 2020.

37. Stephan Ramon Garcia and Bob Lutz. A supercharacter approach to Heilbronn sums. *J. Number Theory*, 186:1–15, 2018.

38. Stephan Ramon Garcia, Bob Lutz, and Dan Timotin. Two remarks about nilpotent operators of order two. *Proc. Amer. Math. Soc.*, 142(5):1749–1756, 2014.

39. Stephan Ramon Garcia and Daniel E. Poore. On the closure of the complex symmetric operators: compact operators and weighted shifts. *J. Funct. Anal.*, 264(3):691–712, 2013.

40. Stephan Ramon Garcia and Daniel E. Poore. On the norm closure problem for complex symmetric operators. *Proc. Amer. Math. Soc.*, 141(2):549, 2013.

41. Stephan Ramon Garcia, Daniel E. Poore, and William T. Ross. Unitary equivalence to a truncated Toeplitz operator: analytic symbols. *Proc. Amer. Math. Soc.*, 140(4):1281–1295, 2012.

42. Stephan Ramon Garcia, Daniel E. Poore, and James E. Tener. Unitary equivalence to a complex symmetric matrix: low dimensions. *Linear Algebra Appl.*, 437(1):271–284, 2012.
43. Stephan Ramon Garcia, Daniel E. Poore, and Madeline K. Wyse. Unitary equivalence to a complex symmetric matrix: a modulus criterion. *Oper. Matrices*, 5(2):273–287, 2011.
44. Stephan Ramon Garcia, Vincent Selhorst-Jones, Daniel E. Poore, and Noah Simon. Quotient sets and Diophantine equations. *Amer. Math. Monthly*, 118(8):704–711, 2011.
45. Stephan Ramon Garcia and Amy L. Shoemaker. On the matrix equation $XA + AX^T = 0$. *Linear Algebra Appl.*, 438(6):2740–2746, 2013.
46. Stephan Ramon Garcia and Amy L. Shoemaker. Wetzel's problem, Paul Erdős, and the continuum hypothesis: a mathematical mystery. *Notices Amer. Math. Soc*, 62(3):243–247, 2015. (part of Erdős retrospective).
47. Stephan Ramon Garcia and George Todd. Supercharacters, elliptic curves, and the sixth moment of Kloosterman sums. *J. Number Theory*, 202:316–331, 2019.
48. Stephan Ramon Garcia and Samuel Yih. Supercharacters and the discrete Fourier, cosine, and sine transforms. *Comm. Algebra*, 46(9):3745–3765, 2018.
49. Rafael C. Gonzalez and Richard E. Woods. *Digital image processing*. Pearson, 2017. Fourth Edition.
50. Fernando Q. Gouvêa. *p-adic numbers*. Universitext. Springer-Verlag, Berlin, second edition, 1997. An introduction.
51. Ben Green and Terence Tao. The primes contain arbitrarily long arithmetic progressions. *Ann. of Math. (2)*, 167(2):481–547, 2008.
52. G. H. Hardy and J. E. Littlewood. Some problems of 'Partitio numerorum'; III: On the expression of a number as a sum of primes. *Acta Math.*, 114(3):215–273, 1923.
53. G. H. Hardy and E. M. Wright. *An introduction to the theory of numbers*. Oxford University Press, Oxford, sixth edition, 2008. Revised by D. R. Heath-Brown and J. H. Silverman, With a foreword by Andrew Wiles.
54. D. R. Heath-Brown. An estimate for Heilbronn's exponential sum. In *Analytic number theory, Vol. 2 (Allerton Park, IL, 1995)*, volume 139 of *Progr. Math.*, pages 451–463. Birkhäuser Boston, Boston, MA, 1996.
55. D. R. Heath-Brown. Heilbronn's exponential sum and transcendence theory. In *A panorama of number theory or the view from Baker's garden (Zürich, 1999)*, pages 353–356. Cambridge Univ. Press, Cambridge, 2002.
56. Shawn Hedman and David Rose. Light subsets of $\mathbb{N}$ with dense quotient sets. *Amer. Math. Monthly*, 116(7):635–641, 2009.
57. Anders Olaf Flasch Hendrickson. *Supercharacter theories of cyclic p-groups*. ProQuest LLC, Ann Arbor, MI, 2008. Thesis (Ph.D.)–The University of Wisconsin - Madison.
58. David Hobby and D. M. Silberger. Quotients of primes. *Amer. Math. Monthly*, 100(1):50–52, 1993.
59. Kenneth Ireland and Michael Rosen. *A classical introduction to modern number theory*, volume 84 of *Graduate Texts in Mathematics*. Springer-Verlag, New York, second edition, 1990.
60. Henryk Iwaniec and Emmanuel Kowalski. *Analytic number theory*, volume 53 of *American Mathematical Society Colloquium Publications*. American Mathematical Society, Providence, RI, 2004.
61. Nicholas M. Katz. *Gauss sums, Kloosterman sums, and monodromy groups*, volume 116 of *Annals of Mathematics Studies*. Princeton University Press, Princeton, NJ, 1988.
62. H. D. Kloosterman. On the representation of numbers in the form $ax^2 + by^2 + cz^2 + dt^2$. *Acta Math.*, 49(3-4):407–464, 1927.
63. Neal Koblitz. *A course in number theory and cryptography*, volume 114 of *Graduate Texts in Mathematics*. Springer-Verlag, New York, second edition, 1994.
64. Philip C. Kutzko. The cyclotomy of finite commutative P.I.R.'s. *Illinois J. Math.*, 19:1–17, 1975.
65. Serge Lang. *Math talks for undergraduates*. Springer-Verlag, New York, 1999.

66. Piotr Miska, Nadir Murru, and Carlo Sanna. On the p-adic denseness of the quotient set of a polynomial image. *J. Number Theory*, 197:218–227, 2019.
67. Piotr Miska and Carlo Sanna. p-adic denseness of members of partitions of $\mathbb{N}$ and their ratio sets. *Bulletin of the Malaysian Mathematical Sciences Society*. (in press) https://arxiv.org/abs/1808.00374.
68. Andrzej Nowicki. Editor's endnotes. *Amer. Math. Monthly*, 117(8):755–756, 2010.
69. Paul Pollack. *Not Always Buried Deep: A Second Course in Elementary Number Theory*. American Mathematical Society, Providence, RI, 2009.
70. Paulo Ribenboim. *The Book of Prime Number Records*. Springer-Verlag, New York, 2nd edition, 1989.
71. T. Šalát. On ratio sets of sets of natural numbers. *Acta Arith.*, 15:273–278, 1968/1969.
72. T. Šalát. Corrigendum to the paper "On ratio sets of sets of natural numbers". *Acta Arith.*, 16:103, 1969/1970.
73. V. Sanchez, P. Garcia, A. Peinado, J. Segura, and Rubio A. Diagonalizing properties of the discrete cosine transforms. *IEEE Transactions on Signal Processing*, 43(11):2631–2641, 1995.
74. Victoria Sanchez, Antonio M. Peinado, Jose C. Segura, Pedro Garcia, and Antonio J. Rubio. Generating matrices for the discrete sine transforms. *IEEE Transactions on Signal Processing*, 44(10):2644–2646, 1996.
75. Carlo Sanna. The quotient set of k-generalised Fibonacci numbers is dense in $\mathbb{Q}_p$. *Bull. Aust. Math. Soc.*, 96(1):24–29, 2017.
76. J.-P. Serre. *A course in arithmetic*. Springer-Verlag, New York-Heidelberg, 1973. Translated from the French, Graduate Texts in Mathematics, No. 7.
77. I. D. Shkredov. On Heilbronn's exponential sum. *Q. J. Math.*, 64(4):1221–1230, 2013.
78. Brian D. Sittinger. Quotients of primes in an algebraic number ring. *Notes Number Theory Disc. Math.*, 24(2):55–62, 2018.
79. Paolo Starni. Answers to two questions concerning quotients of primes. *Amer. Math. Monthly*, 102(4):347–349, 1995.
80. S. A. Stepanov. The number of points of a hyperelliptic curve over a finite prime field. *Izv. Akad. Nauk SSSR Ser. Mat.*, 33:1171–1181, 1969.
81. S. A. Stepanov. Estimation of Kloosterman sums. *Izv. Akad. Nauk SSSR Ser. Mat.*, 35:308–323, 1971.
82. P. Stevenhagen and H. W. Lenstra, Jr. Chebotarëv and his density theorem. *Math. Intelligencer*, 18(2):26–37, 1996.
83. Oto Strauch and János T. Tóth. Asymptotic density of $A \subset \mathbf{N}$ and density of the ratio set $R(A)$. *Acta Arith.*, 87(1):67–78, 1998.
84. Oto Strauch and János T. Tóth. Corrigendum to Theorem 5 of the paper: "Asymptotic density of $A \subset \mathbb{N}$ and density of the ratio set $R(A)$" [Acta Arith. **87** (1998), no. 1, 67–78; MR1659159 (99k:11020)]. *Acta Arith.*, 103(2):191–200, 2002.
85. Terence Tao. An uncertainty principle for cyclic groups of prime order. *Math. Res. Lett.*, 12(1):121–127, 2005.
86. S. Vajda. *Fibonacci & Lucas numbers, and the golden section*. Ellis Horwood Series: Mathematics and its Applications. Ellis Horwood Ltd., Chichester; Halsted Press [John Wiley & Sons, Inc.], New York, 1989. Theory and applications, With chapter XII by B. W. Conolly.
87. André Weil. On some exponential sums. *Proc. Nat. Acad. Sci. U. S. A.*, 34:204–207, 1948.

Projects in *(t, r)* Broadcast Domination

Pamela E. Harris, Erik Insko, and Katie Johnson

Abstract

Domination theory is a subfield within graph theory that aims to describe subsets of the vertices of a graph which satisfy certain distance properties. The original domination problem asked one to find subsets of the vertices of a graph (with minimal cardinality) so that every vertex in the graph was either in the set or adjacent to a vertex in the set. Since its development, thousands of papers on domination theory and its many variants have appeared in the literature. We focus our study on (t, r) broadcast domination, a variant with a connection to the placement of cellphone towers, where some vertices send out a signal to nearby vertices (with the signal decaying linearly along edges according to distance), and where all vertices must receive a minimum predetermined amount of this signal. The overall goal is to minimize the number of tower vertices needed to have all vertices receive the appropriate amount of signal reception. We summarize our past work with students and present many remaining open problems in this field. We end the chapter by providing some advice on how we continue to develop new research projects with and for students; although the mathematical content of the chapter is in domination theory, the suggestions can be implemented in any area.

P. E. Harris (✉)
Williams College, Williamstown, MA, USA
e-mail: peh2@williams.edu

E. Insko · K. Johnson
Florida Gulf Coast University, Fort Myers, FL, USA
e-mail: einsko@fgcu.edu; kjohnson@fgcu.edu

© Springer Nature Switzerland AG 2020
P. E. Harris et al. (eds.), *A Project-Based Guide to Undergraduate Research in Mathematics*, Foundations for Undergraduate Research in Mathematics,
https://doi.org/10.1007/978-3-030-37853-0_8

Suggested Prerequisites Discrete mathematics and a proof writing course. In addition, a first semester abstract algebra course or graph theory is preferable, but not required.

---

1   Introduction

Akira, Luke, and Gemma are working on a math project together and are taking a break in order to attend a dinner, where they and five friends will sit around a large rectangular table that sits four people on each of the longer sides. This is illustrated in Fig. 1. They want to enjoy the dinner and, in order to not be tempted to continue discussing their project, they agree to not sit next to each other nor directly across from each other. Is it possible for Akira, Luke, and Gemma to sit around this table in order to satisfy their constraints?

One example of a seating arrangement that would leave them unable to directly interact with each other has Akira sitting at spot 1, Luke at spot 4, and Gemma at 7. In this case they would neither be next to each other nor directly across from each other. So there is at least one way to sit so that they are able to enjoy time off from their project and interact with others at the dinner.

Let us shift our focus and study this toy problem through graph theory. The configuration of the seats around the table can be thought of as a graph, where the spots to sit (numbered 1 through 8) are the vertices of the graph and the edges connect two of these vertices if they are either next to each other or directly across from each other. This set up yields the graph of Fig. 2.

The seating arrangement where Akira sits at vertex 1, Luke at vertex 4, and Gemma at vertex 7 has an additional interesting property. When Leif, Luke's younger brother and one of the other dinner attendees, arrives and sits in any open seat, he must end up sitting at a distance of 1 from Akira, Luke, or Gemma. In this example, a distance of 1 means that Leif would sit either next to or directly across

Fig. 1 Dining table with sitting arrangement for eight people

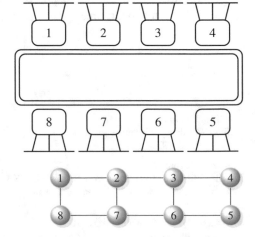

Fig. 2 Graph associated with seating arrangements of Fig. 1

from one of them. In this example, we would then say that the set of vertices (seats) taken by Akira, Luke, and Gemma constitutes a dominating set!

Let us make this mathematically precise. Let G be a (typically finite) graph with vertex set $V(G)$ and edge set $E(G)$. The classical graph domination problem seeks to find a set $S \subset V(G)$ such that every vertex v in $V(G)$ is either contained in S or adjacent to an element of S. Such a set is called a **dominating set** of G. In the case of Fig. 2 notice that the set $\{1, 4, 7\}$ forms a dominating set, as we saw previously, while another dominating set consists of all of the vertices of the graph $\{1, 2, 3, 4, 5, 6, 7, 8\}$.

Of course, it is natural to feel that taking the entire set of vertices to form a dominating set would be a waste of resources. For example, suppose that the vertices of the graph illustrated in Fig. 2 represent street corners, the edges the streets, and we want to place fire hydrants at the vertices so that every street corner either has a fire hydrant or is one block away from a fire hydrant. In this case, placing fire hydrants at every single corner is certainly a wasteful expense. In a resource allocation problem like this we want to minimize the resources used, i.e., the number of vertices in a dominating set. This is what we call the domination number of a graph.

The **domination number** of the graph G, denoted $\gamma(G)$, is the minimum cardinality among all dominating sets of G. For example, in Fig. 3, the set of colored vertices of Fig. 3a, b are both dominating sets, whereas the colored vertices in Fig. 3c do not dominate G. One can verify that the graph in Fig. 3 cannot be dominated with a single vertex. Hence we conclude that $\gamma(G) = 2$ in this instance.

The domination number problem on graphs was introduced by Berge in 1958, but the name "domination number" was coined by Ore in his 1962 book [2]. Since the publication of these books, approximately 2000 research papers have been published on domination in graphs and more than 80 domination related parameters have been defined in these papers. Most of these parameters are listed in the excellent survey text on this subject by Haynes et al. [17].

A classical domination theory problem is to determine the domination number for grid graphs. The domination numbers of $2 \times n$, $3 \times n$, and $4 \times n$ grids were first calculated by Jacobson and Kinch [18]. Chang and Clark found domination numbers for $5 \times n$ and $6 \times n$ in 1993 [7]. Notice that our previous dinner example implies that the domination number of the 2×4 grid graph is no more than 3, but showing that the domination number is not 2 requires us to show that none of the $\binom{8}{2} = 28$ choices of two vertices from eight form a dominating set. In this case a computer implementation would prove helpful; this is a great exercise for a student who is learning to code for the first time.

Fig. 3 Two dominating sets for G and a non-dominating set

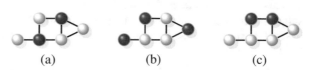

(a) (b) (c)

We continue with a brief history on the domination numbers for the grid graphs as it will motivate some of our future research projects. In his 1992 Ph.D. thesis Chang [6] proved that any large grid $G_{m,n}$ has domination number satisfying

$$\gamma(G_{m,n}) \leq \left\lfloor \frac{(n+2)(m+2)}{5} \right\rfloor - 4, \tag{1}$$

and Chang also conjectured that

$$\gamma(G_{m,n}) = \left\lfloor \frac{(n+2)(m+2)}{5} \right\rfloor - 4 \tag{2}$$

when m and n are sufficiently large. Chang proved this result by constructing a dominating set of the infinite grid $\mathbb{Z} \times \mathbb{Z}$ that uses $\frac{1}{5}$ of the vertices in any row or column. He realized that for sufficiently large m and n an optimal dominating set for an $m \times n$ grid would be close to the one he constructed for the infinite grid; in particular, to dominate the $m \times n$ grid using his set, it sufficed to simply restrict it to the $(m+2) \times (n+2)$ grid. He then showed that in condensing the dominating set of the $(m+2) \times (n+2)$ grid to the $m \times n$ grid, one could eliminate a vertex at each of the four corners. Thus he arrived at his well-known conjecture in (2). The examples below illustrate this procedure.

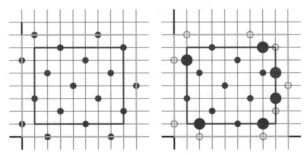

Concurrently, a number of mathematicians and computer scientists worked on dominating grid graphs. Algorithms for computing domination numbers of grid graphs were described by Hare et al. [15], and Fisher [13], but these algorithms were still not efficient to handle grids when $m, n \geq 21$ [14]. However, Chang's conjecture remained an open problem until 2011 when Gonçalves, Pinlou, Rao, and Thomassé finally proved the conjecture for $m, n \geq 16$ with a computer-aided proof [14].

This history illustrates that domination related problems often break down into subproblems that follow a fixed hierarchy of difficulty, and this is one of the reasons we feel domination related problems are great for undergraduate research projects! Students can typically construct a minimal dominating set (pattern) that gives a near optimal upper bound on the domination number of a graph very quickly.

Generalizing that pattern to a larger family of graphs usually takes a little more work, but is also very doable. Finding a combinatorial formula for the number of vertices in the pattern as the number of vertices in the graph grows is a fun challenge, where the students may use recursions, inductions, or even generating functions. Finally, proving that the pattern they have identified is optimal is often very difficult, but can sometimes be achieved using analytic arguments or a computer-aided proof.

The above results are also just a glimpse of this very fruitful area of study for graph theorists and computer scientists. In fact, there are many directions for possible research in this field. Most domination problems fall into one of two categories. The first is to consider variants of the domination parameter or domination number. The second is to consider specific families of graphs and compute domination numbers for those graphs. Our work will do both of these by focusing our study on the following domination parameters:

- domination numbers,
- distance-k domination numbers,
- (t, r) broadcast domination numbers.

We will consider certain families of finite graphs, including grid graphs, triangular matchstick graphs, trees, and even infinite graphs such as tessellations of the Euclidean plane.

We now describe the organization of this chapter. Section 2 contains the necessary background on the domination parameters mentioned above and includes exercises for you, our brave reader, to undertake in order to gather the needed experience to work on the research projects we present. In Sect. 3 we detail our past work on (t, r) broadcast domination, including Insko and Johnson's original work with students on small grid graphs. This is followed by Harris's work with students on efficient (t, r) broadcasts of the infinite square and triangular grid graphs. We will state some of the results precisely and give some of the proofs in detail to illustrate the kinds of techniques that are needed to begin working on the open projects we present. We also scatter challenge problems throughout. These exercises are more difficult, but we provide references to where detailed solutions are found. Throughout the chapter we incorporate open problems after we have presented enough background for students to begin undertaking their study.

There is great value in also discussing how we have developed some of the research project ideas presented in this chapter. Hence, we present Sect. 4, which provides a short guide to faculty and students who would like to develop further research projects. We focus on how to do this for domination theory since these types of problems lie at the heart of discrete mathematics, have many applications, and are of interest to mathematicians and computer scientists alike. However, the ideas presented there can be used to develop research ideas in other fields. Our goal is to leave the reader with the correct impression that mathematics continues to be a vibrant and lively field to which all can contribute.

2   Background

Throughout the remainder of the chapter G will be a finite connected simple graph, $V(G)$ its vertex set and $E(G)$ its edge set. Let us begin this background section with the definitions, examples, and some exercises for the dominating graph parameters we study.

2.1   Distance Domination

A **distance-k dominating set** is a subset $S \subset V(G)$ such that every vertex in $V(G)$ is within distance k of one vertex in S. The minimum cardinality among all distance-k dominating sets of G is called the **distance-k domination number** and is denoted $\gamma_k(G)$. Notice that if $k = 1$, then a distance-k dominating set is just our old friend, a dominating set. So the distance-k domination parameter is a generalization of the domination parameter.

Example 1 The Petersen graph is illustrated in Fig. 4. The red vertices in Fig. 4a–c give a dominating set, a distance-2 dominating set, and a distance-3 dominating set, respectively.

Exercise 1 Prove that the distance dominating sets given in Example 1 are the best possible (of smallest cardinality), or give other sets and prove that they are the best possible.

2.2 (t, r) Broadcast Domination

In 2015, Blessing et al. [3] introduced a generalization of the domination number and distance-k domination number of a graph called the (t, r) **broadcast domination number,** which depends on two fixed nonnegative integer parameters t and r. Their definition of this domination parameter was motivated by the idea of placing

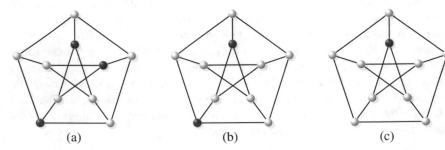

| (a) | (b) | (c) |

Fig. 4 The Petersen graph with a dominating set, a distance-2 dominating set, and a distance-3 dominating set highlighted in red

cellphone towers on the vertices of a grid graph such that each tower sends a signal to all vertices within distance t of the tower, and the signal decays linearly as it traverses an edge. We then ask what is the minimal number of towers of signal strength t needed to guarantee that every vertex in the graph receives a required minimal signal reception, which we denote by r. Notice that tower vertices give themselves signal t.

Example 2 Throughout this example let $r = 2$. We denote the locations of the vertices on the 3×5 grid graph as we would on a matrix, by specifying the row and column position of the vertex. Figure 5 presents the 3×5 grid graph with a $t = 3$ tower placed on the colored vertex located at position $(1, 1)$ on the grid graph along with the signal received by nearby vertices. Our goal is to make sure that every vertex on the graph receives a signal of at least 2 from the tower vertices in the graph. As one can see from Fig. 5 there are many vertices receiving signal 0. So we must place a second tower vertex somewhere within the graph.

Let us introduce an additional $t = 3$ tower at the vertex in position $(3, 3)$. Then the signal sent out from this new tower vertex is added to all of the surrounding vertices, as depicted in Fig. 6. Again some vertices are receiving no signal—just like those pesky bad reception locations in real life!

We still need another $t = 3$ tower. Let us place it on the grid graph at position $(1, 5)$. The result of this is depicted in Fig. 7. With this third tower we now note that every vertex in the graph is receiving a signal strength of at least 2 from the tower vertices.

You may wonder about the vertex in Fig. 7b that is highlighted in blue. While all other non-tower vertices are receiving exactly signal strength $r = 2$, the blue vertex

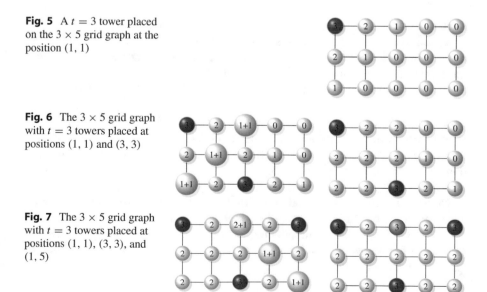

Fig. 5 A $t = 3$ tower placed on the 3×5 grid graph at the position $(1, 1)$

Fig. 6 The 3×5 grid graph with $t = 3$ towers placed at positions $(1, 1)$ and $(3, 3)$

Fig. 7 The 3×5 grid graph with $t = 3$ towers placed at positions $(1, 1)$, $(3, 3)$, and $(1, 5)$

at position $(1, 3)$ is receiving more signal than it needs. Of course if this was our cell phone we would be happy about having a stronger signal! However, if sending out more signal than is needed is costly, or if this is modeling water levels in an irrigation canal system, then we could consider whether there was a better way to arrange the tower vertices to avoid having this waste. Maybe we could even arrange the tower vertices differently so that only two were needed.

Exercise 2 Show that the vertices of the 3×5 grid graph cannot receive at least 2 worth of signal from only two $t = 3$ towers.

We now provide a precise definition of the concepts we illustrated in Example 2.

Throughout we let $t \in \mathbb{N} := \{1, 2, 3, \ldots\}$ and given two vertices $u, v \in V(G)$ the distance between u and v, denoted $d(u, v)$, is the minimum length of the paths connecting u and v. We say that $v \in V(G)$ is a **broadcast vertex** of transmission strength t if it transmits a *signal* of strength $t - d(u, v)$ to every vertex $u \in V(G)$ with $d(u, v) < t$. Given a vertex v and integer t, we define the **distance t neighborhood** of v to be

$$N_t(v) = \{w \in V(G) : d(w, v) < t\}.$$

If v is selected to be a broadcasting vertex, then we call $N_t(V)$ the **broadcasting neighborhood of** v. Given a set of broadcast vertices $S \subseteq V(G)$ each with transmission strength t, we say that the *reception* at vertex $w \in V(G)$ is

$$r(w) = \sum_{v \in S \cap N_t(w)} (t - d(w, v)).$$

That is, the reception $r(w)$ is the sum of the transmissions from its neighboring broadcast vertices in S. Then a set $S \subseteq V(G)$ is called a (t, r) **broadcast dominating set** if every vertex $w \in V$ has a reception strength $r(w) \geq r$. For a finite graph G, the minimal cardinality among all broadcast dominating sets of G is called the (t, r) **broadcast domination number** of G and is denoted $\gamma_{t,r}(G)$.

Example 3 We continue our previous example of the 3×5 grid graph G with $t = 3$ towers at the vertices v_1, v_2, and v_3 located at $(1, 1)$, $(3, 3)$, and $(1, 5)$, respectively (see Fig. 7). In this case we have shown that $S = \{v_1, v_2, v_3\}$ is a $(3, 2)$ broadcast dominating set, and that $\gamma_{3,2}(G) \leq 3$.

Note that the (t, r) broadcast domination parameter generalizes both domination and distance-k domination, since a $(2, 1)$ broadcast dominating set is a dominating set, and a $(k+1, 1)$ broadcast dominating set is a distance-k dominating set. Because of this, we focus now only on the study of (t, r) broadcast domination as doing so will yield general results.

3   Projects in (t, r) Broadcast Domination

In this section we summarize our work on (t, r) broadcast domination, on which we collaborated with students. This work is organized by the type of graph family involved: finite grid graphs, infinite square and triangular grid graphs, and triangular matchstick graphs. We scatter exercises and research problems throughout.

3.1   Grid Graphs

One great way to start thinking about projects in the area of broadcast domination is to attempt to recover patterns of tower placement on small grid graphs. In fact several of the first papers in this area were dedicated to identifying the domination numbers of $2 \times n$, $3 \times n$, and $4 \times n$ grids [8, 18].

Exercise 3 Verify that the domination numbers of the 3×5 and 5×7 grids are 3 and 9, respectively.

Blessing et al. established the (t, r) broadcast domination numbers for small grid graphs with $(t, r) \in \{(2, 2), (3, 1), (3, 2), (3, 3)\}$, and provided upper bounds for these broadcast domination numbers for arbitrarily large grid graphs [3]. We summarize the results below.

Theorem 1 (Theorems 2.1, 2.4, and 2.6 [3]) *If $n \geq 3$, then*

$$\gamma_{2,2}(G_{3,n}) = \left\lceil \frac{4n}{3} \right\rceil, \quad \gamma_{3,1}(G_{3,n}) = \left\lceil \frac{n}{3} \right\rceil, \quad and \quad \gamma_{3,2}(G_{3,n}) = \left\lceil \frac{n+1}{2} \right\rceil.$$

Theorem 2 (Theorem 2.2, 2.5, and 2.7 [3]) *If $n \geq 4$, then*

$$\gamma_{2,2}(G_{4,n}) = 2n - \left\lceil \frac{n-6}{4} \right\rceil,$$

$$\gamma_{3,1}(G_{4,n}) = \left\lfloor \frac{n+5}{7} \right\rfloor + \left\lfloor \frac{n+3}{7} \right\rfloor + \left\lfloor \frac{n+1}{7} \right\rfloor + 1, \; and$$

$$\gamma_{3,2}(G_{4,n}) = \left\lceil \frac{n+4}{5} \right\rceil + \left\lfloor \frac{n+2}{5} \right\rfloor + \left\lfloor \frac{n}{5} \right\rfloor + 1.$$

Theorem 3 (Theorem 2.3 [3]) *If $n \geq 5$, then $\gamma_{2,2}(G_{5,n}) = 2n + \left\lceil \frac{n+2}{7} \right\rceil$.*

We provide the proof to the first part of Theorem 1, as it appeared in [3], in order to illustrate the types of arguments typically used to solve these broadcast domination problems.

Proof (Part 1 of Theorem 2.1 [3]) Let $n \geq 3$. We want to show that the $(2, 2)$ broadcast domination number of the $3 \times n$ grid is

$$\gamma_{2,2}(G_{3,n}) = \left\lceil \frac{4n}{3} \right\rceil .$$

To establish an upper bound, we will describe how to construct a dominating set, denoted D_n, for the grid $G_{3,n}$. For each column of the grid, D_n will contain either the vertex in the second row or the vertices in the first and third rows. When describing a pattern, we simply use the number of vertices included from that column.

For each $n \geq 3$, the dominating set D_n will always start with the pattern 1-2 and end with the pattern 2-1, i.e., $\{(1, 2), (2, 1), (2, 3), (n - 1, 1), (n - 1, 3), (n, 2)\} \subseteq D_n$. Starting with the second column, the middle pattern will repeat -2-1-1- as many times as necessary, until column $n - 1$ is reached. See Figs. 8 and 9 for a variety of examples.

It is clear that the set D_n dominates $G_{3,n}$. It remains to show that D_n is the smallest dominating set for G_n, proving $|D_n| = \gamma_{2,2}(G_{3,n})$. Then we will still need to compute the cardinality of D_n to arrive at the result.

Notice that each D_n uses the pattern 2-1-1 as often as possible, so the only way a dominating set could possibly be more efficient would be if it used the pattern 2-1-1-1. Figure 10 demonstrates why this will not be possible. Any pattern that includes -2-1-1- will have two vertices (circled in red) that only have reception strength 1. Hence, we must dominate them using two vertices from the following column, and a 2-1-1-1 pattern would be insufficient. As an additional check for this and future patterns, our students wrote a computer program in SAGE called MPS (minimal pattern search) that searched over every pattern which begins 2-1-1, and it found that the fourth column must then contain two vertices for a dominating set.[1] Hence,

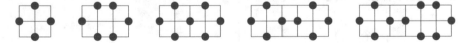

Fig. 8 Dominating sets D_n for $G_{3,n}$

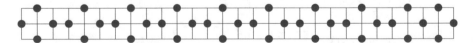

Fig. 9 Dominating set D_{29} for $G_{3,29}$

Fig. 10 Pattern containing 2-1-1

---

[1]The SAGE code for the minimal pattern search algorithm is available as a supplemental document on the Discrete Applied Mathematics website for Blessing et al.'s article [3].

no dominating set of $G_{3,n}$ can contain the pattern 2-1-1-1. Similarly, the program also shows that the patterns 2-1-0 and 1-1-1 are not dominating subpatterns. This proves our construction for D_n is the smallest dominating set for $G_{3,n}$.

Next, we use complete induction to show that $|D_n| = \left\lceil \frac{4n}{3} \right\rceil$. We will need three base cases, consisting of D_3, D_4, and D_5. The reader can (and should!) check that the equality holds, using the graphs in Fig. 8. By the principle of strong mathematical induction, assume that for all $k \leq n$ the cardinality of the set D_k is given by the formula $|D_k| = \left\lceil \frac{4k}{3} \right\rceil$. Now consider the set D_{n+1}. It can be built up from the set D_{n-2} by using D_{n-2} to dominate columns 4 through $n+1$ of $G_{3,n+1}$ and using the pattern 1-2-1, i.e., that of D_3, for the first three columns of $G_{3,n+1}$. Thus the cardinality of D_{n+1} is

$$|D_{n+1}| = |D_{n-2}| + |D_3| = \left\lceil \frac{4(n-2)}{3} \right\rceil + 4 = \left\lceil \frac{4(n+1)}{3} \right\rceil.$$

□

Although Theorems 1–3 are all results that have appeared in the literature, it is worthwhile to challenge yourself and work through some of those proofs. With this in mind we pose the following.

Challenge Problem 1 Construct dominating sets whose cardinalities are the (t, r) broadcast domination numbers for $3 \times n$ grid graphs as given above. That is, provide ways to select the placement of the towers within the grid graph, so that they create a (t, r) broadcast dominating set of the correct cardinality.

It is often straightforward to construct efficient dominating sets that give upper bounds on (t, r) broadcast domination numbers for the family of graphs you are studying, and programming is a useful tool for checking large cases. This leads us to our first open problem.

Research Project 1 Find formulas for the (t, r) broadcast domination number for grid graphs $G_{m,n}$ with $m \in \{2, 3, 4, 5\}$, $n \in \mathbb{N}$, and (t, r) different from $(2, 2)$, $(3, 1)$, $(3, 2)$, $(3, 3)$.

For smaller grids, the (t, r) broadcast domination is not yet well-understood and is much more unpredictable than it is for large grids. It would be interesting if one could develop a web app game to crowd-source the problem of finding optimal broadcast dominating sets for $m \times n$ grids when m and n are relatively small.

As we mentioned, finding closed formulas for (t, r) broadcast domination numbers for large grid graphs can be rather difficult. Another way to discover new results is by focusing on computing upper bounds of (t, r) broadcast domination number for large $m \times n$ grid graphs. Doing a good job will require the power of abstract algebra!

Our students Michael Farina and Armando Grez improved the known upper bounds established by Fata et al. [12] for the distance-k broadcast domination number of large grid graphs [11]. Their main result follows.

Theorem 4 (Theorem 1 [11]) *Assume that m and n are greater than $2(2k^2 + 2k + 1)$. Then the k-distance domination number of the $m \times n$ grid graph $G_{m,n}$ is bounded above by the following formula:*

$$\gamma_k(G_{m,n}) \leq \left\lfloor \frac{(m + 2k)(n + 2k)}{2k^2 + 2k + 1} \right\rfloor - 4.$$

The proof of Theorem 4 described an embedding of the grid graph into the integer lattice $\mathbb{Z}^2$ and the k-distance neighborhood of the graph. By describing a family of efficient dominating sets for $\mathbb{Z}^2$ as the inverse images of a ring homomorphism, and showing that for sufficiently large m and n they could remove at least one vertex from each corner of the grid graph, they were able to show the resulting set remained a dominating set, while having smaller cardinality, thereby providing an improved upper bound.

Using similar algebraic techniques, Tim Randolph, another one of our students, expanded Armando and Michael's work by establishing an asymptotically optimal upper bound for the $(t, 2)$ broadcast domination number of the grid graph when $t \geq 3$.

Theorem 5 (Theorem 1 [21]) *If $G_{m,n}$ is the grid graph with dimensions $m \times n$, and $t \geq 3$, then*

$$\gamma_{t,2}(G_{m,n}) \leq \left\lfloor \frac{(m + 2(t - 2))(n + 2(t - 2))}{2(t - 1)^2} \right\rfloor.$$

In light of this work we now provide an open problem first posed in [21].

Research Project 2 The authors of [3] and [11] systematically improve their bounds for the $(t, 1)$ broadcast domination numbers by a constant value of 4 by adjusting vertices at the corners of the grid graph $G_{m,n}$. Are similar constant improvements possible for the upper bounds on the $(t, 2)$ broadcast domination numbers of Theorem 5?

3.2   The Infinite Square Grid Graph

As we saw at the end of the last section, working with the infinite grid graph, also known as the square lattice or integer lattice, allows us to pull back information about the (t, r) broadcast domination numbers for subgraphs, in this case the finite grid graphs. Thus another direction for research is to consider methods to (t, r) dominate infinite graphs "efficiently," that is, with minimal wasted signal. Recall that we discussed wasted signal in Example 2, where the blue vertex in Fig. 7b received more signal than it required.

This approach was originally explored by Chang for regular domination theory and Blessing et al. and Drews, Harris, and Randolph in (t, r) broadcast domination theory [3, 6, 9]. In order to state those results we first need to define the density of a (t, r) broadcast. The **density** of a (t, r) broadcast is defined intuitively as the proportion of the vertices to lattice points needed in a (t, r) broadcast dominating set of the infinite grid graph, which we denote by G_∞.

Suppose that S_1 and S_2 are (t, r) broadcast dominating sets of the infinite grid graph, with densities d_1 and d_2, respectively. If $d_1 < d_2$, then when bounding the (t, r) broadcast domination number for a finite grid graph we should use the dominating set S_1 since this is "less dense" and so might require fewer vertices to dominate the finite grid graph. This shows that computing the density of (t, r) broadcast dominating sets allows us to determine which sets are more optimal. There are only a few known results on the density of (t, r) broadcasts on the infinite square grid graph. We summarize them below.

Theorem 6 (Theorems 1, 2, and 3 [9])

- *If $t \geq 1$, then $\gamma_{t,1}(G_\infty) = \dfrac{1}{2t^2 - 2t + 1}$.*
- *If $t > 2$, then $\gamma_{t,2}(G_\infty) = \dfrac{1}{2(t - 1)^2}$.*
- *If $t > 2$, then $\gamma_{t,3}(G_\infty) \leq \gamma_{t-1,1}(G_\infty)$.*

First, note that the bounds in Theorem 6 are only for $r = 1, 2, 3$. Second, note that part 3 of Theorem 6 has not been proven to be optimal but only a bound. These observations lead us naturally to the following.

Research Project 3 Determine the optimal $(t, 3)$ broadcast density on the infinite grid graph.

Research Project 4 Determine the optimal (t, r) broadcast density on the infinite grid graph for $r > 3$.

As information about the density of a (t, r) broadcast on the infinite grid graph allows us to provide bounds for the finite grid graphs, we pose the next problem.

Research Project 5 By a method similar to that employed in [21], take the bound for the density of the $(t, 3)$ broadcast as given in Theorem 6 and use it to give an upper bound on the $(t, 3)$ broadcast domination number for finite grid graphs.

Note that since the $t = 3$ bound presented in Theorem 6 (Part 3) has not been proven to be optimal we know that the resulting bounds for finite grids may be off by an amount proportional to the size of the grid. Hence we also ask: can we improve the bound for the finite grid graphs?

3.3   Infinite Triangular Grid Graph

Let us now turn our attention to another infinite graph: the **infinite triangular grid graph**, denoted $\mathbb{T}_\infty$. The vertex set of $\mathbb{T}_\infty$ can be placed in correspondence with the points $(x, y) = \left(\frac{1}{2}a - b, \frac{\sqrt{2}}{3}a\right)$, where $a, b \in \mathbb{Z}$, and we connect two vertices only when their corresponding coordinate entries are a Euclidean unit distance apart.

Figure 11 illustrates the triangular lattice with $t = 3$ towers at the red vertices. The black hexagonal outline denotes the locations where the center red tower sends a signal of 1 to the vertices lying on this outline. Vertices outside of these boundaries do not receive any signal from those particular towers.

In 2018, Pamela Harris and her students Dalia Luque, Claudia Reyes Flores, and Nohemi Sepulveda [16] considered the triangular lattice and presented a complete description of where to place tower vertices in order to efficiently (t, r) broadcast dominate the triangular lattice. **Efficient broadcasts** are those (t, r) broadcasts on the infinite triangular lattice that minimize wasted signal in the sense that every vertex that is far from broadcast towers receives exactly the signal strength needed and no more, while those vertices close to a tower only get their signal from that tower and from no other. To state our condition of an efficient (t, r) broadcast precisely we give the following.

Fig. 11 Triangular lattice
with $t = 3$ tower vertices

Definition 1 A (t, r) broadcast dominating set S for T_∞ is said to be *efficient* if

$$r(u) = \begin{cases} r & \text{if } d(u, v) \geq t - r \text{ for all } v \in S \\ t - d(u, v) & \text{if } 0 \leq d(u, v) < t - r \text{ for exactly one } v \in S. \end{cases} \tag{3}$$

Notice that Fig. 11 provides a placement of towers that efficiently $(3, 1)$ broadcast dominates T_∞. We refer to such a placement of tower vertices as a (t, r) broadcast domination pattern.

Exercise 4 Determine a placement of towers on T_∞ to efficiently $(3, 2)$ and $(3, 3)$ broadcast dominate T_∞. Are there multiple efficient $(3, 2)$ and $(3, 3)$ broadcast domination patterns?

As we mentioned, the main result in [16] provides a concrete description of where one should place towers to efficiently (t, r) broadcast dominate T_∞. This result is very general as it not only holds for all values $t \geq r \geq 1$, but also it does so in a way that is efficient.

Theorem 7 (Theorem 4.1 [16]) *Let $\alpha_1 = (1, 0)$ and $\alpha_2 = \left(\frac{-1}{2}, \frac{\sqrt{3}}{2} \right)$. If $t \geq r \geq 1$, then an efficient (t, r) broadcast domination pattern for T_∞ is given by placing towers at every vertex of the form*

$$[(2t - r)x + (t - r)y]\alpha_1 + [tx + (2t - r)y]\alpha_2 \tag{4}$$

with $x, y \in \mathbb{Z}$.

Challenge Problem 2 Fix values of $t \geq r \geq 1$ and prove that placing tower vertices as described in Theorem 7 efficiently (t, r) broadcast dominates T_∞.

Challenge Problem 3 Prove Theorem 7.

A second efficient (t, r) broadcast domination pattern for $\mathbb{T}_\infty$ is acquired by the reflection of the pattern presented in Theorem 7 across the line through the origin and the point $\alpha_1 + \alpha_2$. This gives the following result.

Corollary 1 *If* $t \geq r \geq 1$, *then a second efficient* (t, r) *broadcast domination pattern is obtained by placing towers at every vertex of the form*

$$[tx + (2t - r)y]\alpha_1 + [(2t - r)x + (t - r)y]\alpha_2 \tag{5}$$

with $x, y \in \mathbb{Z}$.

Figure 12 illustrates efficient (t, r), broadcast dominating patterns for $\mathbb{T}_\infty$ when $t = 4$ and $1 \leq r \leq 4$. Having found two efficient (t, r) broadcast domination patterns for $\mathbb{T}_\infty$ we pose the following problem, first appearing in [16].

Research Project 6 Determine whether or not there are other efficient (t, r) broadcast dominating patterns for $\mathbb{T}_\infty$.

It was interesting to have found these efficient (t, r) broadcast dominating patterns. We wonder whether this is unique to the infinite triangular lattice or if there are other families of graphs which have efficient (t, r) dominating patterns for all $t \geq r \geq 1$.

Research Project 7 Find families of graphs (finite or infinite) that can be efficiently (t, r) broadcast dominated for all $t \geq r \geq 1$.

Theorem 7 allowed Harris, Luque, Reyes Flores, and Sepulveda to compute upper bounds for the (t, r) broadcast domination numbers of **triangular matchstick graphs**, denoted T_n, which are subgraphs of $\mathbb{T}_\infty$ with vertex set

$$\left\{ \left(\frac{1}{2}a - b, \frac{\sqrt{3}}{2}a \right) \ : \ a, b \in \mathbb{Z}, 0 \leq b \leq a \leq n \right\}.$$

Figure 13 illustrates the triangular matchstick graphs T_n for $n = 1, 2, 3, 4$.

To find upper bounds for the (t, r) broadcast domination numbers for T_n we first must figure out when the pattern given in Theorem 7 repeats along a horizontal line in the lattice. In other words, if we have a tower at the origin we want to find

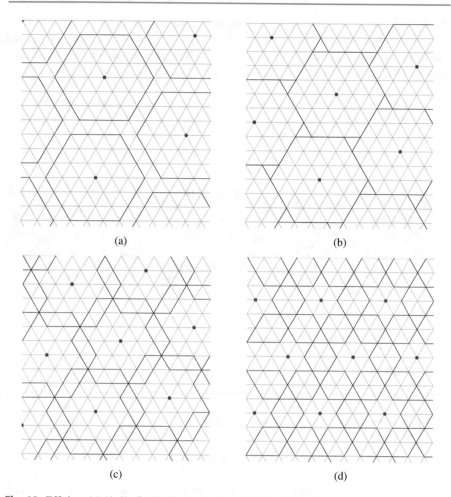

Fig. 12 Efficient (**a**) $(4, 1)$, (**b**) $(4, 2)$, (**c**) $(4, 3)$, and (**d**) $(4, 4)$ broadcast dominating sets on $\mathbb{T}_\infty$

Fig. 13 Triangular
matchstick graphs T_1, T_2, T_3,
and T_4

the closest instance where a tower is placed on a point of the form $m\alpha_1$, that is a
point with coordinates $(m, 0)$. This value then allows us to tile T_n using the smaller
triangles of size T_m (with possibly some portion of T_n unaccounted for as $n = km + b$, where $0 \leq b \leq m - 1$). Then we count how many towers are needed to dominate
T_m and multiply this by the number of occurrences of T_m needed to tile most of, if
not all of T_n. By subtracting some of the double counting along the interior edges of

the T_m's as they tile T_n, and accounting for the number of towers needed to dominate the missing portion of T_n which was not tiled by T_m's we get an upper bound for the (t, r) broadcast domination number of T_n. Using this technique we established bounds for $\gamma_{t,r}(T_n)$, when $(t, r) \in \{(2, 1), (3, 1), (3, 2), (4, 1), (4, 2), (4, 3), (t, t)\}$ [16].

What this technique does not account for is a potential rearrangement of the towers along the outer boundary of T_n that would allow us to decrease the number of towers needed. It is possible that this would give a better bound for the (t, r) broadcast domination number of T_n, which leads to the following open problem.

Research Project 8 Accounting for a rearrangement of the tower vertices on the boundary of T_n will likely give better bounds for $\gamma_{t,r}(T_n)$ for $(t, r) \in \{(2, 1), (3, 1), (3, 2), (4, 1), (4, 2), (4, 3), (t, t)\}$. More generally, by using the technique presented in [16] and by reducing the number of towers needed on the boundary of T_n, find sharper bounds for $\gamma_{t,r}(T_n)$, for $t > r \geq 1$.

3.4   Other Families of Graphs

We now present open (t, r) broadcast domination problems on a variety of graphs.

Research Project 9 Consider the family of graphs called trees and find the (t, r) broadcast domination number of these graphs.

As before, if finding closed formulas is difficult, then one can resort to finding upper bounds for these domination numbers. We also could further specialize the trees. For example, one could restrict to full binary trees, trees in which every vertex other than the leaves (degree one vertices) has two children. See Fig. 14 for an example of a full binary tree. These graphs have a lot of symmetry that one could exploit in order to establish results.

Fig. 14 An example of a full binary tree

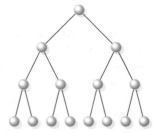

We could also consider (t, r) broadcast domination numbers for hypercubes. Recall that the vertices of an n-dimensional hypercube are the n-tuples of 0's and 1's, and two vertices u and v are connected by an edge if u and v differ only in a single coordinate. For example, the two-dimensional hypercube has vertices $v_1 = (0, 0)$, $v_2 = (1, 0)$, $v_3 = (0, 1)$, $v_4 = (1, 1)$, and there are edges from v_1 to v_2, from v_2 to v_3, from v_3 to v_4, and from v_4 to v_1, as expected since the two-dimensional hypercube is just a square.

Research Project 10 For varying values of t and r, determine (t, r) broadcast domination numbers (or bounds) for the n-dimensional hypercube.

Fun fact: A four-dimensional hypercube is called a tesseract. The previous research problem is asking you to "dominate" the tesseract. We hope you use that power only for good![2] Figure 15 illustrates a tesseract that has been projected from four-dimensional space to three-dimensional space, by using a Schlegel diagram [19].

A famous family of graphs are the generalization of the Petersen graph we presented in Fig. 4. The generalized Petersen graphs are defined by two positive

Fig. 15 A tesseract projected into 3-space as a Schlegel diagram

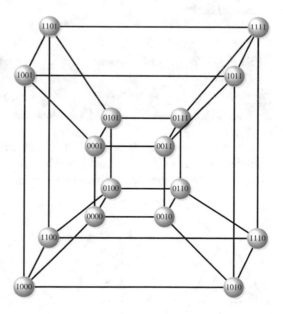

integer parameters $n \geq 3$ and $1 \leq k \leq \lfloor \frac{n-1}{2} \rfloor$ and are denoted $P(n, k)$. The parameter n denotes the number of sides for the outer regular polygon, which is a cycle graph on n vertices, and also the number of vertices of the interior star polygon, while the parameter k denotes how the vertices of the inner star polygon are connected: each vertex in the interior star polygon is connected to the kth vertex to its left and right. We also then identify corresponding vertices in the inner and outer polygons and connect them with edges. Note that the Petersen graph is $P(5, 2)$ since both the exterior and interior polygons have five vertices, and the interior star polygon has edges connecting every other vertex. See Fig. 16 for other examples of generalized Petersen graphs.

Research Project 11 For varying values of t and r, determine (t, r) broadcast domination numbers (or bounds) for the generalized Petersen graphs $P(n, k)$.

For another research direction, let us return to thinking about the infinite square and triangular grid graphs. These graphs are in fact regular tilings of the Euclidean plane. So we could consider other tilings of the Euclidean plane and think of them as graphs! One starting case is to study the remaining regular tiling of the plane: the hexagonal lattice.

Research Project 12 Consider the hexagonal lattice and determine the placement of towers to efficiently (t, r) broadcast dominate the hexagonal lattice for different values of t and r.

However, there is no reason to restrict our work to only regular tilings of the Euclidean plane. In fact we could consider any tessellation: a tiling of a plane using one or more geometric shapes with shapes neither overlapping nor leaving

 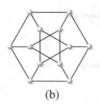 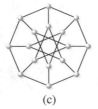 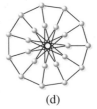

| (a) | (b) | (c) | (d) |

Fig. 16 Examples of generalized Petersen graphs. (a) $P(3, 1)$. (b) $P(6, 2)$. (c) $P(8, 3)$. (d) $P(11, 5)$

any gaps. By interpreting a tessellation as a graph, as we did in the regular square and triangular tessellation of the plane through the infinite square and triangular grid graphs, we can then work on the following project.

Research Project 13 For any tessellation of the plane, determine the placement of towers to efficiently (t, r) broadcast dominate the graph associated to this tessellation for different values of t and r.

For a concrete example take the tessellation of the plane presented in Fig. 17, which is a semiregular tiling of the plane and has one type of vertex (notice all vertices have degree 3), but has more than one type of faces (in this case a square and an octagon). We remark that the many symmetries of these tessellations should allow one to find patterns to exploit in any arguments and supporting proofs. Also the sheer number of tilings of the plane creates numerous research projects for students to undertake. For some of the tilings of the plane we recommend seeing [1].

There is also no reason to restrict ourselves to tessellations of the plane. We can go wild and study any space-filling polyhedron! For concrete projects we state the following.

Research Project 14 Compute densities for (t, r) broadcast dominating sets of $\mathbb{Z}^n$.

Of course the above project may be very difficult, so one could restrict the values of (t, r) or set $n = 3$. Then one could use these densities to give bounds for the three-dimensional grid graphs of dimension $\ell \times m \times n$.

Fig. 17 A semiregular tiling of the plane with regular octagons and squares

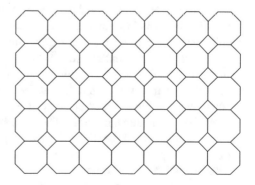

Research Project 15 Find (t, r) broadcast dominating patterns for the three-dimensional triangular lattice. Are these patterns efficient? Use these patterns to give bounds on the (t, r) broadcast domination number for tetrahedrons whose base is the triangular matchstick graph T_n.

We end this section by mentioning a few other possible directions for research:

- Introduce a cost of a tower, which depends on t the strength of the broadcast, with the idea that stronger towers are more expensive than weaker towers. If we fix r and allow towers with varying broadcast strengths, i.e., costs, how do we minimize the cost of providing every vertex with reception r? This would generalize the broadcast domination parameter that was introduced by Dunbar et al. [10].
- Consider directed graphs on which signal only flows in certain directions. What would be the directed (t, r) broadcast domination number for grid graphs? This would generalize the directed domination numbers of graphs which have been previously studied by Caro and Henning [4, 5].
- Mix and match approach: Use multiple types of towers on directed graphs.

The only constraint in finding new research problems in this area is time, which unfortunately is finite.

4 Developing Accessible Research Projects

As faculty members we often have an idea of a project we would like to work on, yet this problem, even if accessible, might not be something the student is interested in working on. When considering a student research project involving domination theory, or any other field for that matter, we recommend you first spend time with students brainstorming original variations of problems in the field. Many of our most interesting projects have arisen in this manner, as the students will have ideas that would never occur to us with our expert blind spots.

One common way that our students have gained experience developing research projects is through an honors contract or a similar coursework modification that delves deeper into the material. One of these course-enriching experiences for mathematics students might be to read extra research papers that go above and beyond the usual course material. The student may also conduct their own research project, which would be limited to a single semester, but this could develop into a senior project or honors thesis. However, at our institutions, as at many others, a mathematics honors thesis often takes the following form: the faculty mentor/research advisor chooses a problem they would like to work on, they provide the student several papers to read to catch them up on background material, and finally the student and faculty member plan meetings to begin working on proving new results, which may require writing code to make some initial conjectures to explore.

While the described system can provide a valuable experience, we suggest considering the following alternative, which is especially relevant for young mathematicians. The professor shows the student a game or puzzle such as the original domination problem. (Even better, the student chooses a problem they found especially interesting!) They may then discuss some simple applications such as cops patrolling a downtown grid of streets, security guards in an art gallery, or radio broadcast towers. The *student* is then asked to brainstorm a question without too much guidance from the mentor. Otherwise, it is too easy to accidentally push a student towards a problem that has already been studied. Naturally, these are the ones we gravitate to as experts, because we know of them.

In this way, our students have developed questions such as minimal patrolling domination (where cops can move) and limited attention span domination (where a guard can only look in k directions at a time). These are both interesting questions that, as far as we know, were previously unstudied. They are described in more detail in Sect. 4.2.

Next, the student does a literature review. This can be tricky because the student will probably develop different notations or use their own names for certain definitions that could already exist in the literature. Hence, this is the time for a research advisor to offer guidance. As the student searches for papers based on keywords that are somewhat related to the problem they have developed, students are learning the background for their problem! They are not reading papers a professor has assigned to them as required. Instead, they are finding the papers, skimming them, maybe choosing to read more in-depth, and then summarizing them to share with the faculty mentor.

Flipping the usual relationship between faculty and student in this manner can be extremely beneficial to both parties. The student feels empowered, as they have control over what to read based on what they find interesting and relevant. They are starting to develop their identity as a researcher and mathematician. On the other side, the faculty mentor is provided with new inspiration. They are introduced to new papers they may not have read before, conveniently summarized by the student.

Throughout, the student should write what they have learned. Using the research they have found and the standard notation they have encountered along the way, they can now state their problem in a clear and mathematically precise way. If, as part of their reading, they discover that their problem has been solved, then they have learned something new, regarding a problem in which they were keenly interested. In fact, this is wonderful! They are already thinking like a real mathematician, proven by the fact that they developed a problem on which other mathematicians have published.

If they have developed a new problem, then this is also wonderful! At this point, they may even get started on working to find a solution. As an alternative, and depending on the situation and time constraints, consider repeating the above process for three or four different problems. Why should this be done rather than working on the new problem? Working on a new problem is always exciting, but the student–faculty pair should consider how much progress can actually be made on the particular problem. Sometimes, this is seen more clearly with a little distance from the problem. Developing, researching, and writing up three or four problems gives the student the opportunity to choose their longer research problem from a selection of interesting problems. If they will be working on this problem for multiple semesters, it is critical that they choose something that can hold their attention, and that contains a lot of different questions that can be attacked. Moreover, developing a set of possible research projects is helpful to both the faculty (as they work with future students) and to the student (perhaps the first project does not work out or they publish and need a new topic!).

Finally, this process more closely imitates what career mathematicians actually do. It is worth reconsidering why we do research with students in the manner that we do. Likely, we want to give them the same (presumably inspiring) experience we had as undergraduates that motivated us to pursue a graduate degree. But we should also work to inspire them as individuals capable of productive research. Giving them the power to develop their own problems allows them to exercise their mathematical creativity in a way that is often missing from standard coursework. Helping them improve their abilities regarding literature reviews and written communication is one of the most versatile skills we can provide them.

This is another reason it is ideal to repeat this process multiple times. It is difficult to learn a new skill by only doing it once. If students can undergo the process in Fig. 18 a few times, then what they have learned will help them develop a skill needed beyond any one academic course.

4.1   Sample Honors Contract

Below we present a copy of part of an Honors Contract at Florida Gulf Coast University, as inspiration to what such a proposal might entail.

[...] Thus, [student] will practice learning to think like a mathematician by

1. developing her own questions that build on material learned in class,
2. researching what is currently known about the question through a literature search, and
3. writing up a summary of findings to share with the class.

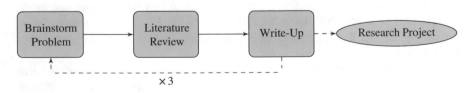

Fig. 18 An alternative method to the standard undergraduate research project

As many math majors do graph theory research for their senior capstone project, this experience will not only help [student] prepare to tackle her own project, but will likely lead to research ideas for other students.

The learning outcomes include navigation of arXiv and research materials, applications of graph theory, a deeper understanding of research methods and how research develops, and the scope of how graph theory applies to the world and other fields.

One unique result of this experience will be developing the skill of "asking the right question," which is not usually covered in college course. Students are always being given a problem to solve, not asked to develop and research one on their own (or in this case, at least five).

4.2   Student Developed Projects

We now summarize some projects developed by students under the honors contract as detailed in Sect. 4.1.

A common application of domination on grids is that of police officers guarding a downtown grid of streets. This is a natural fit for (t, r) broadcast domination, as officers can see a certain distance, and if an officer is farther away from an intersection, we would likely want multiple officers able to respond, which would mean the reception at every vertex, i.e. intersection, should be at some minimum level. Past students wondered what might happen if officers were allowed to patrol/move along the edges of the graph, or in fact if they were required to patrol through the graph.

The simplest approach to add the requirement that the officers must be constantly moving along the edges of the graph would be to find two domination sets that are "adjacent" to each other, i.e., two sets such that there is a bijection between the domination sets that maps vertices to adjacent vertices. This led our students to pose the following.

> **Research Project 16** Find the patrolling domination number for grids and other graphs, i.e., the minimum number of vertices in a dominating set that has an adjacent dominating set of the same cardinality.

The last variation we share is inspired by the same application. Suppose a group of cops is guarding a set of intersections represented by a graph. At any given moment, each cop can only see in a limited number of directions. In light of this, students proposed a "limited attention span" domination problem, where each vertex may only dominate a limited number of other vertices besides itself, say ℓ.

In particular, define a function $s : V \to V$ that maps vertices to vertices so that $s(v) = w$ means w dominates or watches v. Let the security set S be the set of all

vertices that map to themselves, i.e., $S = \{v \in V \mid s(v) = v\}$. (Due to restriction
(2) below, $S = \text{Range}(s)$.) To be a *limited attention span dominating function*, the
function must have the following properties:

1. If $s(v) = w$, then vertices v and w must be adjacent.
2. If $s(v) = w$ for some vertex v, then $s(w) = w$; in other words, if w watches
   some vertex, then it must be part of the security set.
3. Each vertex in the security set watches at most ℓ vertices besides itself, i.e.,
   $|s^{-1}(w)| \le \ell + 1$.

Then, the goal is to minimize the size of the security set over all valid limited
attention span dominating functions. For an example, see the grid in Fig. 19, where
the arrows represent the function s by indicating the directions the security set is
watching.

In addition, one could expand this to broadcast domination by requiring that
each vertex in the dominating set is given strength ℓ, but looking farther in a single
direction requires more strength, as the cop would be focusing. For this variation,
the restriction described in item (1) above would be removed, and restriction (3)
would be changed to:

3. Each vertex in the security set uses total strength at most ℓ, where watching a ver-
   tex at distance d requires a strength of d units; specifically, $\sum_{v \in s^{-1}(w)} d(v, w) \le \ell$ for every vertex $w \in S$.

Note that this inequality also implies that a cop cannot watch an intersection at a
distance greater than ℓ. This leads to the following problem.

> **Research Project 17** Explore the limited attention span domination problem
> by changing the parameter ℓ and determining how the size of the security
> set S increases. Consider both the domination and the broadcast domination
> variants. When $\ell = \Delta(G)$, the maximum degree of any vertex, this will be
> the standard model of domination or broadcast domination. When $\ell = 0$, we
> must include all vertices. What happens in between?

Fig. 19 The 3×4 grid and a
limited attention span
security set with $\ell = 2$. The
arrows indicate the directions
the guards are watching

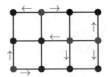

Note that a variation of this problem is described by Mamidisetty, Ghamande, Ferrara, and Sastry in [20] from a topological perspective. They provide fascinating applications to data aggregation and bandwidth limiting, as a network may have connected computers but limited capabilities to transmit data.

Acknowledgements We end by remarking that our students have been the research leaders for these and many other projects. As faculty, we believe that it is our responsibility to guide them and support them, but not to impose on them a problem to solve. Students are inquisitive and have wonderful intuition for asking interesting and challenging mathematical questions. We have continued to leverage their curiosity to develop more research projects than we could solve in a lifetime! We are thankful for having been part of our students' experience in research and for their hard work on developing and researching these problems.

References

1. Euclidean tilings by convex regular polygons. [Online; accessed 20-March-2019].
2. Claude Berge. *The theory of graphs and its applications*. Translated by Alison Doig. Methuen & Co. Ltd., London; John Wiley & Sons Inc., New York, 1962.
3. David Blessing, Katie Johnson, Christie Mauretour, and Erik Insko. On (t, r) broadcast domination numbers of grids. *Discrete Appl. Math.*, 187:19–40, 2015.
4. Yair Caro and Michael A. Henning. A greedy partition lemma for directed domination. *Discrete Optim.*, 8(3):452–458, 2011.
5. Yair Caro and Michael A. Henning. Directed domination in oriented graphs. *Discrete Appl. Math.*, 160(7-8):1053–1063, 2012.
6. Tony Yu Chang. *Domination numbers of grid graphs*. ProQuest LLC, Ann Arbor, MI, 1992. Thesis (Ph.D.)–University of South Florida.
7. Tony Yu Chang and W. Edwin Clark. The domination numbers of the $5 \times n$ and $6 \times n$ grid graphs. *J. Graph Theory*, 17(1):81–107, 1993.
8. Tony Yu Chang and W. Edwin Clark. The domination numbers of the $5 \times n$ and $6 \times n$ grid graphs. *J. Graph Theory*, 17(1):81–107, 1993.
9. Benjamin F. Drews, Pamela E. Harris, and Timothy W. Randolph. Optimal (t,r) broadcasts on the infinite grid. *Discrete Applied Mathematics*, 255:183–197, 28 February 2019.
10. Jean E. Dunbar, David J. Erwin, Teresa W. Haynes, Sandra M. Hedetniemi, and Stephen T. Hedetniemi. Broadcasts in graphs. *Discrete Appl. Math.*, 154(1):59–75, 2006.
11. Michael Farina and Armando Grez. New upper bounds on the distance domination numbers of grids. *Rose-Hulman Undergrad. Math. J.*, 17(2):Art. 7, 133–145, 2016.
12. Elaheh Fata, Stephen L Smith, and Shreyas Sundaram. Distributed dominating sets on grids. In *American Control Conference (ACC), 2013*, pages 211–216. IEEE, 2013.
13. David C. Fisher. The domination number of complete grid graphs. *manuscript*.
14. Daniel Gonçalves, Alexandre Pinlou, Michaël Rao, and Stéphan Thomassé. The domination number of grids. *SIAM J. Discrete Math.*, 25(3):1443–1453, 2011.
15. E. O. Hare, S. T. Hedetniemi, and W. R. Hare. Algorithms for computing the domination number of $k \times n$ complete grid graphs. In *Proceedings of the seventeenth Southeastern international conference on combinatorics, graph theory, and computing (Boca Raton, Fla., 1986)*, volume 55, pages 81–92, 1986.
16. Pamela Harris, Dalia Luque, Claudia Reyes Flores, and Nohemi Sepulveda. Broadcast domination of triangular matchstick graphs and the triangular lattice. *arXiv preprint arXiv:1804.07812*, 2018.
17. Teresa W. Haynes, Stephen T. Hedetniemi, and Peter J. Slater. *Fundamentals of domination in graphs*, volume 208 of *Monographs and Textbooks in Pure and Applied Mathematics*. Marcel Dekker, Inc., New York, 1998.

18. M. S. Jacobson and L. F. Kinch. On the domination number of products of graphs. I. *Ars Combin.*, 18:33–44, 1984.

19. A.L. Loeb. *Schlegel Diagrams*. Space Structures. Design Science Collection. Birkhäuser, Boston, MA, 1991.

20. K. K. Mamidisetty, M. Ghamande, M. Ferrara, and S. Sastry. A domination approach to clustering nodes for data aggregation. In *IEEE GLOBECOM 2008 - 2008 IEEE Global Telecommunications Conference*, pages 1–5, Nov 2008.

21. Timothy W. Randolph. Tight bounds for (t,2) broadcast domination on finite grids. *arXiv preprint arXiv:1805.06058*, 2017.

Squigonometry: Trigonometry in the p-Norm

William E. Wood and Robert D. Poodiack

Abstract

We can define the traditional trigonometric functions in several different ways: via differential equations, via an arclength definition on the unit circle $x^2 + y^2 = 1$, or via an analytic approach. In this project, we adapt these approaches to define analogous functions for a unit *squircle* $|x|^p + |y|^p = 1$, $p \geq 1$. As we develop these functions using only elementary calculus, we will ponder the importance and role of π, and glimpse some very deep ideas in elliptic integrals, special functions, non-Euclidean geometry, number theory, and complex analysis.

Suggested Prerequisites Knowledge of basic trigonometric functions, differential and integral calculus are a must. Some knowledge of differential equations is helpful, but not essential.

1 Introduction

The circle is a "perfect" object, and our familiarity with it can cause us to lose sight of just how remarkable it is. It is one of the first geometric shapes we encounter in school; children can identify it easily. The unit circle can be described as the set of

W. E. Wood
Mathematics Department, University of Northern Iowa, Cedar Falls, IA, USA
e-mail: bill.wood@uni.edu

R. D. Poodiack (✉)
Department of Mathematics, Norwich University, Northfield, VT, USA
e-mail: rpoodiac@norwich.edu

© Springer Nature Switzerland AG 2020
P. E. Harris et al. (eds.), *A Project-Based Guide to Undergraduate Research in Mathematics*, Foundations for Undergraduate Research in Mathematics,
https://doi.org/10.1007/978-3-030-37853-0_9

263

points (x, y) in the plane satisfying the equation $x^2 + y^2 = 1$. The sine and cosine functions emerge as natural functions with which to describe it.

In this project, we look to enhance appreciation of the circle's perfection by introducing some imperfection, by developing an analog of trigonometry for something that is *not quite* a circle.

Our primary model is the *unit p-circle*, or "squircle," a superellipse defined as the set of points (x, y) in the plane satisfying the equation $|x|^p + |y|^p = 1$ for some real $p \geq 1$. This generalizes the Euclidean circle which emerges as the case $p = 2$ (note that we can drop the absolute values when p is a rational number whose reduced fraction has an even numerator). We will examine the $p = 4$ case quite closely for examples.

2 Defining Trigonometric Functions

We often define the sine and cosine functions in one of three ways:

- via differential equations,
- geometrically, as coordinates of points on the unit circle $x^2 + y^2 = 1$,
- analytically, via the use of definite integrals.

Conveniently, all three definitions yield the same functions. Let us examine these three methods separately, as these will be the directions for our generalizations to squigonometric functions.

2.1 Differential Equations Approach

This approach is inspired by methods discussed in [6] and [7] of using initial value problems (IVP's) to develop transcendental functions in a first-year calculus course.

Recall from calculus that $\frac{d}{dt} \cos t = -\sin t$ and $\frac{d}{dt} \sin t = \cos t$. We view these as defining properties for the cosine and sine functions. By choosing to start the parameterization on the x-axis, we further adopt the initial conditions $\cos 0 = 1$ and $\sin 0 = 0$. Thus we can say that cosine and sine are the functions satisfying

$$\begin{cases} x'(t) = -y(t) \\ y'(t) = x(t) \\ x(0) = 1 \\ y(0) = 0, \end{cases} \tag{1}$$

where x corresponds to cosine and y to sine. This is an example of a *coupled initial value problem* (or CIVP). It turns out problems like these always have unique solutions [5]. Therefore, we may *define* cosine and sine to be the unique solution to (1).

Exercise 1 Consider the function $f(t) = u(t)^2 + v(t)^2$, where $u(t) = \cos(t)$ and $v(t) = \sin(t)$. Use the properties (1) to show that $f'(t)$ is identically 0 and that $f(0) = 1$. Conclude that $f(t)$ is identically 1 and thus $\cos^2 t + \sin^2 t = 1$ for all t.

Exercise 2 Use similar techniques to show that cosine is even and sine is odd.

2.2 Unit Circle Approach

The unit circle is the set of points (x, y) that are exactly 1 unit from the origin. Such points satisfy the equation $x^2 + y^2 = 1$.

Any line through the origin intersects the unit circle at a point t units away from $(1, 0)$ as measured along the arc of the circle. We define the *radian* as a measure of angle such that one radian subtends an arc of the unit circle of length 1. Thus the line passing through our point subtends an angle of t radians, and we define the cosine and sine of t to be the x- and y-coordinates, respectively of our intersection point (Fig. 1). Thus we have that $\cos^2 t + \sin^2 t = 1$ for all $t \in \mathbb{R}$.

As a full circuit of the unit circle subtends 2π radians, each point on the unit circle is associated with infinitely many angles, all multiples of 2π radians apart. In particular, this means that the cosine and sine functions are periodic—they repeat their values every 2π radians:

$$\cos(t + 2\pi \cdot k) = \cos t$$

$$\sin(t + 2\pi \cdot k) = \sin t$$

Fig. 1 The first quadrant of the unit circle

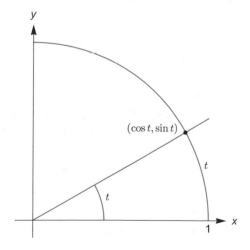

for all $k \in \mathbb{Z}$.

The parameter t in cosine and sine thus simultaneously specifies both the arclength and the angle. It also determines the area of the sector. We see that a sector traced out by t radians has an area of $\frac{t}{2\pi}$ times the area of the whole circle, or $\frac{t}{2\pi} \times \pi = \frac{t}{2}$. Thus the angular, arclength, and areal parameterizations all lead to the same functions. We will come to appreciate this remarkable feature of Euclidean geometry when we lose it in the geometries we will soon explore.

2.3   Analytic Approach

We can also define the sine and cosine functions as solutions to integral equations. The sine function is the function $S(t)$ such that

$$\int_0^{S(t)} \frac{1}{\sqrt{1-u^2}} \, du = t. \tag{2}$$

In a similar manner, we can define the cosine function to be the function $C(t)$ such that

$$\int_{C(t)}^1 \frac{1}{\sqrt{1-u^2}} \, du = t. \tag{3}$$

Suppose that (1) holds. The result of Exercise 1 shows that $\cos^2 t + \sin^2 t = 1$ and we have the sine and cosine functions from our unit circle approach.

Now let $S(t)$ and $C(t)$ be, respectively, the sine and cosine functions defined in Eqs. (2) and (3). We first claim that $C(t) = \sqrt{1 - S^2(t)}$ (and thus $S(t) = \sqrt{1 - C^2(t)}$ and $C^2(t) + S^2(t) = 1$). If we substitute $s = \sqrt{1 - u^2}$ into Eq. (2), then $u = \sqrt{1 - s^2}$, $du = -\frac{s}{\sqrt{1-s^2}} \, ds$ and

$$\int_0^{S(t)} \frac{1}{\sqrt{1-u^2}} \, du = \int_1^{\sqrt{1-S^2(t)}} \frac{1}{s} \cdot \frac{-s}{\sqrt{1-s^2}} \, ds = \int_{\sqrt{1-S^2(t)}}^1 \frac{1}{\sqrt{1-s^2}} \, ds = t.$$

Comparing this last integral to (3), we see that $C(t) = \sqrt{1 - S^2(t)}$. Furthermore, Eqs. (2) and (3) show that $S(0) = 0$ and $C(0) = 1$.

Taking the derivatives across Eqs. (2) and (3), we obtain

$$1 = \frac{1}{\sqrt{1-S^2(t)}} \cdot S'(t) = -\frac{1}{\sqrt{1-C^2(t)}} \cdot C'(t)$$

and so $S'(t) = \sqrt{1 - S^2(t)} = C(t)$ and $C'(t) = -\sqrt{1 - C^2(t)} = -S(t)$. Thus S and C fulfill the CIVP (1) and must be the same cosine and sine functions found there.

Once again, we see these three approaches inevitably yield the same cosine and sine functions.

Exercise 3 Deduce (2) and (3) directly from CIVP (1).

3 Squigonometric Functions

We will now develop an analog of trigonometry for the unit p-circle, the set of points (x, y) in the plane satisfying $|x|^p + |y|^p = 1$, $p \geq 1$, depicted in Fig. 2. The absolute values are necessary in the defining equation, as our next exercise shows.

Exercise 4 Use a computer algebra system to implicitly plot the equation $x^3 + y^3 = 1$ where x ranges from -2 to 2. Do the same for $x^5 + y^5 = 1$.

3.1 Differential Equations Approach

We first generalize our differential equations approach in order to define our squigonometric functions. Consider the function $g(t) = u(t)^p + v(t)^p$ for some $p \geq 1$. We will design a CIVP whose solutions $u(t)$ and $v(t)$ will make $g(t)$ constant. To wit:

$$\begin{cases} x'(t) = -y(t)^{p-1} \\ y'(t) = x(t)^{p-1} \\ x(0) = 1 \\ y(0) = 0. \end{cases} \tag{4}$$

Fig. 2 Unit squircles $|x|^p + |y|^p = 1$ for $p = 1, 2, 5,$ and 10

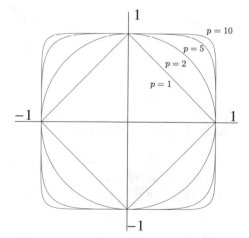

Thus if $u(t)$ and $v(t)$ are the solutions to (4) then

$$g'(t) = p \cdot u(t)^{p-1} u'(t) + p \cdot v(t)^{p-1} v'(t)$$

$$= -p \cdot u(t)^{p-1} v(t)^{p-1} + p \cdot v(t)^{p-1} u(t)^{p-1} = 0.$$

so $g(t)$ must be constant. Since $g(0) = 1$, we then have that $g(t) \equiv 1$, as desired.

We denote $\cos_p(t) = x(t)$ and $\sin_p(t) = y(t)$ to be the unique pair of functions that solve (4). (The *cosquine* and *squine* functions, if we prefer.) Note that $\cos_p^p(t) + \sin_p^p(t) = 1$. For $p = 2$ or any other positive even integer, we get the entire unit squircle from this equation. Otherwise, we can either take absolute values (as Exercise 4 showed, $|\cos_p(t)|^p + |\sin_p(t)|^p = 1$ works), or we can deftly restrict the interval of solution for Eqs. (4) and then extend the solutions by symmetry. We will see how to do this shortly.

We can define the other squigonometric functions through the usual definitions using ratios of the squine and cosquine functions:

$$\tan_p(t) = \frac{\sin_p(t)}{\cos_p(t)}, \qquad\qquad \sec_p(t) = \frac{1}{\cos_p(t)},$$

$$\cot_p(t) = \frac{\cos_p(t)}{\sin_p(t)} = \frac{1}{\tan_p(t)}, \qquad\qquad \csc_p(t) = \frac{1}{\sin_p(t)}.$$

It is here that we abandon whimsy for practicality and refer to these functions by referencing p, so, for example, $\tan_p$ is simply the p-tangent.

Exercise 5 Prove that $1 + \tan_p^p(t) = \sec_p^p(t)$ and $1 + \cot_p^p(t) = \csc_p^p(t)$.

Exercise 6 Use the quotient rule along with our initial derivative formulas for $\sin_p(t)$ and $\cos_p(t)$ to obtain

$$\frac{d}{dt} \tan_p(t) = \sec_p^2(t), \qquad\qquad \frac{d}{dt} \sec_p(t) = \sec_p^2(t) \sin_p^{p-1}(t), \qquad (5)$$

$$\frac{d}{dt} \cot_p(t) = -\csc_p^2(t), \qquad\qquad \frac{d}{dt} \csc_p(t) = -\csc_p^2(t) \cos_p^{p-1}(t)$$

(see [9]).

3.2   The Many Values of π

If we look at a graph of, say, $\sin_4(t)$ and $\cos_4(t)$, we notice that, much like the squircle, the graphs resemble those of our usual sine and cosine functions, but flattened out a bit (Fig. 3).

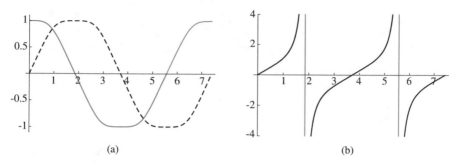

Fig. 3 Plots of $y = \sin_4(t)$ (**a**, dashed), $y = \cos_4(t)$ (**a**, solid), and $y = \tan_4(t)$ (**b**)

We also notice that the period of these sine and cosine functions is longer than the usual 2π, and that the first vertical asymptote in the tangent graph is far beyond the classical $t = \pi/2 \approx 1.57$. We can find this new value of π by first defining inverse functions for our squine and cosquine functions.

Let $x = \sin_p(y)$. Then $dx/dy = \cos_p^{p-1}(y) = (1 - x^p)^{(p-1)/p}$. This is a separable differential equation and we can solve for y to obtain

$$y = \arcsin_p(x) = \int_0^x \frac{1}{(1 - t^p)^{(p-1)/p}} \, dt.$$

In a similar way, since $\cos_p(0) = 1$, we have

$$\arccos_p(x) = \int_x^1 \frac{1}{(1 - t^p)^{(p-1)/p}} \, dt.$$

This leads organically to being able to define π as a function of p (see Fig. 4):

$$\pi_p = 2 \int_0^1 \frac{1}{(1 - t^p)^{(p-1)/p}} \, dt = 2 \arcsin_p(1) = 2 \arccos_p(0). \tag{6}$$

In particular,

$$\pi_4 = 2 \int_0^1 \frac{dt}{(1 - t^4)^{3/4}}.$$

Exercise 7 Use a computer algebra system to find an approximate value for π_4.

Though not expressible in terms of elementary functions, π_p can be expressed in terms of the beta and gamma functions [2]. The *beta function* is a function of two variables defined in terms of an integral:

$$B(x, y) = \int_0^1 u^{x-1}(1 - u)^{y-1} \, du. \tag{7}$$

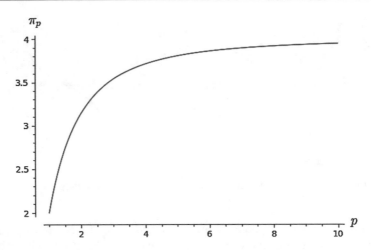

Fig. 4 Graph of π_p for $p \geq 1$

The *gamma function* is defined as $\Gamma(z) = \int_0^\infty t^{z-1}e^{-t}\,dt$. If n is a positive integer, then $\Gamma(n) = (n-1)!$. Thus the gamma function is a generalization of the factorial. Its relationship to the gamma function is via the identity $B(x, y) = \frac{\Gamma(x)\Gamma(y)}{\Gamma(x+y)}$.

Exercise 8 Verify these properties of the beta and gamma functions.

If we make the change of variable $u = t^p$ in Eq. (6), we get

$$\pi_p = \frac{2}{p}\int_0^1 u^{1/p-1}(1-u)^{1/p-1}\,du = \frac{2}{p}B(\tfrac{1}{p}, \tfrac{1}{p}) = \frac{2}{p}\frac{\Gamma^2(\frac{1}{p})}{\Gamma(\frac{2}{p})}. \qquad (8)$$

We can use the well-known gamma function identities

$$\Gamma(x+1) = x\Gamma(x) \quad \text{and} \quad \Gamma(x)\Gamma(1-x) = \frac{\pi}{\sin(\pi x)}. \qquad (9)$$

to derive some facts and, in particular, values for the gamma function and for some π_p's.

Exercise 9 Show that $\Gamma(\tfrac{1}{2}) = \sqrt{\pi}$.

Challenge Problem 1 (Legendre Duplication Formula) We will use the identity $B(x, y) = \frac{\Gamma(x)\Gamma(y)}{\Gamma(x+y)}$ to derive the Legendre duplication formula, which gets the value of $\Gamma(2z)$ from the value of $\Gamma(z)$ and $\Gamma(z + \tfrac{1}{2})$.

(a) Use the above identity with $x = y = z$ to show

$$\frac{\Gamma^2(z)}{\Gamma(2z)} = \int_0^1 u^{z-1}(1-u)^{z-1}\,du.$$

(b) Substitute $u = (1+x)/2$ in the previous integral and show that

$$\frac{\Gamma^2(z)}{\Gamma(2z)} = 2^{2-2z}\int_0^1 (1-x^2)^{z-1}\,dx.$$

(c) Show that

$$B(m,n) = 2\int_0^1 x^{2m-1}(1-x^2)^{n-1}\,dx.$$

(Hint: Set $u = x^2$ in formula (7).)

(d) Use the result of the previous problem to show

$$\frac{\Gamma^2(z)}{\Gamma(2z)} = 2^{1-2z}\,B(\tfrac{1}{2}, z) = 2^{1-2z}\frac{\Gamma(\tfrac{1}{2})\Gamma(z)}{\Gamma(z+\tfrac{1}{2})}.$$

(e) Use the result of part (d) above and Exercise 9 to conclude

$$\Gamma(2z) = \frac{\Gamma(z)\Gamma(z+\tfrac{1}{2})}{2^{1-2z}\sqrt{\pi}}. \tag{10}$$

Exercise 10 Use the Legendre duplication formula (10) and the first identity from (9) to show that $\lim_{p\to\infty} \pi_p = 4$.

Using the formulas (9) leads to

$$\pi_3 = \frac{2[\Gamma(\tfrac{1}{3})]^2}{3\Gamma(\tfrac{2}{3})} = \frac{[\Gamma(\tfrac{1}{3})]^3}{\sqrt{3}\,\pi} \quad \text{and} \quad \pi_4 = \frac{[\Gamma(\tfrac{1}{4})]^2}{2\Gamma(\tfrac{1}{2})} = \frac{[\Gamma(\tfrac{1}{4})]^2}{2\sqrt{\pi}}. \tag{11}$$

This leads to the formulas

$$\Gamma(\tfrac{1}{3}) = \sqrt[3]{\pi_3\pi\sqrt{3}} \quad \text{and} \quad \Gamma(\tfrac{1}{4}) = \sqrt{2\pi_4\sqrt{\pi}}.$$

Since these are considered to be "unknown" values of the gamma function, a good source of classroom discussion could encompass whether these are now solved. (Both values have been proved to be transcendental [4].)

Exercise 11 Use either formula (9) or (10) to find exact values of $\Gamma(\tfrac{3}{4})$ and $\Gamma(\tfrac{2}{3})$ in terms of π_4 and π_3, respectively. Then use formula (8) to find formulas for $\Gamma(\tfrac{1}{2^n})$ (see [19]) and $\Gamma(\tfrac{1}{3\cdot2^n})$ for $n \in \mathbb{N}$.

Research Project 1 The integral that defines π_4 is elliptic in nature, and several relationships exist between gamma function values and elliptic integral singular values, or between various gamma function values. Consult with [16, 21] and [22], and rewrite some of the listed relationships in terms of various π_p's.

The solutions for the coupled initial value problem (4) can be restricted to hold for $0 \leq t \leq \pi_p/2$. We can then extend the function definitions via symmetry, first to $0 \leq t \leq \pi_p$, then to $0 \leq t < 2\pi_p$, and finally, via periodicity, to the entire real line. To wit:

$$\sin_p(t) = \begin{cases} \sin_p(\pi_p - t) & \pi_p/2 < t \leq \pi_p, \\ -\sin_p(2\pi_p - t) & \pi_p < t < 2\pi_p \end{cases}$$

$$\cos_p(t) = \begin{cases} -\cos_p(\pi_p - t) & \pi_p/2 < t \leq \pi_p, \\ \cos_p(2\pi_p - t) & \pi_p < t < 2\pi_p. \end{cases}$$

We finish by setting $\sin_p(t + 2\pi_p k) = \sin_p(t)$ for $k \in \mathbb{Z}$ and do the same for our cosquine function.

Exercise 12 Show that the squine functions are odd and the cosquine functions are even for all p. (cf. Exercise 2).

Exercise 13 Previously, we were able to find an integral definition for $\arcsin_p(x)$. Use a similar technique along with formula (5) and the result of Exercise 5 to derive an integral formula for $\arctan_p(x)$. Use that formula to state another integral definition for π_p. Then consult [18].

4 The Geometry of p-Circles

We generalize the unit circle approach by studying a geometry in which the squircles really are circles, i.e., they represent the set of points whose distance from a center point is a constant. To do that we will need to rethink what distance means.

4.1 The Planar p-Norm

A *metric* on $\mathbb{R}^2$ is a function $d : \mathbb{R}^2 \to \mathbb{R}$ for which the following properties hold for any x, y, z in $\mathbb{R}^2$.

Positive Definite: $d(x, y) \geq 0$ with equality if and only if $x = y$

Symmetric: $d(x, y) = d(y, x)$

Triangle Inequality: $d(x, y) \leq d(x, z) + d(z, y)$

The metrics we will study are translation-invariant, meaning for any $b \in \mathbb{R}^2$, $d(x, y) = d(x + b, y + b)$. In this case, we can understand the metric by the norm, $\|x\|_d = d(x, 0)$. The Euclidean metric $d_2(x, y) = ((x_1 - y_1)^2 + (x_2 - y_2)^2)^{1/2}$ satisfies these properties and is of course the most familiar. We will study a generalization of this metric called the p-metric, wherein we turn the 2's into p's and add absolute values as before, giving, $d_p(x, y) = (|x_1 - y_1|^p + |x_2 - y_2|^p)^{1/p}$ with norm $\|x\|_p = |x_1|^p + |x_2|^p$. We also refer to this metric as the p-norm, a reference to the fact that because it is translation-invariant, the geometry can be fully explored by assuming one point is the origin.

Theorem 1 *The p-metric is a translation-invariant metric on $\mathbb{R}^2$.*

Exercise 14 Verify that the p-metric is positive definite, symmetric, and translation-invariant.

Exercise 15 Show that if $p \neq 2$ then the p-metric is not rotation-invariant, meaning that if points A and B are rotated about the origin, the distance between the points may change.

The triangle inequality for the p-metric is a consequence of Minkowski's inequality, a discussion of which would take us a bit astray (see, e.g., [10]).

Of course, $p = 2$ is the usual Euclidean metric. The special case of $p = 1$ is also well-studied. It is often called *taxicab geometry* or the *Manhattan metric* because it models the distance a car would travel through a gridded street. See [12] for a full exploration of this geometry. It is useful to study as the extreme case of the p-metric satisfying the triangle inequality. The other extreme is $p = \infty$, which is defined as $d(x, y) = \max(|x|, |y|)$. This is the limiting case of the definition for finite p. In explorations, we find $p = 4$ to be a good case study because the 4-squircle is not extremal but still asymmetric, and $p = 4$ has some features that simplify some calculations.

4.2   Sines Everywhere

There are plenty of reasonable ways to parameterize a p-circle. We saw in the $p = 2$ case that all of these reasonable ways turn out to be the same, making the classical definitions of the trigonometric functions extremely natural. The choice is much less obvious for other values of p, and we will explore some of this variation.

We started with $\arcsin_p(x) = \int_0^x \frac{1}{(1-t^p)^{(p-1)/p}} \, dt$ which we will now show corresponds to an areal parameterization of the p-circle, one in which the area of the unit p-circle is π_p square units. This argument is due to Levin [14].

Fig. 5 Area of a p-circular
sector

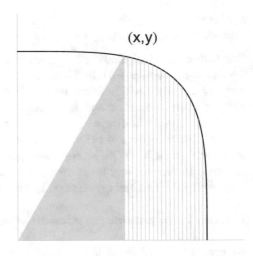

(x,y)

To mimic the classical case, we want the parameter α to be twice the area of the sector from the positive x-axis to the segment connecting the origin to $(\cos_p(\alpha), \sin_p(\alpha))$. See Fig. 5. For example, if $p = 2$ then $\alpha = \pi$ gives the semicircle with area $\frac{\pi}{2}$. This sector splits into a triangle and half of a p-circular segment whose combined area is

$$\frac{\alpha(y)}{2} = \frac{1}{2}y(1-y^p)^{1/p} + \int_{(1-y^p)^{1/p}}^{1} (1-t^p)^{1/p}\, dt.$$

By expressing the area in terms of y, we have found an expression for the inverse of the areal version of sine. A more explicit formula is found by differentiating and integrating:

$$\alpha'(y) = (1-y^p)^{1/p} + y(1-y^p)^{\frac{1}{p}-1}(-y^{p-1}) + 2y(1-y^p)^{\frac{1}{p}-1}(-y^{p-1}) = (1-y^p)^{\frac{1}{p}-1}.$$

Then $\alpha(y) = \int_0^y (1-t^p)^{\frac{1}{p}-1}\, dt = \arcsin_p(y)$. In particular, when $y = 1$, α returns a value of $\pi_p/2$, twice the area of the first quadrant portion of the p-circle. Thus the area of the unit p-circle is π_p square units.

Exercise 16 Repeat this calculation in terms of x to show $\alpha(x) = \arccos_p(x)$.

Exercise 17 Use polar coordinates to find explicit formulas in terms of classical trigonometric functions for the p-squigonometric functions parameterized by angle.

Note on Exercise 17 The polar formulas can be useful for numerical estimates and drawing pictures because they are expressed in terms of well-known and well-implemented functions. However, the non-Euclidean p-norms do not have a natural

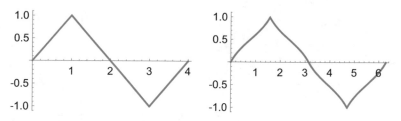

Fig. 6 Graphs of the areal and angular versions of the 1-sine function

notion of angle so this parameterization is not as useful intrinsically. The next exercise illustrates this.

Exercise 18 Show that the areal 1-squine function is piecewise linear but the angular version of the 1-squine from Exercise 17 is not. See Fig. 6

4.3 Transcendental Functions

With our generalized trigonometric functions in place, we may now use them to build new p-norm versions of other transcendental functions. We start by adapting the defining differential equations (4) to define the hyperbolic p-trigonometric functions.

$$
\begin{cases}
x'(t) = y(t)^{p-1} \\
y'(t) = x(t)^{p-1} \\
x(0) = 1 \\
y(0) = 0
\end{cases}
\tag{12}
$$

This CIVP has a unique solution $x(t) = \cosh_p(t)$ and $y(t) = \sinh_p(t)$ which we define to be the hyperbolic p-cosine and p-sine, respectively. Just as $(\cos_p(t), \sin_p(t))$ parameterize the p-circle, $(\cosh_p(t), \sinh_p(t))$ parameterize a p-hyperbola. See Fig. 7.

Exercise 19 Find integral expressions for the inverse hyperbolic p-sine and p-cosine functions.

Exercise 20 Show that if $\omega^p = -1$, then $\omega \sinh_p(t) = \sin_p(\omega t)$ and $\cosh_p(t) = \cos_p(\omega t)$.

Exercise 21 Show that $|\cosh_p(t)|^p - |\sinh_p(t)|^p = 1$ for all t.

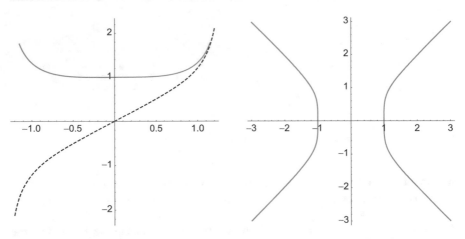

Fig. 7 Graphs of $y = \sinh_4(t)$ (left, dashed), $y = \cosh_4(t)$ (left, solid), and $x^4 - y^4 = 1$ (right)

Exercise 21 brings us back to the geometry. It seems reasonable to define the curve $|x|^p - |y|^p = 1$ to be a rectangular p-hyperbola, and it is clearly parameterized by $(\cosh_p(t), \sinh_p(t))$.

Exercise 22 Continuing with the methods of Exercises 2 and 12, prove the symmetries of the hyperbolic p-sine and p-cosine functions.

Exercise 23 Try to replicate Fig. 7 in your favorite computer algebra system. More generally, try to implement and graph as many of our special functions as you can.

Exercise 24 How might we define more general p-hyperbolas and p-ellipses?

Exercise 25 Use a computer algebra system to plot families of p-ellipses and p-hyperbolas and describe how they change shape as p varies from 1 to ∞.

Research Project 2 (p-Conics) Conic sections have a few different analytic and geometric definitions that are equivalent in Euclidean geometry but not in the p-metric. Explore p-conics in general. To what extent do classical geometric properties of these curves carry over to the p-norm? (See [12] for a delightful exploration of conics when $p = 1$).

Research Project 3 (Geometric Curves) There are plenty of other interesting geometric curves besides conics. For example, a lemniscate is the set of

(continued)

points such that the product of the distances to two foci is constant (as opposed to the sum or difference in conics). Levin shows in [14] that the circumference of a standard Euclidean lemniscate is equal to the area of the 4-squircle. Can this relationship be generalized in any way? More broadly, explore geometric properties of curves and how they generalize to different p.

Taking a cue from the identity $e^t = \sinh t + \cosh t$, our generalized hyperbolic functions offer a candidate for the p-exponential function $\exp_p(t) = \sinh_p(t) + \cosh_p(t)$.

Exercise 26 Show that $\exp_1(t) = \begin{cases} 1 & t \leq 0 \\ 2t + 1 & t > 0 \end{cases}$

Research Project 4 (p-Exponential Growth) Exercise 26 shows that $\exp_p(t)$ grows linearly in t if $p = 1$. It is obviously exponential if $p = 2$. What is the growth rate for general p?

Research Project 5 (General Transcendentals) Keep going. We have generalized trigonometric and exponential functions, and other functions can be expressed in terms of those. Explore how such functions as a p-logarithm mirror their classical counterparts.

4.4 π Redux

We have defined $\pi_p = 2 \int_0^1 \frac{1}{(1-t^p)^{(p-1)/p}}\, dt$. It is the half-period of $\sin_p$ and also represents the area of the unit p-circle. Another definition could be expressed in terms of p-arclength. Recall that the Euclidean arclength of a parameterized curve $(x(t), y(t))$ is obtained by considering a small arc of the curve to be the hypotenuse of an infinitesimal Euclidean triangle with side lengths dx and dy. We integrate over some interval $a \leq t \leq b$ to obtain $L = \int_a^b \sqrt{(dx)^2 + (dy)^2}\, dt$. If we want to change to the p-arclength, we simply adjust our measure of the length of this hypotenuse to obtain $L = \int_a^b \sqrt[p]{|dx|^p + |dy|^p}\, dt$.

Exercise 27 Show that if $f(x)$ is a continuous function on $a \leq x \leq b$, then the p-arclength of the graph of $y = f(x)$ over the interval $a \leq x \leq b$ is

$\int_a^b \sqrt[p]{1 + |f'(x)|^p} \, dx$. Experiment with some of your favorite familiar functions to get a sense of how their "lengths" change with p, and which such integrals can be calculated exactly.

Exercise 28 Calculate the p-arclength of the segment connecting the origin to the point $(1, 0)$. Compare with the p-arclength of the segment connecting the origin to the point $(1, 1)$.

Define

$$\pi_p^{arc} = 4 \int_0^{2^{-1/p}} \left(1 + (t^{-p} - 1)^{1-p}\right)^{1/p} dt = \frac{2}{p} \int_0^1 \left(t^{1-p} + (1-t)^{1-p}\right)^{1/p} dt.$$

Notice that π_p^{arc} has different dynamics from π_p. It takes a minimum value at $p = 2$, where $\pi_2^{arc} = \pi_2 = \pi$ [1].

Another candidate for defining π is to simply use the Euclidean arclength, $\pi_p^{euc} = 4 \int_0^1 \sqrt{1 + (t^{p-1}(1 - t^p))^2} \, dt$. Although this does not consider arclength

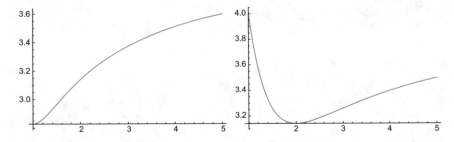

Fig. 8 Graphs of π_p^{euc} and π_p^{arc}

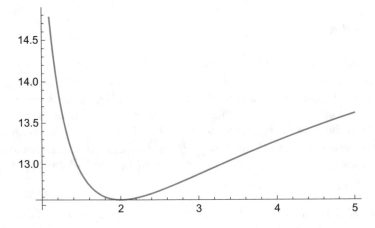

Fig. 9 Graph of I_p

in the circle's natural geometry unless $p = 2$, it offers some insight into the geometry of the p-circle. The *isoperimetric ratio* of simple closed curve is the square of the length of the curve divided by its area. For the unit p-circle, this is $I_p = \frac{\text{length}^2}{\text{area}} = \frac{(2\pi_p^{euc})^2}{\pi_p} = \frac{4(\pi_p^{euc})^2}{\pi_p}$. (See Figs. 8 and 9.) For $p = 2$, where all of our definitions of π are equivalent, this is 4π. The circle is the solution to the *isoperimetric problem*, meaning that 4π is the minimum possible value of the isoperimetric ratio. That is, the circle is the most efficient shape to enclose the most area with the smallest perimeter [3].

Exercise 29 Calculate directly the values of π_p^{arc} and π_p^{euc} for $p = 1$ and $p = \infty$.

Research Project 6 (Flavors of π) Analyze π_p^{euc} and π_p^{arc} as functions of p. Is there any significance to the inflection point of π_p^{arc} near $p = 2.9$?

Research Project 7 (p-Isoperimetric Ratios) Explore the isoperimetric properties of p-circles. What can be said about the isoperimetric ratio if we use p-arclength instead of Euclidean? We may define $\hat{I}_p = \frac{4(\pi_p^{arc})^2}{\pi_p}$ and we see that the minimum is not attained at $p = 2$. I_p and $\hat{I}_p$ also have inflection points perhaps worth exploring.

5 Widening the Scope

We are seeing that for $p \neq 2$, we get different definitions of π depending not just on the value of p, but on what geometric features we wish to capture. We will look closer at this variety and offer some tools to try to find connections among them.

5.1 Duality

Two real numbers $p, p' > 0$ are *conjugate* if $\frac{1}{p} + \frac{1}{p'} = 1$. There are well-documented connections between the p-norm and p'-norm when p and p' are conjugate, in which case the metric spaces are called *dual*. We explore how these connections are reflected in the squigonometric functions, beginning with an observation of Lindqvist and Peetre [15].

Exercise 30 Confirm the following notable relationships for conjugate numbers $p, p' > 1$.

- $pp' = p + p'$
- $p = p'(p - 1)$
- $p = \frac{p'}{p'-1}$
- $p = p'$ if and only if $p = p' = 2$

Note that these expressions are all valid if p and p' are exchanged. We also extend the definition to $p = 1$ by defining its conjugate to be $p' = \infty$.

Theorem 2 *Let p and p' be conjugate. Then the p'-circumference of the unit p-circle is $2\pi_p$.*

Proof By symmetry, we may take the circumference to be four times the arclength of the quarter-circle. We use our p-trigonometric parameterization and calculate the circumference C directly to be

$$C = 4 \int_0^{\pi_p/2} \left(\left| \frac{dx}{dt} \right|^{p'} + \left| \frac{dy}{dt} \right|^{p'} \right)^{1/p'} dt$$

$$= 4 \int_0^{\pi_p/2} \left(\left| \frac{d}{dt} \sin_p(t) \right|^{p'} + \left| \frac{d}{dt} \cos_p(t))^{p'} \right| \right)^{1/p'} dt$$

$$= 4 \int_0^{\pi_p/2} \left(|\cos_p(t)^{p-1}|^{p'} + |-\sin_p(t)^{p-1}|^{p'} \right)^{1/p'} dt$$

$$= 4 \int_0^{\pi_p/2} \left(\cos_p(t)^{p'(p-1)} + \sin_p(t)^{p'(p-1)} \right)^{1/p'} dt$$

$$= 4 \int_0^{\pi_p/2} \left(\cos_p(t)^p + \sin_p(t)^p \right)^{1/p'} dt = 4 \int_0^{\pi_p/2} dt = 2\pi_p.$$

One phenomenon to watch for is seeing a "nice" property of Euclidean geometry as an artifact of the self-conjugacy of $p = 2$. For example, consider a particle moving according to the standard Euclidean circle parameterization $(\cos t, \sin t)$. The velocity of the particle $(-\sin t, \cos t)$ also parameterizes a unit circle. When we look at the generalized setting, we find that the velocity is actually parameterizing the conjugate circle, which is obscured in Euclidean geometry where the circle and its dual are indistinguishable.

Exercise 31 The graph of π_p^{arc} in Fig. 8 suggests a symmetry related to conjugacy. Explore π_p^{arc} and $\pi_{p'}^{arc}$ for conjugate p', p and make a conjecture. Then see [11].

Exercise 32 Parameterize the unit p-circle by $r(t) = (\cos_p(t), \sin_p(t))$. Show that $r'(t)$ parameterizes the unit p'-circle, where p and p' are conjugate. Find a related result for a p-hyperbola.

Research Project 8 (Duality Relationships) Theorem 2 suggests we might look for other relationships among geometric quantities in dual p-norms. Compare quantifiable features of various geometric objects in conjugate norms and try to establish some new connections.

5.2   Arclength Trigonometric Functions

We can use the definition of π_p^{arc} to analytically present an arclength parameterization of the unit squircle. Define our new sine function $S(t)$ to be the function such that

$$t = \int_0^{S(t)} \left(1 + (u^{-p} - 1)^{1-p}\right)^{1/p} du, \tag{13}$$

and our new cosine function $C(t)$ to be the function such that

$$t = \int_{C(t)}^1 \left(1 + (u^{-p} - 1)^{1-p}\right)^{1/p} du, \tag{14}$$

with $0 \le t \le \pi_p^{arc}/2$. We first claim that $S^p(t) + C^p(t) = 1$. In Eq. (13), set $s = (1 - u^p)^{1/p}$. Then $u = (1 - s^p)^{1/p}$ and $du = -\frac{s^{p-1}}{(1-s^p)^{(p-1)/p}} ds$. Thus Eq. (13) becomes

$$t = \int_1^{(1-S^p(t))^{1/p}} \left(1 + \left(\frac{1}{1-s^p} - 1\right)^{1-p}\right)^{1/p} \cdot \frac{-s^{p-1}}{(1-s^p)^{(p-1)/p}} ds$$

$$= \int_{(1-S^p(t))^{1/p}}^1 \left(1 + \left(\frac{s^p}{1-s^p}\right)^{1-p}\right)^{1/p} \cdot \frac{s^{p-1}}{(1-s^p)^{(p-1)/p}} ds$$

$$= \int_{(1-S^p(t))^{1/p}}^1 \left[\left(1 + \left(\frac{s^p}{1-s^p}\right)^{1-p}\right) \cdot \frac{s^{p(p-1)}}{(1-s^p)^{p-1}}\right]^{1/p} ds$$

$$= \int_{(1-S^p(t))^{1/p}}^1 \left[\frac{s^{p(p-1)}}{(1-s^p)^{p-1}} + 1\right]^{1/p} ds$$

$$= \int_{(1-S^p(t))^{1/p}}^1 \left[\left(\frac{1-s^p}{s^p}\right)^{1-p} + 1\right]^{1/p} ds$$

$$= \int_{(1-S^p(t))^{1/p}}^{1} \left((s^{-p} - 1)^{1-p} + 1 \right)^{1/p} ds,$$

which is (14) if $C(t) = (1 - S^p(t))^{1/p}$, and the claim follows.

If we differentiate (13) with respect to t, we get

$$1 = (1 + (S(t)^{-p} - 1)^{1-p})^{1/p} \cdot S'(t)$$

$$S'(t) = \frac{1}{\left[1 + \left(\frac{1-S^p(t)}{S^p(t)} \right)^{1-p} \right]^{1/p}}$$

$$S'(t) = \frac{1}{\left[1 + \left(\frac{C^p(t)}{S^p(t)} \right)^{1-p} \right]^{1/p}}$$

$$S'(t) = \frac{1}{\left[1 + \left(\frac{S^p(t)}{C^p(t)} \right)^{p-1} \right]^{1/p}}$$

$$S'(t) = \frac{C(t)^{p-1}}{\left[C(t)^{p(p-1)} + S(t)^{p(p-1)} \right]^{1/p}}.$$

Through similar means, we can derive

$$C'(t) = \frac{-S(t)^{p-1}}{\left[C(t)^{p(p-1)} + S(t)^{p(p-1)} \right]^{1/p}}.$$

From (13) and (14), we have that $S(0) = 0$, and $C(0) = 1$. We will call S and C the *arclength sine* and *arclength cosine* functions, respectively, notate them as $\sin_p^{arc}$ and $\cos_p^{arc}$, and note that they are the unique solutions on $0 \le t \le \pi_p^{arc}/2$ of the CIVP

$$\begin{cases} x'(t) = -y(t)^{p-1} \cdot [x(t)^{p(p-1)} + y(t)^{p(p-1)}]^{-1/p} \\ y'(t) = x(t)^{p-1} \cdot [x(t)^{p(p-1)} + y(t)^{p(p-1)}]^{-1/p} \\ x(0) = 1 \\ y(0) = 0, \end{cases} \tag{15}$$

with x corresponding to the arclength cosine and y corresponding to the arclength sine. We can then extend the functions to the entire real line in the usual manner via symmetry. A graph of $y = \sin_4^{arc}(t)$ and its intimidating derivative

$$y' = \frac{(\cos_4^{arc}(t))^3}{[(\cos_4^{arc}(t))^{12} + (\sin_4^{arc}(t))^{12}]^{1/4}}$$

can be seen in Fig. 10.

Fig. 10 Graphs of
$y = \sin_4^{arc}(x)$ (solid) and its
derivative (dashed)

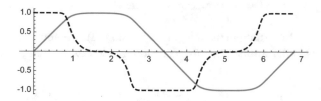

This graph was rendered in Mathematica by first defining arclength p-sine and p-cosine functions in terms of the Euclidean angle t, then another for the "p-angle" φ. We then produced a parametric plot of the sine function and the cosine function against φ. (Try this in your favorite computer algebra system.)

5.3   Other Parameterizations

We have seen two different parameterizations of the unit p-circle leading to quite different versions of trigonometric functions, as well as different versions of π_p. With (4), π_p gave the correct area of the unit p-circle. With (15), $2\pi_p^{arc}$ gave the correct arclength. This points to a key difference between the circle and the p-circle, which relates to the existence of π. The trigonometric functions are arranged so that their periods reflect the arclength and the areal and the angular parameterizations of the circle. No parameterization of the p-circle can do all three at the same time. We have values that play some of the roles of π, but none that can do it all.

We can, in fact, choose any CIVP of the form

$$\begin{cases} x'(t) = -y(t)^{P-1}\zeta(x(t), y(t), t) \\ y'(t) = x(t)^{P-1}\zeta(x(t), y(t), t) \\ x(0) = 1 \\ y(0) = 0 \end{cases} \tag{16}$$

to parameterize the unit p-circle, where $\zeta(x(t), y(t), t)$ is any appropriate differentiable function [23], and the solutions are taken over a suitable interval. In general, each such parameterization will occasion a new definition of π_p, which may or may not correspond to a geometric trait of the p-circle, and result in generalized trigonometric functions which may not do much beyond parameterizing the unit p-circle.

Exercise 33 Find a function $\zeta(x(t), y(t), t)$ so that (16) results in the angular parameterization; that is, one in which the angle function $\theta(t) = \arctan(\frac{y(t)}{x(t)})$ is the identity map. Compare with Exercise 17.

Exercise 34 We saw why the solution curves to a CIVP of the form (4) must lie on the unit p-circle, but we want to get the whole p-circle. What goes wrong if we let $\zeta(x, y, t) = x$? How can we fix it?

Returning to the integral expression (2), we notice that we are seeing integrals of similar forms. This invites a more general two-parameter definition

$$\arcsin^*_{p,q}(x) = \int_0^x (1 - t^q)^{-1/p} \, dt$$

and define $\sin^*_{p,q}$ as the inverse function of $\arcsin^*_{p,q}$.

Exercise 35 Verify that $\sin^*_{p,p'}(x) = \sin_p(x)$ where p and p' are conjugate. We may then define an alternate version of our generalized trigonometric functions by $\sin^*_p(x) = \sin^*_{p,p}(x)$ and $\cos^*_p(x) = \frac{d}{dx} \sin^*_p(x)$. Show that $(\cos^*_p(x), \sin^*_p(x))$ parameterizes the unit p-circle. Define yet another version of π in this context.

Exercise 35 illustrates an expository inconvenience in this topic. Depending on your point of view, there are multiple definitions of generalized trigonometric functions that may seem natural, and they are all related by duality or other geometric properties. The definition in Exercise 35 is used in [13], which details many of the properties and applications of generalized trigonometry. That definition sets cosine as the derivative of sine, which is certainly convenient, but the parameter is no longer the area. It is mostly a matter of preference but one must take care to keep straight which versions have which properties. Unless $p = 2$, you will have multiple reasonable options.

Challenge Problem 2 (General Parameterizations) Revisit the previous exercises in this chapter using different parameterizations, such as $\sin^*_p$ and $\sin^{arc}_p$. For example, how would the derivative and integral tables change?

Research Project 9 (Unification) We have cut across the different types of squigonometric functions both by varying p and varying what geometric information they encode. The previous Challenge Problem is not so challenging if $p = 2$ because the same functions encode everything. These lead to different functions for each p, but they are related by geometric properties and duality. In practice this makes for a somewhat disorganized theory. Are there any unifying principles we might establish to more generally work with the various versions of squigonometric functions?

6 Onward

In addition to the specific projects we suggested, we should also observe a variety of projects available by exploring ways the study of generalized trigonometric functions cuts across other fields.

For example, consider the following algebraic property of p-circles [24].

Theorem 3 *For any integer $p > 2$, the unit p-circle passes through no rational points except on the axes. That is, if (x, y) is on the unit p-circle and $x, y \in \mathbb{Q}$, then either $x = 0$ or $y = 0$.*

Proof Suppose $(\frac{m_1}{n_1}, \frac{m_2}{n_2})$ is a rational point on the unit p-circle, where m_1, m_2, n_1, n_2, p are positive integers and $p > 2$. Then $\left(\frac{m_1}{n_1}\right)^p + \left(\frac{m_2}{n_2}\right)^p = 1$, or $(m_1 n_2)^p + (m_2 n_1)^p = (n_1 n_2)^p$. But this contradicts Fermat's Last Theorem, a storied result [20] that says $x^p + y^p = z^p$ has no solutions in integers for $p > 2$.

This suggests looking at curves defined geometrically in the planar p-norm as algebraic curves, bringing in a new set of tools to understand their properties.

In the case $p = 4$, the integrals defining the inverse trigonometric functions are elliptic. Elliptic integrals have a wealth of special properties that can lead to special identities in the 4-norm, including a (messy) formula for $\sin_4(t + u)$ [14]. Could this be pushed further?

The squigonometric functions offer a generalization of Fourier analysis, the foundations of which are laid out in [13]. The geometries we are studying are special cases of Minkowski geometry, which offers a more general lens through which to see these problems. Also, duality is a cornerstone of the function-analytic view in that the space of linear functionals on a vector space with the p-norm is isometric to the conjugate p'-norm. This offers yet another set of tools. Studies of higher dimensions may also open new territory.

Overall, this is a subject with a long but fractured history (cf. [1, 8, 17–19, 23] and the references therein) and there are many connections among the special functions associated with p-norm geometry yet to explore.

References

1. Adler, C.L., Tanton, J.: π is the minimum value of Pi. College Mathematics Journal **31**(2), 102–106 (2000)
2. Andrews, G., Askey, R., Roy, R.: Special Functions (Encyclopedia of Mathematics and its Applications). Cambridge University Press (1999). https://doi.org/10.1017/CBO9781107325937
3. Blåjö, V.: The Isoperimetric Problem. The American Mathematical Monthly **112** (6), 526–566 (2005)
4. Borwein, J., Bailey, D.: Mathematics by Experiment: Plausible Reasoning in the 21st Century. A K Peters (2003)

5. Boyce, W., DiPrima, R.: Elementary Differential Equations, 9th edition. Wiley (2008)
6. Callahan, J., Cox, D., Hoffman, K., O'Shea, D., Pollatsek, H., Senechal, L.: Calculus in Context: The Five College Calculus Project. W.H. Freeman (1995)
7. Cha, B.: Transcendental Functions and Initial Value Problems: A Different Approach to Calculus II. College Mathematics Journal **38**(4), 288–296 (2007)
8. Euler, R., Sadek, J.: The π's go full circle, Mathematics Magazine **72**, 59–63 (1999)
9. R. Grammel: Eine Verallgemeinerung der Kreis- und Hyperbelfunktionen, (German) Ing.-Arch. **16**, 188–200 (1948)
10. Hardy, G., Littlewood, J., Pólya, G., Inequalities, Cambridge Univ. Press (1934)
11. Keller, J., Vakil, R.: π_p, the Value of π in ℓ_p. The American Mathematical Monthly, **116**(10), 931–935 (2009). https://doi.org/10.4169/000298909X477069
12. Krause, E.: Taxicab Geometry: Adventures in Non-Euclidean Geometry. Dover (1987)
13. Lang, J., Edmunds, D.: Eigenvalues, Embeddings and Generalised Trigonometric Functions, Springer Lecture Notes in Mathematics (2016)
14. Levin, A.: A Geometric Interpretation of an Infinite Product for the Lemniscate Constant. The American Mathematical Monthly, **113**(6), 510–520 (2006). https://doi.org/10.2307/27641976
15. Lindqvist, P., Peetre, J.: p-arclength of the q-circle, Math. Student **72** (1–4) 139–145(2003)
16. Maican, C.C: Integral Evaluations Using the Gamma and Beta Functions and Elliptic Integrals in Engineering: A Self-Study Approach, International Press, 2005.
17. Markushevich, A.I.: The Remarkable Sine Functions, Elsevier (1966)
18. Poodiack, R.: Squigonometry, Hyperellipses, and Supereggs, Mathematics Magazine, **89**(2) 92–102 (2016)
19. Shelupsky, D.: A Generalization of the Trigonometric Functions, The American Mathematical Monthly, **66**(10), 879–884 (1959). https://doi.org/10.1080/00029890.1959.11989425
20. Singh, S.: Fermat's last theorem: the story of a riddle that confounded the world's greatest minds for 358 years. Harper Perennial (2007)
21. Weisstein, E.W.: Gamma Function. From *Mathworld* – A Wolfram Web Resource. http://mathworld.wolfram.com/GammaFunction.html
22. Weisstein, E.W.: Elliptic Integral Singular Value. From *Mathworld* – A Wolfram Web Resource. http://mathworld.wolfram.com/EllipticIntegralSingularValue.html
23. Wood, W.: Squigonometry, Mathematics Magazine, **84**(4): 257–265 (2011)
24. Young, R.M. Excursions in Calculus: An Interplay of the Continuous and the Discrete, Cambridge Univ. Press (1992)

Researching in Undergraduate Mathematics Education: Possible Directions for Both Undergraduate Students and Faculty

Milos Savic

Abstract

Research in Undergraduate Mathematics Education (RUME) is a new field to both mathematics and mathematics education. It borrows theory and methodology from other disciplines including psychology, sociology, and neurology. At its core, RUME is attempting to find out about the teaching and learning of undergraduate mathematics education in order to improve it. In this book chapter, I attempt to give a quick overview on how to conduct RUME with undergraduate students. I pull from my experiences as a mentor of ten undergraduate projects. There is also a suggested timeline of RUME in a semester, some ways to generate RUME open questions, and a large amount of open questions conjectured by others. My hope is that this book chapter has information for both mentors and undergraduates alike.

Suggested Prerequisites For research in undergraduate mathematics education, a pre-requisite may be the mathematical knowledge of whatever topic you would like to research. For example, if an undergraduate student wants to research in the teaching of real analysis, they must have some knowledge of real analysis topics in order to understand the mathematics in the education. The rest of the preparation before research can be accomplished during a semester or year.

M. Savic (✉)
University of Oklahoma, Norman, OK, USA
e-mail: savic@ou.edu

© Springer Nature Switzerland AG 2020
P. E. Harris et al. (eds.), *A Project-Based Guide to Undergraduate Research in Mathematics*, Foundations for Undergraduate Research in Mathematics,
https://doi.org/10.1007/978-3-030-37853-0_10

287

1 Introduction

Research in undergraduate mathematics education (RUME) is a relatively new field, with researchers beginning to publish and discuss findings in the 1980s [23]. Much of the work in RUME is focused on humans at the undergraduate level and how they teach or learn mathematics. RUME has links to many other disciplines, whether it be psychology, neurology, social sciences, ethnography, or (and perhaps most importantly) mathematics education at the K-12 level. Therefore, the possibilities of an undergraduate researching in RUME are endless.

In this chapter, I detail what *I* deem is necessary and sufficient for undergraduate research in RUME and offer both open questions and ways to generate more. The audience of this book chapter is new(ish) faculty that have students interested in RUME. One caveat is that the faculty themselves must be somewhat comfortable with RUME or be open to learning new RUME topics. For example, I have mentored ten undergraduate students in research of some type, and every mentoring process I have learned new ways of approaching research with undergraduate students. However, I think that I would have liked an outline or more suggestions to mentoring than the word-of-mouth or experiences I received as a graduate student. This book chapter is an aggregation of those first-hand experiences. Another audience is undergraduate students themselves. I hope that an undergraduate can pick up some ideas about how to generate research questions or common structure of a thesis from this chapter. In fact, Sect. 4 is written directly for undergraduate students.

This book chapter will not include everything on RUME. It is anecdotal in nature. Many of the citations here are ones that have been influential for me as a student and a researcher. For another approach to RUME, please see what Selden and Selden [24] have written. Intrinsic to data analysis is the filtering of what the researcher attends to and how they make sense of what they see. Some of these theoretical perspectives are also presented by Selden and Selden. Here, I offer a cursory view of theoretical perspectives for two reasons: one is that I am gaining knowledge about theoretical perspectives every year, but do not feel expert enough to explain each or all of the perspectives. The second reason is with the short timeline that an undergraduate student has, I have often guided them towards more open questions that can cite articles which utilize other RUME theoretical perspectives (see [24] for more on this subject).

2 Beginning RUME with a Question

There are possibly two routes that a faculty member could go with undergraduates who want to do RUME. The first route is to incorporate the student in to one of the faculty member's projects. This route has some positives and negatives for the student and the research supervisor. On the one hand, there is likely a developed research plan already in place. This may include a research proposal, a collection of related research literature, and perhaps data that has already been collected. Thus, when bringing a student into an existing project, there is a pretty clear path to follow: familiarize the student with the research plan, have the student read the related research literature, and then have them reflect on that literature (and possibly the already collected data) to develop a research question that fits into the existing project. On the other hand, however, this route may not offer the student the same level of motivation since they are assuming another person's project. It is for this reason alone that I have yet to go this route for my own mentorship, although I can recognize how much time and effort would have been reduced by incorporating an undergraduate into my own projects.

The second route is to have the student generate the question. As an undergraduate student at the University of Oklahoma, Katherine (Kaki) Simmons approached me about doing a directed research project. I think she assumed that she was going to join me on my research, but I asked her what she was interested in. She stated that she started an American Sign Language club at the university and was interested in Deaf and hard-of-hearing (D/HH) students' experiences in undergraduate mathematics courses. Therefore, I asked her to look at the previous literature for this intersection of D/HH and undergraduate mathematics education. Kaki did not find much literature in undergraduate math education, but did find some K-12 math education literature about D/HH students. So her research question ended up being: "What roadblocks and successes happened in D/HH students' undergraduate mathematics education?" While being an example of a first encounter with advising an undergraduate student, this is also an example of how to generate a RUME open question.

Therefore, if a student emails me with a request to research, I ask the question, "Why RUME? What part of RUME interests you, either as a student or a future teacher?" I believe having the student ask the research question first may allow or sustain intrinsic motivation [19] that the student has for the study (see Sect. 4.1 for more). In this same first meeting, I show how I would approach research in a certain topic, including utilizing Google Scholar and other internet resources. I may also bring up a couple of foundational articles or learning researchers and show them how to do "reverse citations." Describing a research question into keywords to search is critical because it can be difficult in RUME to have a synonymous vocabulary about teaching or learning. This might be due to the field of RUME being so "young." I suggest articulating and being specific about one's research question.

Critical thinking about an article can also be a way of generating new research questions. For example, a research question of mine came from a reading of Sio and Ormerod [25], where the authors stated that all 117 psychological studies analyzed included an incubation period in their study from 1 to 60 minutes. An incubation period is a break in problem solving, which precedes an AHA! moment [12]. I kept saying to myself, "When I do math proofs in graduate school, I usually take days to incubate and think about approaches." So, I attempted to investigate what I felt was incubation in mathematics [21] using methods that could allow for incubation over a few days or weeks. It is up to the advisor to figure out how to mold an undergraduate's research question into something "new" for the field. There are a large amount of suggestions for research questions towards the end of this book chapter (see Sect. 6).

After having a meeting with the student, I always suggest that they take 2 weeks and try to read quite a bit about their interest. Reading previous research has multiple benefits for undergraduate students:

1. **Students will need to situate their research within the field's work for any publication.** Whether their research question is extremely novel or well-researched, more reading of a similar topic builds more knowledge of the subject.
2. **Reading helps refine research questions.** By reading many articles on the topic of interest, students may find that their research question has been explored in previous studies. Also, the student can attempt to position their question as different from those same previous studies.
3. **Reading provides models of frameworks, methods, and interview tasks that can be used.** Some studies provide the exact surveys or questions in appendices, and those methodological tools can assist in their research. If one aspect of the research question is more developed, borrowing methods is very good for establishing their perspective on the study (see Sect. 4.2 for more).
4. **Reading is important for examples of conducting and writing research.** A student can see a template and examples of what RUME research looks like (see Sect. 4.5.1 for a suggested template). The background literature sections of any article, in particular, could be useful to find more literature on the same or a similar subject.

Once you have conjectured a research question and started to read some research, then one must figure out which theoretical way can one answer the question. For example, suppose that one wants to research how students learn partial derivatives. There are many perspectives on how students learn a concept: is it their engagement with the material, or their peers, or what beliefs arise with learning partial derivatives? These different perspectives require different data collection techniques, and each one of these perspectives is associated to some educational (or other) theory.

In RUME, one usually "grounds" their work in a philosophical theory about education. This is similar to mathematics research; one can assume the axiom of choice or not, and because a mathematician assumed an axiom, it influences how they look at or create their subsequent mathematics. Influential authors of educational psychology theory include Piaget, Vygotsky, and Bandura (for an overview, please see [30]). These theories may sometimes be difficult to understand, so if you do include a theoretical background with undergraduates, please approach this with thoughtfulness and care. These perspectives usually align with certain methodologies. In the example above, if I were mentoring an undergraduate on investigating the beliefs of students as they pertain to partial derivatives, I would suggest reading up on Pajares [18], belief and affect theory (e.g., [8]), and any article about students and partial derivatives (e.g., [7] which is in physics education). The reading, which is already situated in a theoretical perspective, now allows me to figure out which method of research I should suggest.

3 Common Methods of RUME Research

There are many ways of approaching an undergraduate mathematics education research study. In this section, I explain some of the common methods that I have seen. I also detail some procedures of each method, and attempt to give ample examples of the RUME research that influenced me in that method.

3.1 Qualitative Research

Qualitative research[1] is another method of conducting RUME, and one that I personally use most often with undergraduate students due to my own comfort, sample sizes, and time. I believe the reason why one would use qualitative methods is to explore the *why* and *how* of certain mathematics education phenomena. It is to "dig deeper" into interesting situations. For example, Ellis et al. [6] concluded, using quantitative data, that women are 1.5 times more likely to leave STEM majors after Calculus compared to men. However, this result generates more research questions, including "*why* are women leaving?". Ellis and Cooper [5], using open-ended prompts in the quantitative study and qualitative methods, attempted to answer this question. They coded 454 open-ended responses and found "the proportion of affective statements made by male and female Switchers [out of a STEM major] was significantly greater than the proportion for Persisters [staying in a STEM major]" [5, p. 133], and also that females specifically reported lower self-efficacy, which may come from more negative experiences with the instructor.

Data can come from many other sources. One may want to answer a research question about teaching ("How does a teacher introduce the Fundamental Theorem of Calculus?"), so data collected may be video observations of multiple teachers teaching the theorem. Depending on the research question, other data collection can include written work or video.

When done with the data collection, data analysis is another aspect of qualitative research that researchers must be careful to conduct and document. For example, one could take all the videos for recording the Fundamental Theorem of Calculus and rewatch them to look for actions that the teacher conducted during the class period. Then, after cataloging the actions, the researcher could categorize the actions into larger groups that could describe many actions. Saldaña [20] has a wonderful resource for coding qualitative data which explains what all the qualitative coding techniques are and in what situations one might use them. When mentoring students in qualitative research, I have them read excerpts from this handbook that I believe will suit their needs.

---

[1] A broad scope of definitions of qualitative research can be found in Chesebro and Borisoff [2].

Larnell [10] is an article that has some of the most robust coding details I have seen, and is, in my mind, one of the most influential qualitative articles in our field. He observed 10 weeks of a course, wrote field notes, audio-recorded post-observation notes, and conducted 5–6 semi-structured interviews of 1–2 hours each with two participants. The author then went through two phases of identity coding with two participants, Vanessa and Cedric, and re-analyzed for a "confluence of themes. . . or the confluence of socialization forces and themes" [10, p. 246]. Larnell concluded that "the institution itself can present barriers to Black learners and their identities" and that the two students are constantly "negotiating" their social, racial, and mathematical identities [10, p. 261]. The author ended the article with appendices that detailed the coding he did for sample data.

One of the most common ways that qualitative research is conducted is through clinical interviews. However, there are many other different methods, including conducting a literature review, teaching experiment, or textbook analysis. I detail each of these below.

3.1.1 Literature Review

A literature review is usually qualitative research where the author tries to aggregate the books, articles, and any other published items of a certain topic in order to find common themes. For example, Leyva [11] did a general search on sex and gender in mathematics education on Google scholar. The author then filtered through "peer-reviewed articles published in the top 100 journals across the fields of education, mathematics (with a focus on education), and gender studies" [11, p. 399]. Finally, Leyva reduced the amount of articles to 56 based on citation amounts and years published. With those 56 articles, Leyva found that there were mainly "two perspectives on studying gender in mathematics education—namely, achievement and participation" [11, p. 425], while also arguing for more detailed research on gender in the form of intersectionality theory.

Literature reviews do not required an Institutional Review Board application, since there are no human subjects in this research. This might be an advantage for some undergraduates, especially those that do not much time to collect data for qualitative or quantitative research. However, literature reviews require use of time differently; a critical reading of the material is required, as well as deep thought about the themes and ideas that emerge.

3.1.2 Teaching Experiment

A teaching experiment is qualitative research where a researcher recruits a handful of students and sets up a series of tasks. These tasks and meetings are separate from their enrolled courses. While doing these tasks, students' mathematics emerges, and the researcher is the person that constructs models of those emergent actions. Swinyard and Lockwood [28, p. 11] stated that "the researcher's central purpose in a teaching experiment is to construct a model of student thinking or reasoning in relation to a particular concept or idea." The authors cited Steffe and Thompson [27] who wrote an essential treatise on why and how to conduct a teaching experiment. For example, Cook [4] examined how two students came to understand properties of algebraic rings through four sessions each lasting 75–90 minutes long and centered around solving equations. One of the results is that utilizing secondary algebra is a starting point can be effective if student thinking can be leveraged productively.

3.1.3 Textbook/Material Analysis

A textbook or material analysis is one method of doing RUME that can be a good way of incorporating undergraduates into research. It also requires no human subjects, so there is no Institutional Review Board application needed (see Sect. 4.3). What it frequently entails is an examination of the constructs or communication of some materials, often with quantitative variables or statistics. One could ask: "What are the kinds of math problems posed written in a textbook?" For example, Lithner [13] examined the exercises in a Calculus textbook using three constructs of how a student could approach the exercise: *identification of similarities* or solving using earlier methods described in the book in solved examples, *local plausible reasoning* or slight modifications in problem solving from what has been previously done in the book, and *global plausible reasoning* or analyzing and considering intrinsic mathematical properties of the components in the exercise. Lithner concluded that this textbook had 85% of the exercises as identification of similarities. Mkhatshwa and Doerr [16] did a similar study on business calculus textbooks.

Tallman and Carlson [29, p. 105] investigated 150 post-secondary Calculus I exams and found that "exams generally require low levels of cognitive demand, seldom contain problems stated in a real-world context, rarely elicit explanation, and do not require students to demonstrate or apply their understanding of the course's central ideas." One critical thinking aspect to consider is based off of Lithner [13]: what beliefs of mathematics are the students acquiring by looking at textbook (or exam) tasks? This question seems to be important especially for undergraduates that would like to be in graduate school or other industry leadership roles, since they may be in charge of training or teaching others.

3.2 Quantitative Research

From many qualitative studies, categories are either used or created. Then, one can either discuss what the categories mean for undergraduate learning and teaching, relationships between the categories, or create a new framework that encapsulates many of the categories. There is another approach to use the categories; with any of these approaches and codings, frequencies appear and quantitative research can be conducted in order to discuss results or answer research questions further.

Quantitative research, particularly for undergraduate students, usually entails accumulating data through surveys or other metrics and using statistics to answer research questions. For example, a student of mine, Ben Gochanour, wanted to investigate the relationship between mathematics anxiety and mathematical motivation, since both constructs have been primarily investigated with mathematical performance (i.e., grades or test scores). He created an online survey with both math anxiety and math motivation questions from previous surveys [9, 14]. The subsequent validation, correlation, and/or causation statistics studied of that online survey would be examples of questions to answer using quantitative research.

Statistical measures are common, so a large sample size would be important. Also, a recommendation would be to have the student take a statistics course or be very fluent with some statistical software (i.e., R or SPSS). Examples of quantitative studies include the previously discussed Ellis et al. [6] article as well as Melhuish [15]. Melhuish took many articles on group theory in abstract algebra that had small-scale qualitative studies and replicated them quantitatively. Her results showed that replication with large-scale sample sizes can "validate theories by reproducing a number of the original results... [using] the general population of undergraduate students taking an introductory abstract algebra course in the United States" [15, p. 31].

One of the biggest questions with a quantitative study is one of reason of correlation or causation. For example, why did some statistically significant result happen with the population one was studying? Making explicit the discussion of statistical results and limitations will go a long way to helping undergraduate students be critical and mindful for their future work.

Mixed-methods research may be another (perhaps strong) approach to research since it combines the quantitative and qualitative methods. It gives both perspectives of research and can answer questions with more information than one method alone. However, mixed methods research can be time-consuming, so I would personally suggest that undergraduates may be guided to one method or another.

4 A Sample Timeline for Mentoring Undergraduate Students in RUME

This section is concerned with a (non-linear) timeline for RUME research that I have experienced with undergraduates. Again, some of this will vary due to the choice of the mentor, as well as if the undergraduate is being incorporated into an existing research project.

4.1 Research Question

A research question is much like a conjecture in mathematics. One is interested with a self-observed phenomenon and wants to see if this is a commonality. Some RUME researchers have undergraduate students work on projects that have been already started, and the research question is a sub-investigation of the larger study. Others let the student generate a research question and negotiate how one could research in that topic. Understandably, both approaches have limits, especially due to the amount of time and energy of both parties, along with the professor's background on methods or topics. There are many research questions aggregated in Sect. 6.

4.2   Background Literature

An undergraduate student comes to you with the research question. What do you do? First, an examination of what has been done previously would be a good step. Perhaps the research question has been already examined, but not in the setting or environment that the undergraduate student has proposed. Some institutions require undergraduate students to have at least five citations before they even submit an Institutional Review Board (IRB) application (see Sect. 4.3). Reading literature is in itself a very difficult task.[2] Mentors should read some literature with the student and discuss in order to reach some mutual understanding of the topics. Much like mathematics, mathematics education has terms and definitions. However, as Selden and Selden [24, p. 432] stated, "Definitions do not have, perhaps cannot have, mathematical precision. Concepts may only be approachable rather than precisely definable." Some literature not only has definitions but the methodological approaches for how they are examining those definitions, and one can look deeper into those methodologies in order to figure out their own. One suggestion is to also properly define each of the terms used in the student's research question. This will often bring up theoretical constructs that need to be addressed.

4.3   Institutional Review Board Application (If Needed)

Almost every institution has an institutional review board (IRB). This review board makes sure that any human-based research projects are within the bounds and scopes of ethics. Prior to applying for permission to research using human subjects, each researcher must be trained in ethics and compliance. This is usually done online through the Collaborative Institutional Training Initiative program [3], and linked to your university. One must plan ahead for this process; the CITI training and the IRB process can both be lengthy. If your project is to be completed in one semester, and the data collected has to also be conducted in the same semester, I would suggest to start writing pieces of the article or thesis that can be done while you wait for the approval.

4.4   Data Collection and Analysis

If you are using human subjects in your research, once you have received permission from your IRB, you can start the data-collection process. Depending on the research conducted, this could take up to 1 month. The data collection should be in accordance with the methodology that you are choosing.

Data analysis is also different depending on methodology chosen, so make sure that you check multiple sources about data analysis before performing it. One of the

---

[2]How one reads mathematics education literature is, in itself, an open RUME question!

ways to alleviate difficulties in writing is to document, either any journal or set of notes, your data collection and data analysis. This way, when you get to the writing process some aspects of the thesis are going to be easier to write out.

4.5 Writing or Presenting Findings

Writing and/or presenting research might be the most difficult aspect of the research process. Once data analysis has ended, one has to interpret what the results mean. This may involve more questions than answers. In this subsection, I will detail the aspects of a typical RUME paper/article/thesis. I will write a bit about what each section includes in order to give guidance. There is no requirement for any of these sections, and authors may combine sections together if appropriate.

4.5.1 Typical RUME Paper Sections

1. **Introduction**. An introduction is typically where one introduces the setting for the research. This may involve stating the math topic or motivation to examine a certain math process. It may also be a place to state what sections are coming in the paper, along with a short description of what each section will discuss.
2. **Background Literature**. This section should be where the majority of your citations of previous research should be. A thorough background literature section should provide both what has been done in the area of the research question while also providing rationale for why your paper is adding to or filling in the previous research. My advice is that a background literature section, in both a paper and a presentation, should be a communication to the audience that you have done your "homework" as well as you could. When doing my dissertation, I had difficulty with not knowing if there were any journal articles that I was missing. Over time, I have been better with my conscience about being fine with not knowing all of the literature. I also know that there are reviewers that will cite other literature that I might have missed.
3. **Theoretical Framework**. This section may be included if one has a theoretical background that they are coming from. Please see Sect. 2 or [24] for more details.
4. **Research Question(s)**. Here is where one writes the question(s) that they wanted to investigate. Many of the previous sections build up to this inquiry, and the rest of the article or thesis deals with how to (try to) answer this question(s).
5. **Methods for Data Collection and Analysis**. I would suggest that this section be as detailed as possible. I also suggest that students include their timeline of both data collection and analysis. If these details are available, please include demographic data with your participants. If possible, and if the participants did not provide their own pseudonyms, I would suggest pseudonyms that encode important information. An example of this pseudonym choice is in Sect. 5.
6. **Results**. The results section is where one presents the data collected or results found. I believe this section should be with as little interpretation or opinion as possible, although some authors combine their results and discussion together. The advice I give to students is to just present the data in the results section.

7. **Discussion**. Here is where the author can discuss and offer the interpretation for the previous section. This is where interpretation and conjecture can happen for the author. It can also be the section that addresses the question, "So what?" or "Why does this research matter?"
8. **Conclusion**. The conclusion should recap the whole article. I believe that we, as humans, only remember a few ideas (for lack of a better word) for any interaction, including reading articles. Therefore, I usually advise that authors include the two big ideas, results, or thoughts that any reader should come away from this article in the conclusion.

5 Paxton's Undergraduate Research Experience

For this section, I will give an example of an experience I had with an honors student, Paxton Martin Clark, doing her thesis in RUME [22].[3] All of the research including reading, IRB, data collection, and writing was done over winter and the spring semester.

Paxton was a pre-med as well as a math major, and she knew that calculus was a requirement for her and many of her peers in pre-med. This requirement of calculus was brought up in her search of articles that mention pre-med and mathematics [17]. However, she also recognized that there was not much literature about pre-med and math. In order to take a first step in this intersection of fields, her (purposely vague) research question was: "what are pre-med majors and physicians' perceptions of undergraduate mathematics education, and what are differences of perceptions between the two groups?"

For Paxton's study, she wanted to know about the intersection of pre-med students and mathematics education. One article by Nusbaum [17] about levels of calculus and pre-med students was the start of the research, and the next step was to see what this article cited or what authors cited in Nusbaum's article. We also looked at pre-nursing or nursing students and mathematics [1], since there was little pre-med and math literature. We did not gain much in terms of methodology and opted instead to do semi-structured interviews with an open qualitative analysis (see Sect. 4.4).

Paxton went through the CITI training immediately after her initial meeting with me. This meeting was arranged in Fall, before her honors thesis work had begun in the Spring. Once done with CITI training, she and I discussed her research question. Since there was little in the literature about this research question, we decided on doing a qualitative study in order to gather more information about mathematics pertaining to Pre-Medicine students and physicians. We then discussed what questions she would like to ask both sets of participants and figure out how we would recruit each set of participants. In our IRB application, we were required to

---

[3] After Paxton submitted her honors thesis, I was so overwhelmed by both the novelty and results of her study that I offered to help her restructure, rewrite, and publish the paper.

detail all of the processes and all of the questions that we used with the participants. This included how we collected the data and what we did with the original data. We also had to make a consent form that allowed the participants to see what the experiment is all about and give their permission to participate.

For example, in Paxton's study, she did a qualitative study that involved interviewing six participants. Each interview was audio-recorded and transcribed, and she gave each student a pseudonym S1–S6 while the physicians were PH1–PH3. Paxton used a process of coding and categorizing, where she initially coded based upon the answers that the interviewees gave and gathered the codes into categories (see [20] and Sect. 3.1 for details). In her results section, she reported on what the participants said in their interviews, presenting the interview data in the same categories she created after her second coding. For example, in her thesis, she gathered quotes into a section labeled "Technological Advances and how that affects math," and one of the quotes that she presented was from a practicing physician talking about more complicated mathematics: "It's actually still pretty integral. I don't personally do the calculations, but I review what the computers say. I have to make sure that they do make sense. The computers help you, and aid you, but they don't replace you." More quotes from the participants were cited in her thesis under that 'technology" category.

Paxton, in her thesis, offered what she believed were some reasons for why the physicians and the pre-med students differed in their beliefs about the relevance of undergraduate mathematics education. She finished her thesis by conjecturing how mathematics educators may use the categories from the interview data in their classrooms.

6   Suggested Projects

Since RUME uses many human-based research questions, and RUME is relatively young, there are infinite ways to generate research questions in RUME. When an undergraduate approaches with an interest in researching with me, the first question I have for them is what *they* are interested when it comes to teaching and learning mathematics. The second question is what they think that they will want to do after this research project; I would steer them towards some research plan that they will enjoy semi-intrinsically.

Perhaps in order to motivate readers interested in undergraduate students and RUME, I offer two collections of open questions in RUME. In 2016, Gulden Karakok (University of Northern Colorado) and I hosted a conference, RUME with a View, that was supposed to be for fairly new researchers to RUME to generate open questions in RUME. 93 participants gathered into five breakout groups in order to both situate themselves with the literature and conjecture questions for the future of RUME to research. Some questions that the participants of the conference generated are listed below:

Research Project 1 Questions generated by the RUME with a View Conference participants

- What does students' mathematical journal entries, notebooks, assignments (or written work), and specifically their use of language reveal about their mathematical understanding?
- How do students' mathematical identities (or affect generally) influence their engagement in pre/business/regular calculus?
- How does collaboration affect the ability to mathematical problem solving or the mathematical problem-solving process?
- What forms do microaggressions take in undergraduate mathematics classrooms? How do students perceive these microaggressions? And, what are the impacts on the students' sense of belongingness?
- What do introductory statistics courses look like across the country at different schools? Why are they offered (for different purposes) and what topics are discussed?

There was also a recent literature review (see Sect. 4.2) on content topics that were under-researched in RUME. This was conducted by Speer and Kung [26], who wanted to investigate the "complement" of RUME research. They [26, p. 1292] stated that:

Knowing where theory development and findings are scarce or plentiful can help researchers (and those who advise them) to know whether their chosen topic is apt to take them into well-understood territory or whether they will encounter few studies and perhaps only limited theoretical frameworks to guide their efforts.

The authors created a list of topics in calculus that have and have not been researched. I cite the topics that have not been researched below:

Research Project 2 Content topics in calculus (from Speer and Kung [26, pp. 1291–1292])

- implicit differentiation, in particular examinations of what sense students make in the transition from df/dx to d/dx and the idea of differentiation as an operator;
- student thinking and sense making about linear approximation and differentials;
- connections between trigonometric functions (as ratios of lengths of triangle sides) and the calculus of them;

(continued)

- Newton's method, in particular what sense students make of the process.
- integration techniques involving the very commonly used method of substitution;
- other integration techniques and what sense, if any, students make of this topic;
- volumes of revolution
- power series, especially the question of what sense students make of the overarching idea of approximating one function with other functions;
- Taylor and Maclaurin Series and what students think the core ideas are behind the computations we ask of them.

7   Conclusion

RUME is a difficult endeavor for anyone. There are many aspects of research that even I am still learning about. Therefore, when an undergraduate student embarks on a semester (or year(s)) journey of RUME, I believe that constant reflection of the journey is necessary. Going through CITI training, then an IRB proposal, then conducting the data collection, then analysis, and then writing and presenting the research is gigantic. In this book chapter, I have tried to make this process explicit for new RUME researchers.

More often than not, what an undergraduate researches in RUME is new to the field. Therefore, I want mentors and mentees to take stock of what they have done throughout their research. I suggest mentors constantly reflect on their advice and how they have nurtured student growth. I suggest mentees to look back at how much they have accomplished themselves. I truly believe research by undergraduates is a formative experience that may ultimately change their lives.

Acknowledgements Thank you to the editors for even considering me; it was an honor. Thank you to Emily Cilli-Turner and Estrella Johnson for reading and making comments prior to submission while always being supportive. I am always indebted to my advisors for their support and care for my professional well-being, while allowing me to be myself throughout this academic journey. Finally, to my family; they are my energy, life, and love.

References

1. Bagnasco, A., Galaverna, L., Aleo, G., Grugnetti, A. M., Rosa, F., Sasso, L.: Mathematical calculation skills required for drug administration in undergraduate nursing students to ensure patient safety: A descriptive study: Drug calculation skills in nursing students. Nurse Ed. in Prac., **16**(1), 33–39. (2016)
2. Chesebro, J. W., Borisoff, D. J.: What makes qualitative research qualitative? Qual. Res. Rep. in Comm., **8**(1), 3–14. (2007)
3. Collaborative Institutional Training Initiative. Available at https://about.citiprogram.org/en/homepage/

4. Cook, J. P.: Monster-Barring as a Catalyst for Bridging Secondary Algebra to Abstract Algebra. In N. Wasserman (ed.), *Connecting Abstract Algebra to Secondary Mathematics for Secondary Mathematics Teachers*, (pp. 47–70). Springer, Cham. (2018)

5. Ellis, J. Cooper, R.: Gender, switching, and student perceptions of Calculus I. In T. Fukawa-Connelly, N. Infante, M. Wawro, & S. Brown (eds.), *Proceedings of the 19th Annual Conference on Research in Undergraduate Mathematics Education* (pp. 125–135). Pittsburgh, PA (2016)

6. Ellis, J., Fosdick, B. K., Rasmussen, C.: Women 1.5 times more likely to leave STEM pipeline after calculus compared to men: Lack of mathematical confidence a potential culprit. PLoS One, **11**(7), e0157447 (2016)

7. Emigh, P. J., Manogue, C. A.: Student Interpretations of Partial Derivatives. In *Proceedings of the 2017 Physics Education Research Conference*. ComPADRE (2017)

8. Grootenboer, P., Marshman, M.: Mathematics, Affect and Learning: Middle School Students? Beliefs and Attitudes About Mathematics Education. Springer (2015)

9. Hopko, D.R., Mahadevan, R., Bare, R.L., Hunt, M.K.: The abbreviated math anxiety scale (AMAS) construction, validity, and reliability. Assessment, **10**(2), 178–182. (2003)

10. Larnell, G. V.: More than just skill: Examining mathematics identities, racialized narratives, and remediation among black undergraduates. J. for Res. in Math. Ed., **47**(3), 233–269. (2016)

11. Leyva, L. A.: Unpacking the male superiority myth and masculinization of mathematics at the intersections: A review of research on gender in mathematics education. J. for Res. in Math. Ed., **48**(4), 397–433. (2017)

12. Liljedahl, P.: The AHA! experience: Mathematical contents, pedagogical implications, Doctoral dissertation. Simon Frasier University, Vancouver (2004)

13. Lithner, J.: Mathematical reasoning in calculus textbook exercises. J. of Math. Beh., **23**(4), 405–427. (2004)

14. Liu, Y., Ferrell, B., Barbera, J., Lewis, J.E.: Development and evaluation of a chemistry-specific version of the academic motivation scale (AMS-Chemistry). Chem. Ed. Res. and Prac., **18**(1), 191–213. (2017)

15. Melhuish, K.: Three conceptual replication studies in group theory. J. for Res. in Math. Ed., **49**(1), 9–38. (2018)

16. Mkhatshwa, T., Doerr, H. M.: Opportunity to learn solving context-based tasks provided by business calculus textbooks: An exploratory study. In T. Fukawa-Connelly, N. Infante, M. Wawro, & S. Brown (eds.), *Proceedings of the 19th Annual Conference on Research in Undergraduate Mathematics Education* (pp. 1124–1132). Pittsburgh, PA (2016)

17. Nusbaum, N. J.: Perspectives: mathematics preparation for medical school: do all premedical students need calculus? Teach. Learn. Med., **18**, 165–168. (2006).

18. Pajares, M. F.: Teachers' beliefs and educational research: Cleaning up a messy construct. Rev. of Ed. Res., **62**(3), 307–332. (1992)

19. Ryan, R., Deci. E.: Self-Determination Theory and the Facilitation of Intrinsic Motivation, Social Development, and Well-Being. Am. Psyc., **55**, 68–78. (2000)

20. Saldaña, J.: The coding manual for qualitative researchers. Sage (2015)

21. Savic, M.: The incubation effect: How mathematicians recover from proving impasses. J. of Math. Beh., **39**, 67–78. (2015)

22. Savic, M., Martin, P.: The perceived vs. actual use of mathematics in medicine according to pre-medicine students and practicing physicians. Tea. Math. and Its App. (2017) doi: https://doi.org/10.1093/teamat/hrx011

23. Selden, A.: A home for RUME: The story of the formation of the Mathematical Association of America's Special Interest Group on Research in Mathematics Education. Technical Report 2012-6, Tennessee Technological University, Cookeville, TN, USA. (2012)

24. Selden, A., Selden, J.: Collegiate mathematics education research: What would that be like? The Col. Math. J., **24**(5), 431–445. (1993)

25. Sio, U. N., Ormerod, T. C.: Does incubation enhance problem solving? A meta-analytic review. Psyc. Bull., **35**, 94–120. (2009)

26. Speer, N., Kung, D.: The complement of RUME: What's missing from our research? In T. Fukawa-Connelly, N. Infante, M. Wawro, & S. Brown (Eds.), *Proceedings of the 19th Annual Conference on Research in Undergraduate Mathematics Education* (pp. 1288–1295). Pittsburgh, PA (2016)
27. Steffe, L. P., Thompson, P. W.: Teaching experiment methodology: Underlying principles and essential elements. In A.E. Kelly, R.A. Lesh (eds.) *Handbook of research design in mathematics and science education*, (pp. 267–306). Routledge. (2000)
28. Swinyard, C. A., Lockwood, E.: Research on students' reasoning about the formal definition of limit: An evolving conceptual analysis. In *Proceedings of the 10th Conference on Research in Undergraduate Mathematics Education* (CRUME2007). (2007)
29. Tallman, M. A., Carlson, M. P., Bressoud, D. M., Pearson, M.: A characterization of calculus I final exams in US colleges and universities. Int. J. of Res. in Und. Math. Ed., **2**(1), 105–133. (2016)
30. Tudge, J. R., Winterhoff, P. A.: Vygotsky, Piaget, and Bandura: Perspectives on the relations between the social world and cognitive development. Human Dev., **36**(2), 61–81. (1993)

Undergraduate Research in Mathematical Epidemiology

Selenne Bañuelos, Mathew Bush, Marco V. Martinez, and Alicia Prieto-Langarica

Abstract

The spread of diseases remains an important issue in public health. The use of mathematics in predicting and understanding epidemics is not new, but still relevant and useful. In this chapter we provide relevant resources and useful exercises for undergraduate students and their mentors. We describe two different modeling techniques which require different backgrounds. For agent based modeling, we suggest students who are either comfortable with programming or willing to learn and who have basic knowledge of probability. For the differential equation approach, we suggest students who have taken at least Calculus 2. Students with a differential equation background will advance faster and can do a more theoretical analysis of the system. A student who might be willing to spend more time working on this topic can model the same disease outbreak using different modeling techniques which will allow for comparison and much deeper analysis of both the mathematics and the biological and public policy implications. Additionally, we include a sample project, developed and written by an undergraduate student who co-authors this chapter. Finally, we provide four different projects that students and their mentors can work on.

S. Bañuelos
California State University Channel Islands, Camarillo, CA, USA
e-mail: selenne.banuelos@csuci.edu

M. Bush · A. Prieto-Langarica (✉)
Youngstown State University, Youngstown, OH, USA
e-mail: mrbush@student.ysu.edu; aprietolangarica@ysu.edu

M. V. Martinez
North Central College, Naperville, IL, USA
e-mail: mvmartinez@noctrl.edu

© Springer Nature Switzerland AG 2020
P. E. Harris et al. (eds.), *A Project-Based Guide to Undergraduate Research in Mathematics*, Foundations for Undergraduate Research in Mathematics,
https://doi.org/10.1007/978-3-030-37853-0_11

Suggested Prerequisites We will describe two different modeling techniques which require different backgrounds. For agent based modeling, we suggest students who are either comfortable with programming or willing to learn and who have basic knowledge of probability. For the differential equation approach, we suggest students who have taken at least Calculus 2. Students with a differential equation background will advance faster and can do a more theoretical analysis of the system of differential equations. A student who might be willing to spend more time working on this topic can model the same disease outbreak using different modeling techniques which will allow for comparison and much deeper analysis of both the mathematics and the biological and public policy implications.

1 Introduction

We begin by discussing our preferred philosophy in leading undergraduate research, as it is with this philosophy in mind that the following projects and exercises are written. The three professors authoring this chapter have a relatively long history of guiding undergraduate students in research activities. All the while prioritizing students' professional goals and acquisition of skills above research topic and research outcomes.

When engaging in a research project students are encouraged to focus on skills, not in the question being asked. In general, working on a research problem involves disseminating the results to the research community. While working with a faculty mentor, students will learn how to best present their results orally either through a poster or a talk (which may include slides). Also, through multiple drafts, students will learn how to write a mathematical paper. Most important, students will experience the difference between answering exercises pertaining to course material, where there is a clear set of instructions in the content, and exploring a topic through inquiry. Research involves getting stuck, persisting, and realizing that not every problem has a single solution, if any at all.

In addition, we take advantage of research activities to enhance students' sometimes "secondary" skills, such as the ability to communicate effectively about research to very diverse audiences, the capacity to work efficiently and harmoniously in diverse teams, and the capability of engaging in interdisciplinary teams. These are all important skills that will be of great benefit to students as they continue their chosen career paths.

In a research investigation involving mathematical modeling, students will learn about different types of mathematical models (e.g., discrete, continuous, stochastic, etc.), what scenarios are best explored with a certain type of model given the conditions of the proposed problem, the limitations of the chosen modeling method, how to run numerical simulations and interpret the results, and the theory on the analysis of such models. We recommend working with a team of biology and mathematics majors, and partnering with biology (or public policy) faculty. This will expose students to different ways of thinking and to knowledge, skills, and processes important in different disciplines. In addition, faculty in these departments may offer great research problems and data.

When research is led by undergraduate students, they develop a sense of ownership of their learning and of the research outcomes. Their curiosity will lead them to a deeper understanding of the problem and of mathematics in general. Through their readings and exploration in the problems below, students will feel empowered to create their own mathematical models. The skills student and mentors will gain by engaging in this chapter will help them sustain further research, for example, modeling in networks.

There is liberty in applied mathematics to explore a research exercise through inquiry. This method will prepare students to approach problems in their future work, where delineations and step by step instructions do not exist. In particular, students will gain the ability to mathematically model physical phenomenon, and more important, to understand the real world implications of their models. These are skills that many undergraduate mathematics programs have not developed in their curriculum. This is of particular importance for those seeking employment outside academia.

This chapter is organized as follows. Section 2 provides a background with plenty of references on mathematical epidemiology. The large amount of references might seem overwhelming, so we recommend students browse through them and then pick for further exploration those which they like the most. Section 4 provides different examples of projects students and mentors can explore. Section 3 offers a set of inquiry based exercises for students of all backgrounds in two different areas, classical ODE modeling, and agent based modeling applied to graphs and networks. Lastly, Sect. 5 offers a particular example, written by an undergraduate research student, Mathew Bush. After some literature review, Bush formulated the problem, the model and did most of the analysis of the model. His project has spanned a couple of years and he is currently working with incoming sophomores who will further advance the project.

1.1 Note to Students

The problems in this chapter are meant to be taken as exploratory exercises. They are not meant to be homework problems, which are usually completed and understood in a couple of days to a week. The exercises presented here aim to be your (maybe) first introduction to research. As opposed to a classroom exercise, these problems might not have a unique solution or the closed, clear solutions that you are accustomed to. The problems in this chapter are open ended in order for you to explore, learn, and question.

1.2 Note to Mentors

This chapter is structured as an inquiry based learning exercise. Students are expected and encouraged to lead the research and make decisions on the direction they want to take. As mentors ourselves, we understand that this can be challenging.

Many times we can predict that the approach students are taking might not yield optimal results. We still encourage you to let them try and to not shield them from possible failure, as we believe learning happens when we allow ourselves to explore. Several times, by letting our students take the lead, we have been pleasantly surprised of how much better their approach is than the one we had in mind.

2   Epidemiology Background

This section will be kept short and will mainly offer bibliography that students can access and discuss with their research mentor.

To Get Started and for Students with a Calculus Background
The following are recommended places to begin. Otto and Day's book on mathematical modeling [35] is a perfect start for any freshmen or sophomore student with knowledge of Calculus 1 or Calculus 2. In particular, Sect. 3.5 on Mathematical Epidemiology is a great first read for potential researchers. A short self-contained guide on the SIR model with a few exercises can also be found in [38].

For Students with Some Exposure to Upper Division Mathematics
A good resource for advanced undergraduate students and graduate students are the following books [1, 9, 15, 22, 28] and the following review articles [2, 27, 37]. For statistical approaches, see [13].

For Examples of Publications by Undergraduates in Mathematical Epidemiology
For ideas on standard things that can be done in epidemiology with advanced undergraduate students, see [12, 19, 21, 30, 32, 47]. See [17, 31] for examples in interesting and non-standard applications of epidemiological models.

3   Mathematical Modeling Problems

3.1   Simple ODE Models

In this section, students will explore simple exercises in ODE modeling in epidemics. Students are asked to add assumptions to each scenario as they consider each exercise. We strongly recommend for students to read the introductory material suggested above, in particular the one presented in [38] before attempting the following exercises:

Exercise 1 Assume there is an isolated population of 10,000 people on an island where for the given time, t, there are no births, deaths, nor migration. An individual infected with an airborne disease lands on this island. This disease is transmitted from an infected person to a non-infected person with transmission probability

$p = 0.75$. An infected individual never recovers (nor dies of the disease). In addition, once an individual contracts the disease, they will no longer be susceptible to the disease. This situation can be described with a simple SI model in which the population is divided into Susceptible and Infected classes. The rate in which a Susceptible individual transitions into the Infected class is given by the transmission probability p times the encounter rate q. One can think of q as a measurement of how sociable the population is. If most people keep to themselves, then q is really low. Otherwise, q can be closer to 1.

1. Create a flow diagram that describes this situation.
2. Write down equations that represent the dynamics of your flow diagram.
3. Solve your equations numerically.
4. Experiment with different values of q.
5. Describe what your results show and how the results depend on q.

Exercise 2 Consider the same scenario as above, however, now assume that infectious individuals can recover from the disease (and gain permanent immunity) after γ days. The individuals that recover from the disease are no longer able to infect others. The model will now need a new group in the population, the *Recovered* class. How will you incorporate γ into this model?

1. Create a flow diagram that describes this situation.
2. Write down equations that represent the dynamics of your flow diagram.
3. Solve your equations numerically.
4. Experiment with different values of γ.
5. Describe what your results show and how the results depend on γ.

Exercise 3 Same scenario as above, however, along with the assumption that infectious individuals can recover from the disease after γ days consider that after spending d days with immunity, they become susceptible again. How would you incorporate d into this model?

1. Create a flow diagram that describes this situation.
2. Write down equations that represent the dynamics of your flow diagram.
3. Solve your equations numerically.
4. Experiment with different values of d.
5. Describe what your results show and how the results depend on d.

Exercise 4 Same scenario as above, however, now assume that humans have a short period of time t_e in which they have been exposed to the disease but are not yet able to spread the disease to others. You will have to add a new population of *Exposed* individuals.

1. Create a flow diagram that describes this situation.
2. Write down equations that represent the dynamics of your flow diagram.

3. Solve your equations numerically.
4. Experiment with different values of t_e.
5. Describe what your results show and how the results depend on t_e.

Exercise 5 So far the exercises above do not included birth into the population, death of natural causes, nor death from the disease. For this exercise, consider the scenario from Problem 4 and add birth and deaths.

1. Where will you include births? Think about different diseases and scenarios.
2. Include a natural death rate μ and a death rate for infected individuals μ_i such that $\mu < \mu_i$. Experiment with different values of μ_i.

Challenge Problem 1 Before attempting this exercise, see the articles and books referenced in Sect. 2 and read about the reproductive number, r_0. For the above model investigate the following:

1. What kinds of diseases behave in the manner described by the model?
2. Choose a disease that can be simulated by the above model and find possible parameter values. Compute r_0 for that particular disease.
3. Keep all the parameters fixed, except for the transmission rate r. Plot r_o vs r. For what values of r can the spread of the disease be contained?
4. What public policies would you recommend for the region to adopt in order to decrease the reproductive number, r_0? (i.e. to contain the epidemic)
5. Keep all parameters fixed, except for the γ (the average time an individual spends in the Infected class). Plot r_0 vs γ. What do you observe? Can we control γ?

3.2   Vector-Borne Diseases

The following set of exercises address diseases that are transmitted via a vector, e.g., mosquitoes, ticks, or fleas. Examples of vector-borne diseases are Zika, Malaria, and Chagas.

For most of these diseases, transmission does not occur from human to human contact. Instead, an infected vector has to bite a susceptible human (or a susceptible vector has to bite an infected human) to transmit the disease. The first exercise begins with a simple scenario.

Exercise 6 Begin with a population of N_h susceptible humans and $N_v - 1$ susceptible vectors where one vector is infected. Assume vectors can transmit the disease to humans via a bite and that neither humans nor vectors will die or recover from the disease (i.e., once a human or vector becomes infected, they remain infected and infectious).

1. Create a flow diagram that describes this situation.
2. Write down equations that represent the dynamics of your flow diagram.

3. Solve your equations numerically for different values of the ratio:

$$y = \frac{N_h}{N_v}.$$

4. Change your initial conditions to start with one infected human and no infected vector and repeat the previous exercise.
5. Describe what your results show and how the results depend on the initial condition and on y.

Exercise 7 Consider the previous scenario and now assume that humans can recover from the disease and gain immunity (they will not transition to the Susceptible class again). In other words, the model will now have a *Recovered* class.

1. Create a flow diagram that describes this situation.
2. Write down equations that represent the dynamics of your flow diagram.
3. Solve your equations numerically for different values of the recovery rate r and different values of the ratio:

$$y = \frac{N_h}{N_v}$$

4. Change your initial conditions to start with one infected human and no infected vector and repeat the previous exercise.
5. Describe what your results show and how the results depend on the initial condition and on y.

Exercise 8 Consider the previous scenario and assume that the immunity gained by humans is only temporary. Let q be the rate at which humans lose immunity.

1. Create a flow diagram that describes this situation.
2. Write down equations that represent the dynamics of your flow diagram.
3. Solve your equations numerically for different values of the recovery rate q and different values of the ratio:

$$y = \frac{N_h}{N_v}$$

4. Change your initial conditions to start with one infected human and no infected vector and repeat the previous exercise.
5. Describe what your results show and how the results depend on the initial condition and on y.

Exercise 9 In addition to the assumptions in the previous exercise, now assume that humans have a short period of time t_e in which they have been exposed to the disease but are not yet able to transmit the disease through a vector. This requires a new group in the population, the *Exposed* class.

1. Create a flow diagram that describes this situation.
2. Write down equations that represent the dynamics of your flow diagram.
3. Solve your equations numerically for different values of t_e and different values of the ratio:

$$y = \frac{N_h}{N_v}.$$

4. Change your initial conditions to start with one infected human and no infected vector and repeat the previous exercise.
5. Describe what your results show and how the results depend on the initial condition and on y.

Challenge Problem 2 For the model that was created above investigate the following:

1. What kinds of diseases behave in the manner described by the model?
2. Choose a disease that can be simulated by the above model and find possible parameter values. Compute r_0 for that particular disease.
3. Keep all the parameters fixed, except for the transmission rate r. Plot r_o vs r. For what values of r can the spread of the disease be contained?
4. In real life, what public policies can be put in place in order to decrease r? (i.e., to contain the epidemic.) Think about the typical intervention strategies:

   • bed nets
   • vector population control
   • use of mosquito, tick, or flea repellent
   • removal of still water (affects mosquito reproduction)

   How do each of these strategies affect the model (which parameters are modified)?
5. Keep all parameters fixed, except for γ (the average time an individual spends in the Infected class). Plot r_0 vs γ. What do you observe? Can we control γ?

3.3   ABM and Epidemics

The following series of exercises aim to familiarize students with working with agent based modeling (ABM) in graphs. We recommend for novice programmers to use introductory programming language such as NetLogo [44]. A great introduction to NetLogo can be found in [7, 14]. For more experienced programmers, MatLab and C++ are excellent tools.

For all these exercises, students will need to be familiar with the adjacency matrix associated with a graph and with some basic concepts in graph theory. However, students are not required to have deep knowledge of these concepts, just the elemental understanding of the definitions. The goal is that through these problems, students will gain a deeper understanding of graphs. We suggest using the following books [11, 20, 42] as a reference for definitions and as a resource for students who wish to learn further about graphs.

Exercise 10 (Randomly Creating Graphs) For $n = 10, 100, 1000$ create a random undirected graph with n vertices and $v = n$, $2n$, n^2 randomly placed edges. For each graph calculate

- The vertex degree
- The average vertex degree
- The graph distance matrix
- The vertex connectivity
- The closeness centrality
- The betweenness centrality

Now assign directions to your graph. Which of the above measurements change? Can you display the graphs above? (Note: this might depend on the programming language being used).

Exercise 11 (Movement in a Graph) Start with a random graph with $n = 100$ vertices and $v = 100, 250, 500, 1000$ directed edges. Randomly pick a vertex for a particle to start (in a visualization this can be displayed with a different color). Following the direction of the edges, let the particle move randomly throughout the graph for $T = 50, 100, 150$ time steps. For the same starting point, record how many vertices were visited and how many times they were visited. What do you observe? Do patterns emerge? Do cycles emerge? Do the same exercise for different initial conditions.

Exercise 12 (Movement in a Graph 2) Randomly create a graph with $n = 100$ vertices and $v = 100, 250, 500, 1000$ directed edges. Randomly place a particle in one of the vertex. Each time step, the particle can move to any other connected vertex. This time the particle can only visit each vertex once. For the same graph and different initial placement, record how many vertices were visited. What do you observe? Do patterns emerge? Do cycles emerge? Do the same exercise for different initial conditions.

Exercise 13 (Weighted Graphs) Randomly create a graph with 100 vertices and 1000 edges. Make sure your graph is completely connected. Randomly assign a weight $w_{i,j} \in [0, 1]$ to the edge connecting vertex i and vertex j. Randomly select a vertex to place a particle on. The particle can move to an adjacent vertex randomly, however, it will move through heavier edges with greater probability.

Run this experiment for $T = 5000$ time steps. Experiment with different ways of assigning this probability. What do you observe? Do patterns emerge? Do cycles emerge? Do the same exercise for different initial conditions.

Challenge Problem 3 (Challenges on Graphs) Randomly create a graph with 100 vertices and 1000 edges. Make sure your graph is completely connected. Randomly assign a weight $w_{i,j} \in [0,1]$ to the edge connecting vertex i and vertex j. Randomly select a vertex to place a particle on. The particle can move to an adjacent vertex randomly, however, it will move through heavier edges with greater probability. Once a vertex has been visited, it cannot be visited again. What do you observe? Do patterns emerge? Do cycles emerge? Do the same exercise for different initial conditions.

Challenge Problem 4 (Challenges on Graphs 2) Randomly create a graph with 100 vertices and 1000 edges. Make sure your graph is completely connected. Randomly assign a weight $w_{i,j} \in [0,1]$ to the edge connecting vertex i and vertex j. Randomly select a vertex to place a particle on. The particle can move to adjacent vertex randomly, however, it will move through heavier edges with greater probability. Once a vertex has been visited, it cannot be visited **for the next 5 time steps**. What do you observe? Do patterns emerge? Do cycles emerge? Do the same exercise for different initial conditions.

4   Projects

Research Project 1 Simple Compartmental Model

For this project, students should benefit from being familiar with the AMC Network show "The Walking Dead." Students interested in familiarizing with the series should watch it or read the comic book series [18,40]. In the world of "The Walking Dead," a disease is being spread among the living, killing them and turning them into *zombie*-like creatures. Before we delve into the complicated world the series (or comic book) depicts, let us start with simpler assumptions:

- Zombies bite humans (but will not eat humans) and thus humans will become zombies
- Humans do not "kill" zombies

Construct a simple SI model that incorporates these assumptions and state all additional assumptions clearly. Experiment with different bite rates. Numerically solve the system of ODEs.

Questions of Interest
1. If you start with one zombie, what percentage of the population becomes a zombie for different bite rates? Look for limiting behavior by running your program for a long time.
2. Can you get rid of all the humans? How or why not?

On the series (or comic book), things are a bit more complicated. The spread of the zombie virus can happen through the following mechanisms:

- Being beaten by a zombie.
- Dying of natural causes.

Once a living human becomes a zombie, their only drive is to finding living humans to devour. Based on this, every time there is a human-zombie interaction one of the following things can happen:

- Human "kills" zombie
- Zombie bites human, but human escapes before being completely eaten.
- Zombie eats human.

Construct an SEI Model that describes the world of "The Walking Dead." Run simulations of your model with different transmission rates. For good references and for examples of zombie epidemic models, see [10, 34, 45].

Questions of Interest
1. Are you adding any assumptions besides the ones provided?
2. What aspects of the show are believable? (i.e., can you find a set of parameters that will (give) scenarios similar to the ones portrayed on the show? If you find a set of parameters that work, are those values plausible?
3. What aspects of the show are not "realistic" based on your results? Why?
4. Are there assumptions you can add or remove in order to get similar results as the ones depicted on the show?
5. Based on your model:
    (a) If a vaccine was available, what percentage of the population would need to be vaccinated to prevent the post-apocalyptic world?
    (b) What percentage of zombies would need to be killed in order to "control" the spread of the disease?

Construct an ABM to simulate this scenario. What assumptions are you making? Does this change the results of your previous investigation?

Research Project 2 Multi-patch model example and analogous ABM

This project is divided into two smaller projects. One project will focus on epidemiological multi-patch ODE models and the second project will model the same system using ABMs. If the same student/group of students wants to work on both projects, the ABM and the ODE model can be run in tandem and the ABM can be used to estimate parameters for the ODE model.

Another way to introduce space in a mathematical model is by using multi-patch metapopulation models. There are a variety of examples of ecological models that use this approach [22, 25, 36]. For epidemiological examples, see [3–5].

One can think about political ideology as something that evolves in time and that can be influenced by people one interacts with. With this in mind, we can model the evolution of political ideologies using epidemiological models. Similar models have been constructed in [6, 33, 46]. Consider four basic ideologies represented in Fig. 1 and assume that any individual's ideology can lie in only one of the quadrants. In reality, most of our political thinking lies in a continuum but for now, this shall suffice.

Consider each quadrant in Fig. 1, which was obtained from the website in [39], as one class in our model. Movement is allowed between any two of the classes at one time, however, the rates of change between classes are highly dependent on the individual's geographical region. Indeed, a person's ideology is highly dependent on their location as can be seen with voters who live in rural or urban communities [26, 41].

Fig. 1 This figure was created by the authors with data from The Political Compass [39]

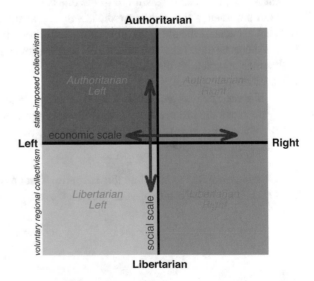

1. Start with a particular location, (say, the rural Midwest). Create a flow diagram that represents the movement of individual among ideologies. You will need 5 classes: one for each ideology plus a class of undecided individuals. Keep in mind that individuals with one particular ideology can influence individuals from all other classes. Individuals with a particular ideology that are influenced by others will move into the undecided class and then to another ideology class.
2. Write a system of ODE's for the model above and write a program to solve the system. What do you notice?
3. In reality, as stated above, an individual's ideology and the rate in which they change from one to the other is highly dependent on geographical region and setting. Use the above model to create a multi-patch model in which:

   - Each patch represents a particular region (e.g. the urban midwest, or the rural south).
   - Individuals interact and are influenced by people in their region at higher rates than by people outside their region.
   - There is interaction among regions (think national media, the internet, migration, or students moving for college and returning home), how would you account for these interactions?

4. Create an ABM that will model the above described situation, with several patches and individuals in 5 different classes. This can be done in several ways. However, looking at this from a network perspective might be the clearest and most easily implementable.
5. Run both models and describe the difference and similarities in your results. How do you think you can use the ABM model to help experiment with different parameters and use that information to help inform the ODE model.

Research Project 3 Spread of Diseases on a Network

In this project we will investigate the spread of information on networks. A basic understanding of Graphs will be assumed. The authors recommend the following books to learn the basics and understand the nomenclature [11, 20, 42]. We also suggest working on Problems 10–12. This project was inspired by the work of Julien Arino [23, 24].

Create a network with all the mathematics majors in at your institution. Each student will be represented by a vertex in the graph, so if there are n mathematics majors, there should be n vertices in your graph. Connect two vertices with an edge if the students are taking a class together. The edges in the graph will be weighted and the weight of each edge will be the number of classes those two students share. Many diseases can be transmitted from student to student by simply sharing the same space. Representing this network in a computer simulation can help us experiment with this idea.

Suppose one of the mathematics majors gets the flu. A question of interest to you might be what is the probability of you contracting that disease. Pick patient zero, P_0, at random from all the mathematics majors, with each vertex having the same probability of being chosen. The disease can then travel from one vertex to the next with a small probability $p_{ij} << 1$ which depends on the weight of edge (i, j) (the more classes you have with a certain mathematics major, the greater the probability of you contracting the disease if they are infected). Run this program until YOUR own vertex contracts the disease or for T time steps, fixing patient zero. For each run, record t_i = the number of time steps it took for you to get infected on simulation i. If you did not contract the disease during simulation i, then $t_i = 0$. Report a histogram of vector t.

Questions of Interest
1. Experiment with a different formulas for $p_{i,j}$. Think about a virus like the cold or the flu. How many times have you been in class with someone that is suffering with that disease and how many times did you get infected?
2. Experiment with a different "patient zero." What do you observe?
3. In reality, diseases are not permanent and people recover from them and many times gain immunity. How can you implement that in this particular network?
4. Assume each time step represents a school day. Most students do not take the same classes each day. How would you accommodate for that? (the weights in your network might need to change every time step)
5. In reality, not all students make it to all the classes all the time (shocker!). Introduce a small probability q_i^j for each vertex i of skipping a class (or a few classes) each time step j.

In reality, you take classes with more than just mathematics majors. A very interesting and more complicated problem will be to get information on the schedule of all students in your college or university and repeat this project for this much larger network.

Questions of Interest
1. For which kind of "patient zero" is the disease more likely to reach you? mathematics majors? STEM majors? students in your same graduating class?
2. How can you incorporate the difference between a resident student or a commuter student in the model? Should there be a difference in who is more likely to contract the disease? How would you implement this?

Research Project 4 Hybrid Model: spread of bacteria/ocean coral reef

Coral reefs are underwater communities of reef-building corals [43] which are in rapid decline worldwide. Reasons for this decline are in part due to bleaching (expulsion of photosynthetic symbionts) and outbreaks of infectious disease [8, 29]. Given that many of these communities behave as entities and can be considered as discrete individuals, a hybrid ABM/PDE model might be an ideal way to model this particular situation. A recommended ODE model can be found in [16].

Start with a grid that represents an underwater region. Each square in your grid can be occupied by a coral reef or it can be empty. As corals reproduce and build more reefs, each time step, the coral can expand to neighboring squares with some probability. Assume bacteria travels in the water (a continuum) inside underwater currents. Model that behavior using PDEs. For every time step in the ABM model, numerically solve the PDE model and update the states of your ABM.

Questions of Interest
1. How would you model bacteria traveling inside underwater currents? What partial differential equation models are in the literature that can help you achieve this?
2. Let r be the reproduction rate of corals and let p be the average bacteria concentration in the fluid. What can you say about the probability of survival of the coral reef in relation to the fraction $\frac{r}{p}$
3. What happens to the fraction $\frac{r}{p}$ as the ocean water warms up?
4. Can you think about good intervention strategies that might help save the corals?

5 An Example: Gendered Zika Model

With a more extensive background in mathematics more challenging questions and problems can be explored. The Gendered Zika model presented here was done as an undergraduate project, by a Junior student, Mathew Bush. He held a full time job as an EMT and worked on this project for almost 2 years. The authors guided the student through this project with their preferred philosophy as described in the introduction—inquiry based research. The following section was written by Mathew.

The challenges that need to be considered when modeling Zika are in its route of transmission. Unlike other vector-borne diseases such as Dengue and Yellow Fever, Zika can be transmitted by humans as well as by mosquitoes. These different transmission routes and their effects on the model are discussed in this section.

The gendered Zika Virus SEIR model is one that presented itself following developments in how the virus is spread. Zika, initially, was thought to be a purely vector-borne illness, meaning it is spread through the bites of mosquitoes. Recent developments, however, have shown other possible routes of infection, including, through blood transfusions, lab acquisition, and sexual contact. Because transmission through blood transfusions and lab acquisition occurs at such low rates, the science community has been focusing on studying the effects of sexual transmission. These new routes of transmission created a need to adapt the most widely accepted models.

Previous models of Zika were created by grouping male and females into one human population. With this single grouping the models do not take into account the different sexual transmission rates of males and females. This new Gendered Zika model allows us to account for male to male transmission and female to female transmission (Fig. 2 and Tables 1 and 2).

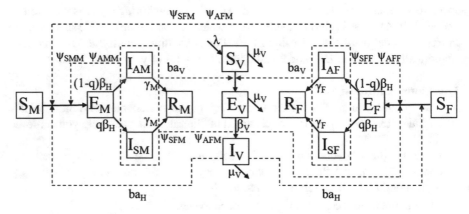

Fig. 2 Diagram of the Gendered Zika model where the dynamics of the female and male populations follow their own SEIR models

Table 1 A list of all thirteen variables in the Gendered Zika model

Variable	Description
N_H	Total human population
S_M	Population of susceptible males
E_M	Population of exposed males
I_{AM}	Population of asymptomatic males
I_{SM}	Population of symptomatic males
R_M	Population of recovered males
S_F	Population of susceptible females
E_F	Population of exposed females
I_{AF}	Population of asymptomatic females
I_{SF}	Population of symptomatic females
R_F	Population of recovered females
N_V	Total vector population
S_V	Population of susceptible vectors
E_V	Population of exposed vectors
I_V	Population of infected vectors

The total population of humans, N_H, and vectors, N_V, is assumed to be constant

Table 2 The parameters govern how human males and females, and mosquitoes move through the compartmental model

Parameter	Description
β_V	Latency rate of Zika in vectors
λ	Birthrate of vectors
μ_V	Lifespan of vectors
a_H	Transmission probability from vector to human
b	Number of bites per mosquito per day
a_V	Transmission probability from human to vector
ψ_{SFM}	Transmission probability from symptomatic female to male
ψ_{SFF}	Transmission probability from symptomatic female to female
ψ_{SMF}	Transmission probability from symptomatic male to female
ψ_{SMM}	Transmission probability from symptomatic male to male
ψ_{AFM}	Transmission probability from asymptomatic female to male
ψ_{AFF}	Transmission probability from asymptomatic female to female
ψ_{AMF}	Transmission probability from asymptomatic male to female
ψ_{AMM}	Transmission probability from asymptomatic male to male
q	Proportion of humans who are symptomatic
β_H	Incubation period in humans (days)
γ_M	Recovery rate of human males
γ_F	Recovery rate of human females

Following the steps described in the exercises above, once the flow diagram was created, we set up the system of differential equations that describes the dynamics of the model.

$$\frac{dS_M}{dt} = \frac{S_M}{N_H}(-a_H b I_V - I_{AM}\psi_{AMM} - I_{SM}\psi_{SMM} - I_{AF}\psi_{AFM} - I_{SF}\psi_{SFM})$$

$$\frac{dE_M}{dt} = \frac{S_M}{N_H}(a_H b I_V + I_{AM}\psi_{AMM} + I_{SM}\psi_{SMM} + I_{AF}\psi_{AFM} + I_{SF}\psi_{SFM})$$
$$-\beta_H E_M$$

$$\frac{dI_{SM}}{dt} = q\beta_H E_M - \gamma_M I_{SM}$$

$$\frac{dI_{AM}}{dt} = (1-q)\beta_H E_M - \gamma_M I_{AM}$$

$$\frac{dR_M}{dt} = \gamma_M(I_{SM} + I_{AM})$$

$$\frac{dS_F}{dt} = \frac{S_F}{N_H}(-a_H b I_V - I_{AF}\psi_{AFF} - I_{SF}\psi_{SFF} - I_{AM}\psi_{AMF} - I_{SM}\psi_{SMF})$$

$$\frac{dE_F}{dt} = \frac{S_F}{N_H}(a_H b I_V + I_{AF}\psi_{AFF} + I_{SF}\psi_{SFF} + I_{AM}\psi_{AMF} + I_{SM}\psi_{SMF})$$
$$-\beta_H E_F$$

$$\frac{dI_{SF}}{dt} = q\beta_H E_M - \gamma_F I_{SF}$$

$$\frac{dI_{AF}}{dt} = (1-q)\beta_H E_M - \gamma_F I_{AF}$$

$$\frac{dR_F}{dt} = \gamma_F (I_{SF} + I_{AF})$$

$$\frac{dS_V}{dt} = \lambda N_V - a_V b(I_{AM} + I_{SM} + I_{AF} + I_{SF})\frac{S_V}{N_V} - \mu_V S_V$$

$$\frac{dE_V}{dt} = a_V b(I_{AM} + I_{SM} + I_{AF} + I_{SF})\frac{S_V}{N_V} - \beta_V E_V - \mu_V E_V$$

$$\frac{dI_V}{dt} = \beta_V E_v - \mu_V I_V$$

Our next step was to solve for the equilibrium states of the system, or when $dx/dt = 0$ for all variables x. Doing so requires that we solve for each variable while setting the equations in the system equal to zero. The total population of humans and mosquitoes is assumed to be constant and

$$N_H = S_M + E_M + I_{AM} + I_{SM} + R_M + S_F + E_F + I_{AF} + I_{SF} + R_F,$$

$$N_V = S_V + E_V + I_V.$$

After some simple computation, we see that at equilibrium

$$E_M = I_{AM} = I_{SM} = E_F = I_{AF} = I_{SF} = 0, \text{ and}$$

$$E_V = I_V = 0.$$

Thus,

$$N_H = S_M + R_M + S_F + R_F, \text{ and}$$

$$N_V = S_V.$$

We can now calculate our basic reproductive number, R_0, by the next generation method. This is done by first setting n equal to the total number of compartments, here $n = 13$. The number of compartments in which infection is present is denoted by m, here $m = 8$. A column vector is formed by reordering the variables of the model and where the m compartments described above are listed first. That is,

$$x = (x_1, x_2, x_3, \ldots, x_n)$$

$$= (E_M, I_{SM}, I_{AM}, E_F, I_{SF}, I_{AF}, E_V, I_V, S_M, S_F, R_M, R_F, S_V).$$

Now we create two matrices, $F_i(x)$ which is the rate of new infections in compartment i, and $V_i(x)$ which is the difference in the rate of transfer out of compartment i and the rate of transfer into compartment i. For our model we found

$$F_1(x) = \frac{S_M}{N_H}(a_H b I_V + I_{AM}\psi_{AMM} + I_{SM}\psi_{SMM} + I_{AF}\psi_{AFM} + I_{SF}\psi_{SFM})$$
$$-\beta_H E_M$$

$$F_4(x) = \frac{S_F}{N_H}(-a_H b I_V \frac{S_F}{N_H} - I_{AF}\frac{S_F}{N_H}(\psi_{AMF} + \psi_{AFF})$$
$$-I_{SF}\frac{S_F}{N_H}(\psi_{SMF} + \psi_{SFF})$$

$$F_7(x) = a_V b(I_{AM} + I_{SM} + I_{AF} + I_{SF})\frac{S_V}{N_V} - \beta_V E_V - \mu_V E_V$$

$$F_2 = F_3 = F_5 = F_6 = F_8 = F_9 = F_{10} = F_{11} = F_{12} = F_{13} = 0$$

This allows us to create our matrix $F_i(x)$.

To create our $V_i(x)$ we first split it into two sets of equations, $V_i^+(x)$ which is our rate of transfer into compartment i, and $V_i^-(x)$ which is our rate of transfer out of compartment i.

$$V_1^+(x) = 0 \qquad\qquad V_1^-(x) = \beta_H E_M$$
$$V_2^+(x) = q\beta_H E_M \qquad\qquad V_2^-(x) = \gamma_M I_{SM}$$
$$V_3^+(x) = (1-q)\beta_H E_M \qquad\qquad V_3^-(x) = \gamma_M I_{AM}$$
$$V_4^+(x) = 0 \qquad\qquad V_4^-(x) = \beta_H E_F$$
$$V_5^+(x) = q\beta_H E_F \qquad\qquad V_5^-(x) = \gamma_F I_{SF}$$
$$V_6^+(x) = (1-q)\beta_H E_F \qquad\qquad V_6^-(x) = \gamma_F I_{AF}$$
$$V_7^+(x) = 0 \qquad\qquad V_7^-(x) = \beta_V E_V + \mu_V E_V$$
$$V_8^+(x) = \beta_V E_V \qquad\qquad V_8^-(x) = \mu_V I_V$$

$$V_9^+(x) = 0 \qquad\qquad V_9^-(x) = \frac{S_M}{N_H}(a_H b I_V + I_{AM}\psi_{AMM}$$
$$+ I_{SM}\psi_{SMM}$$
$$+ I_{AF}\psi_{AFM}I_{SF}\psi_{SFM})$$

$$V_{10}^+(x) = 0 \qquad\qquad V_{10}^-(x) = \frac{S_F}{N_H}(a_H b I_V + I_{AM}\psi_{AMF}$$
$$+ I_{SM}\psi_{SMF} + I_{AF}\psi_{AFF}$$
$$+ I_{SF}\psi_{SFF})$$

$$V_{11}^+(x) = \gamma_M(I_{SM} + I_{AM}) \qquad V_{11}^-(x) = 0$$
$$V_{12}^+(x) = \gamma_F(I_{SF} + I_{AF}) \qquad V_{12}^-(x) = 0$$

$$V_{13}^+(x) = \lambda N_V \qquad\qquad V_{13}^-(x) = a_V b(I_{AM} + I_{SM} + I_{AF} + I_{SF})\frac{S_V}{N_V}$$
$$+ \mu_V S_V$$

Now we let $V_i(x) = V_i^-(x) - V_i^+(x)$.

Now that we have our $V_i(x)$, using Gauss–Jordan elimination we want to find $V_i(x)^{-1}$. Then the dominant eigenvalue of the product, $F_i(x)V_i(x)^{-1}$, is the basic reproduction number. A closed form for the expression of R_0 is difficult to find. Because of this, we use computational software such as MatLab or Maple to find these eigenvalues.

This is an example of a real world application of mathematical modeling in epidemiology. This model is a work in progress and provides many opportunities for future work. We would like to progress this research by including sensitivity analysis, which will allow us to see the driving parameters within this model. Also, future work may introduce seasonality or stochasticity into the model.

Acknowledgement The authors would like to thank the Office of Research at Youngstown State University for their support.

References

1. L. J. Allen, F. Brauer, P. Van den Driessche, and J. Wu. *Mathematical epidemiology*, volume 1945. Springer, 2008.
2. M. Andraud, N. Hens, C. Marais, and P. Beutels. Dynamic epidemiological models for dengue transmission: a systematic review of structural approaches. *PLoS One*, 7(11):e49085, 2012.
3. J. Arino, J. R. Davis, D. Hartley, R. Jordan, J. M. Miller, and P. Van Den Driessche. A multi-species epidemic model with spatial dynamics. *Mathematical Medicine and Biology*, 22(2):129–142, 2005.
4. J. Arino and P. Van den Driessche. A multi-city epidemic model. *Mathematical Population Studies*, 10(3):175–193, 2003.
5. J. Arino and P. Van den Driessche. Disease spread in metapopulations. *Fields Institute Communications*, 48(1):1–13, 2006.
6. S. Bañuelos, T. Danet, C. Flores, and A. Ramos. An epidemiological math model approach to a political system with three parties. *CODEE Journal*, 12(1), 2019.
7. M. J. Berryman and S. D. Angus. *Tutorials on agent-based modelling with NetLogo and network analysis with Pajek*. World Scientific, 2010.
8. H. V. Boyett, D. G. Bourne, and B. L. Willis. Elevated temperature and light enhance progression and spread of black band disease on staghorn corals of the great barrier reef. *Marine Biology*, 151(5):1711–1720, 2007.
9. F. Brauer and C. Castillo-Chavez. *Mathematical models in population biology and epidemiology*, volume 40. Springer, 2012.
10. B. Calderhead, M. Girolami, and D. J. Higham. Is it safe to go out yet? Statistical inference in a zombie outbreak model, 2010.
11. G. Chartrand and P. Zhang. *A first course in graph theory*. Courier Corporation, 2013.
12. M. Clauson, A. Harrison, L. Shuman, M. Shillor, and A. Spagnuolo. Analysis of the steady states of a mathematical model for Chagas disease. *Involve, A Journal of Mathematics*, 5(3):237–246, 2013.
13. D. Clayton, M. Hills, and A. Pickles. *Statistical models in epidemiology*, volume 161. Oxford University Press, Oxford, 1993.
14. M. Dickerson. Multi-agent simulation and NetLogo in the introductory computer science curriculum. *Journal of Computing Sciences in Colleges*, 27(1):102–104, 2011.
15. O. Diekmann, H. Heesterbeek, and T. Britton. *Mathematical tools for understanding infectious disease dynamics*. Princeton University Press, 2013.

16. S. P. Ellner, L. E. Jones, L. D. Mydlarz, and C. D. Harvell. Within-host disease ecology in the sea fan Gorgonia ventalina: modeling the spatial immunodynamics of a coral-pathogen interaction. *The American Naturalist*, 170(6):E143–E161, 2007.

17. S. M. L. Emily K Kelting, Brittany E Bannish. Toxoplasma gondii: A mathematical model of its transfer between cats and the environment. *Siuro*, 11, 2018.

18. Fandom. The walking dead wiki, 2019. https://walkingdead.fandom.com/wiki/Zombies Last accessed on 2019-01-06.

19. K. R. Fister, H. Gaff, E. Schaefer, G. Buford, and B. Norris. Investigating cholera using an SIR model with age-class structure and optimal control. *Involve, A Journal of Mathematics*, 9(1):83–100, 2015.

20. N. Hartsfield and G. Ringel. *Pearls in graph theory: a comprehensive introduction*. Courier Corporation, 2013.

21. G. B. Jiechen Chen. Realistic modeling and simulation of influenza transmission over an urban community. *Siuro*, 8, 2015.

22. M. J. Keeling and P. Rohani. *Modeling infectious diseases in humans and animals*. Princeton University Press, 2011.

23. K. Khan, J. Arino, F. Calderon, A. Chan, M. Gardam, C. Heidebrecht, W. Hu, D. Janes, M. MacDonald, J. Sears, et al. An analysis of Canada's vulnerability to emerging infectious disease threats via the global airline transportation network. *Technical report*, 2011.

24. K. Khan, J. Arino, W. Hu, P. Raposo, J. Sears, F. Calderon, C. Heidebrecht, M. Macdonald, J. Liauw, A. Chan, et al. Spread of a novel influenza a (H1N1) virus via global airline transportation. *New England Journal of Medicine*, 361(2):212–214, 2009.

25. S. Levin, T. Powell, and J. Steele. Patch dynamics (lecture notes in biomathematics 96), 1993.

26. R. L. Lineberry and I. Sharkansky. *Urban politics and public policy*. Harper & Row New York, 1978.

27. S. Mandal, R. R. Sarkar, and S. Sinha. Mathematical models of malaria-a review. *Malaria Journal*, 10(1):202, 2011.

28. P. Manfredi and A. D'Onofrio. *Modeling the interplay between human behavior and the spread of infectious diseases*. Springer Science & Business Media, 2013.

29. J. Mao-Jones, K. B. Ritchie, L. E. Jones, and S. P. Ellner. How microbial community composition regulates coral disease development. *PLoS Biology*, 8(3):1000345, 2010.

30. R. Margevicius and H. Joshi. The influence of education in reducing the HIV epidemic. *Involve, A Journal of Mathematics*, 6(2):127–135, 2013.

31. R. Martin, M. Sauer, E. Olawsky, and M. Marinello. Stochastic models for HIV transmission as a vector-host disease. *Minnesota Journal of Undergraduate Mathematics*, 2(1), 2017.

32. D. Maxin, T. Olson, and A. Shull. Vertical transmission in epidemic models of sexually transmitted diseases with isolation from reproduction. *Involve, a Journal of Mathematics*, 4(1):13–26, 2011.

33. A. K. Misra. A simple mathematical model for the spread of two political parties. *Nonlinear Analysis: Modelling and Control*, 17(3):343–354, 2012.

34. P. Munz, I. Hudea, J. Imad, and R. J. Smith. When zombies attack!: mathematical modelling of an outbreak of zombie infection. *Infectious Disease Modelling Research Progress*, 4:133–150, 2009.

35. S. P. Otto and T. Day. *A biologist's guide to mathematical modeling in ecology and evolution*. Princeton University Press, 2011.

36. C. Ray and M. Hoopes. Metapopulation biology: ecology, genetics, and evolution. *Ecology*, 78(7):2270–2272, 1997.

37. R. C. Reiner Jr, T. A. Perkins, C. M. Barker, T. Niu, L. F. Chaves, A. M. Ellis, D. B. George, A. Le Menach, J. R. Pulliam, D. Bisanzio, et al. A systematic review of mathematical models of mosquito-borne pathogen transmission: 1970–2010. *Journal of The Royal Society Interface*, 10(81):20120921, 2013.

38. D. Smith and L. Moore. the SIR model for spread of disease - introduction. *JOMA, Convergence*, 2004.

39. The Political Compass. Political compass, 2019. https://www.politicalcompass.org/.

40. The Walking Dead Website. The walking dead comic book series official website, 2019. https://www.skybound.com/the-walking-dead/walking-dead-comics-story/ Last accessed on 2019-01-06.

41. United State Census Bureau. New census data show differences between urban and rural populations, 2016. https://www.census.gov/newsroom/press-releases/2016/cb16-210.html.

42. D. B. West et al. *Introduction to graph theory*, volume 2. Prentice Hall Upper Saddle River, 2001.

43. Wikipedia. Coral reef, 2019. https://en.wikipedia.org/wiki/Coral_reef Last accessed on 2019-01-06.

44. U. Wilensky. Netlogo, 1999.

45. C. Witkowski and B. Blais. Bayesian analysis of epidemics-zombies, influenza, and other diseases. *arXiv preprint arXiv:1311.6376*, 2013.

46. J. Woo and H. Chen. Epidemic model for information diffusion in web forums: experiments in marketing exchange and political dialog. *SpringerPlus*, 5(1):66, 2016.

47. K. Yokley, J. T. Lee, A. Brown, M. Minor, and G. Mader. A simple agent-based model of malaria transmission investigating intervention methods and acquired immunity. *Involve, A Journal of Mathematics*, 7(1):15–40, 2013.

Printed in the United States
by Baker & Taylor Publisher Services